Praise for

THE SEDIMENTS OF TIME

"In a field of celebrity scientists, nobody shines brighter than Meave Leakey . . . Extraordinary . . . This inspirational autobiography stands among the finest scientist memoirs."

— *New York Times Book Review,* Editors' Choice

"An excellent overview of how we know what we know about human evolution . . . Meave Leakey is a real-life Indiana Jones. Her life has been filled with adventure, struggle, and discovery after amazing discovery that are detailed in her riveting autobiography, *The Sediments of Time.*"
— *Forbes*

"A fascinating glimpse into our origins. Meave Leakey is a great storyteller, and she presents new information about the far-off time when we emerged from our apelike ancestors to start the long journey that has led to our becoming the dominant species on Earth. That story, woven into her own journey of research and discovery, gives us a book that is informative and captivating, one that you will not forget."

— Jane Goodall, PhD, DBE, founder of
the Jane Goodall Institute

"An engaging memoir . . . A marvelous account of what it is like for a celebrated scientist to take on some of the most vital and vexing questions regarding human origins and to come up with biocultural answers."
— *Science*

"For over fifty years, British-born palaeoanthropologist Meave Leakey has been unearthing fossils of our early ancestors in Kenya's Turkana Basin. Her discoveries have changed how we think about our origins . . . [*The Sediments of Time*] reflects on her life in science and pieces together what we now understand about the climate-driven evolution of our species." —*Guardian*

"An exciting and richly informative scientist's autobiography . . . This major work of scientific dedication and original insight illuminates both our distant past and our current, serious, human-caused planetary challenges." —*Booklist,* starred review

"Masterful . . . Drawing on field notes, interviews, and research papers, Meave recounts the work that led to some of her and her team's greatest discoveries . . . Meave and her cowriter, her youngest daughter Samira, write clearly and compellingly about what these discoveries mean . . . A thrilling account." —*BookPage,* starred review

"Meave Leakey confronts extraordinary challenges that ultimately yield unimaginably rich rewards. *The Sediments of Time* offers everything for everyone: exciting fossil finds, courageous expeditions, seminal palaeoanthropological contributions to science, and convincing scientific evidence to support climate change. Leakey is a researcher extraordinaire." —Gilbert Grosvenor, former president and chairman of the National Geographic Society

"Involved for five decades in collecting, describing, and interpreting an extraordinary range of fossils critical to understanding human evolution, Meave Leakey and her daughter Samira present us here with a welcome and accomplished example of accessible science writing in this engaging and deeply informed book."
—David Pilbeam, PhD, Henry Ford II Research Professor of Human Evolution, Harvard University

THE SEDIMENTS OF TIME

THE SEDIMENTS OF TIME

MY LIFELONG SEARCH FOR THE PAST

MEAVE LEAKEY

with

SAMIRA LEAKEY

MARINER BOOKS
An Imprint of HarperCollinsPublishers
BOSTON   NEW YORK

For our daughters

——————

First Mariner Books edition 2021
Copyright © 2020 by Meave Leakey and Samira Leakey

marinerbooks.com

Library of Congress Cataloging-in-Publication Data
Names: Leakey, Meave G., author. | Leakey, Samira, author.
Title: The sediments of time : my lifelong search for the past /
Meave Leakey with Samira Leakey.
Description: Boston : Houghton Mifflin Harcourt, 2020. |
Includes bibliographical references and index.
Identifiers: LCCN 2019050624 (print) | LCCN 2019050625 (ebook) |
ISBN 9780358206675 (hardcover) | ISBN 9780358308942 | ISBN 9780358311799 |
ISBN 9780358171911 (ebook) | ISBN 9780358629221 (pbk.)
Subjects: LCHS: Leakey, Meave G. | Paleoanthropologists — Great Britain —
Biography. | Women anthropologists — Great Britain — Biography. |
Paleoanthropology — History.
Classification: LCC GN50.6.L43 L43 2007 (print) | LCC GN50.6.L43 (ebook)
| DDC 599.9092 [B] — dc23
LC record available at https://lccn.loc.gov/2019050624
LC ebook record available at https://lccn.loc.gov/2019050625

Book design by Helene Berinsky
Illustrations by Patricia J. Wynne © 2020 by Houghton Mifflin Harcourt

Printed in the United States of America
1 2021
4500837346

Contents

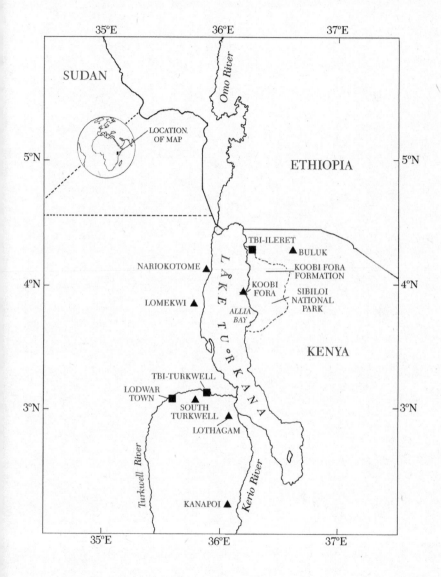

Prologue

NEVER BEFORE HAS A SPECIES BEEN AS INTELLIGENT AS WE ARE or as able to achieve the technological innovations that we have. We can plumb the depths of the oceans and penetrate far into space. We can understand the smallest particles that make up matter and probe the farthest planets in the galaxies. New innovations, new theories, and new breakthroughs are reported every day. We have better communication tools than ever before. With the advent of the Internet and the power of our smartphones and social media, we can communicate with almost every individual on the planet, and we know what is happening everywhere in the world almost as soon as it happens in real time. This is truly extraordinary.

The earliest stone tools were first fashioned more than three million years ago, but the wheel was not invented until around 3500 BC. Today, new innovations appear with increasing rapidity and complexity; the first landing on the moon was in 1969, now we can land on Mars and explore the far reaches of space. In 2017, the international collaboration between two hundred scientists and the combined magnification power of seven telescopes spanning the globe culminated in the seemingly impossible: photographing the event horizon of a supermassive black hole at the heart of the Messier 87 galaxy, nearly fifty-five million light-years away from Earth, thus putting proof to what had previously been an abstract theory. And our technological capacity continues to increase at an exponential pace. This potential to achieve such mind-boggling accomplishments—relying on the combination of technical knowledge, our capacity for abstract thought, and our ability to work cooperatively—represents the pinnacle of our human capabilities.

Our past is full of twists and turns that ultimately led to the species we are today. Our extraordinary accomplishments can be traced back in time through the major milestones of our evolution. The body plan and complex hierarchical social structure of primates provided the latent potential for all the key adaptations that now separate us from other animals. Swings in global climate provided the main impetus for crucial adaptations to occur. Driven by the drying trend towards more open savannah, we left the safety of the trees to begin our bipedal journey, thus freeing our hands for more manipulative tasks. Changes in diet provided the calories necessary to grow our enormously expensive brains to the size they are today.

No part of our evolution could have been anticipated as an obvious outcome of a prior development. Nevertheless, the sequence of adaptive changes our ancestors underwent were the essential stepping-stones that define the species we ultimately became. This process was in no way preset. The course of our evolution could have been very different, and we might have become a very dissimilar being.

I have been fortunate to spend my career studying and exploring human evolutionary developments, and I have come to fully appreciate the importance of the past in charting a direction for our future. But harder to predict and foresee is the impact of our extraordinary technological developments. The rapidity with which human technology is increasingly driving our lives today is a potential threat and, at the same time, seemingly the best and only hope for our species. We are able to better predict the negative consequences of a future in which we continue to willfully destroy the environment on which we all depend. For our species to survive, we must protect the natural resources that sustain us.

As I write this, it feels like we are hurtling ever closer to some sort of tipping point. The threat of man-made climate change is growing inexorably more immediate and urgent, and the dizzying speed at which information can now spread across the world has upended the traditional political and media culture that has prevailed through most of my life. Science and truth, the bedrocks of rational thought ever since the Enlightenment, are now once again contested arenas: a seismic shift whose consequences we cannot predict and are only starting to bear witness to.

This is a book about personal—and collective—discovery. It recounts the events that have led to our current understanding of our past, which was very minimal when I began my career. We have a much more detailed record today and can fill in many of the gaps. Species come and go and are replaced over time by newly evolved, better-adapted forms. This is the ever-changing face of evolution.

Although we cannot determine physical evolutionary changes, we are able to direct technological ones. While technology can be put to many beneficial and productive uses, it can be extremely dangerous if misused. As our capabilities expand, so does the capacity to do irreparable harm. This is much closer to home for many of us than ever before, with many new questions arising each year about how best to regulate research in artificial intelligence, genetic engineering, drone use, and other cutting-edge fields. Alongside these is the greatest challenge of all: what we are to do to address the threats posed by the warming of our planet?

Our challenge is to ensure that we direct our extraordinary technological abilities along channels beneficial to all humanity and the planet. I have no doubt in my mind that we have the mental and technological capacity to prevail. The question is whether we can summon the collective will to do so. To me, these facets—our ability to solve intractable problems and overcome insurmountable odds in opposition to the impulsive, willful, and selfish parts of our human nature—are the dichotomy that defines our uniquely human character. Our continued success as a species depends on our choices about which of these evolutionary traits we will allow to dominate our collective future. I have written this book in the hope that readers develop this essential perspective based on a better understanding of our past.

—Meave Leakey, October 2019

PART I

1

BEGINNINGS

IT WAS THE END OF A LONG HOT DAY SCOURING FOR FOSSILS NOT far from the eastern shore of Lake Turkana, known in 1969 as Lake Rudolf. We were clustered around a small dying cooking fire. Against the rich black darkness, the stars crowded the sky, looming so large and close that it felt as though you could reach out and touch them. A cool breeze rustled the thorny bare branches of nearby wait-a-bit bushes, bringing welcome relief from the searing heat. Our camels were gathered together, legs hobbled. They gurgled and groaned in the manner peculiar to their species, and their chorus carried to where we sat on the stony ground.

As the slender crescent of the moon sank slowly toward the horizon, an animated discussion erupted. Just two weeks earlier, we had gathered together under the same starry sky at our base camp to listen to a crackling voice proclaim over the short-wave radio: "That's one small step for man, one giant leap for mankind!"

Peter Nzube Mutiwa, a spry young man with extremely sharp eyes and an equally sharp sense of humor, was still in total disbelief. "How could there possibly be a man up there in the sky, standing on the moon?" he demanded, yet again adamant that this Neil Armstrong character simply had to be an American fabrication. Pure propaganda, designed to convey US superiority in a cold-war era, which had cast its pall even in this far-flung corner of Kenya.

No wonder Nzube, who preferred to use his Kamba name, was so disbelieving—the contrast between our surroundings and the technology that led to this breakthrough could not have been greater. A wild and remote area, the Lake Turkana region of Kenya is inhabited by the Dassenetch, a local tribe who, at the time, lived by the minimum of technology. Their pastoral lives still depend largely on their sheep and goats to this day, reflecting a major revolution in the way humans relate to the natural world, one that saw us control the number and destiny of animals and plants. Yet in many respects, their pastoral ways were continuous with a deeper, foraging past and were not dissimilar to those of humans living hundreds of thousands of years ago. In 1969, small groups of men habitually followed the lakeshore in canoes crudely crafted from gouged-out tree trunks to supplement their diet with fish and turtles. They foraged for whatever edible plants they could find in the sparsely vegetated desert landscape. If they needed to cut something up, they fashioned a stone tool. To cook their food, they made a fire on the lakeshore or in the bottom of a hardwood canoe using two sticks and some tinder from a ball of dry, fibrous herbivore dung. The meal might be eaten out of a tortoise shell that served as a bowl. Carrying few possessions and apparently feeling no need for them, the Dassenetch could find everything they required readily available from their barren surroundings.

Surrounded by the simplicity of this life in the desert, it was even more difficult to conceive just how big a leap mankind has taken to have ended up not only on the moon but also as the supremely dominant species here on Earth. Both achievements are thanks to the advantages conferred by technology, which continues to evolve at a dizzying pace. Gazing at the glowing embers of our dying campfire, it occurred to me that the same impulse that landed a man on the moon had found us gathered together on the stony ground that night: a burning desire to know where we came from. What makes us special? How did we come to be here? And what happens after we die? These questions would go on to drive my career. They are the same questions that have compelled people for thousands of years.

MY COMPANIONS were just the sort you would choose for an expedition in the remote world of Lake Turkana in 1969. Peter Nzube had worked at Olduvai Gorge in Tanzania with Louis and Mary Leakey for many years. Always the source of much entertainment and a great storyteller with a mischievous sense of humour, he loved to regale us with tales of excavations under Mary's strict and unforgiving supervision. A perfectionist to her core, Mary simply would not tolerate sloppy work in the excavations.

Kamoya Kimeu, like Nzube, was from the Kamba tribe, a farming people who live on fertile terraced lands south of Nairobi. He too had worked with Mary at Olduvai Gorge and knew her son Richard as a teenager. Over the years, Nzube and Kamoya had built a strong friendship with and loyalty to Richard. Kamoya took life more seriously than Nzube; reliable, and solid, he was the master of difficult logistics and never happier than when looking for fossils in a new place. Both men had eagle eyes, and between them, they had already at this early point in their careers found some of the most important fossils from East Africa.

Across the fire from me was Richard, the handsome, bold, and daring Leakey who had already made strides in carrying forward the palaeoanthropological work started by his parents. This expedition was his brainchild, and he had invited me along for my newfound expertise in monkeys. I was in Kenya at the behest of his father, Louis, who had tasked me with helping run a monkey breeding and research centre at Tigoni in the highlands near Nairobi.

Beneath this magical moon in the rugged, expansive northern desert of Kenya, I was blissfully happy. Just twenty-seven years old, I was thrilled—if surprised—to have this rare opportunity to be counted among this circle of talented individuals. The desert of Africa was not at all where I had expected to find myself. Newly graduated from the University of Bangor in Wales, I was technically a marine biologist. Or so I had planned, with my freshly minted joint degree in zoology and marine biology. But behold the world in the late 1960s. *Homo sapiens* had not, apparently, evolved so far as to allow female marine biologists onto ships. "We cannot employ you because we don't have the facilities for you," I had been told time and again by the men running the expeditions. When I had chosen my degree, I didn't know about this important technicality, and it is testament to the open-mindedness of my parents that I had never once questioned my birthright to be a scientist alongside these men. But in the sea of gently undulating desert plains, this obstacle was easily overcome because there were no "facilities" at all!

I WAS BORN at Guy's Hospital in the middle of the London Blitz in 1942. I don't remember very much about my early years, but I am told I was a fractious child, anxious and often crying, presumably because of the highly unsettling incessant bombings during my childhood. When I was eleven months old, my father was deployed to Corsica, leaving my mother and me alone in a bungalow in Orpington just outside of London, plum in the middle of London's deterrent "balloon barrage." This strategy to protect London against bombing during the war consisted of a great number of inflated airships that were kept circling above by a ground operator. The German bombers could not fly below the altitude of the airships without becoming ensnared in the multitude of steel cables anchoring them to the ground. Orpington was also home to bright searchlights and antiaircraft guns positioned to catch the German bombers on their way in to London. It was a noisy and grim place to live. Our days were constantly interrupted by wailing sirens, which would pierce the air, startling everyone. During air raids, we slept in a Morrison shelter—a strong iron structure that doubled as a kitchen table and could withstand the weight of a small house collapsing on it, the theory being that you would be safe inside its metal walls until somebody could dig you out of the rubble. But I don't believe much sleeping was done in our shelter.

"Noise, Mummy, noise," I'd cry, drawing myself up rigid and screaming at the top of my lungs. Nothing and no one could console me. My exhausted mother eventually decided to get me away from all the raids, and we moved to a much more rural and rustic setting near Tunbridge Wells. Robingate Cottage was very isolated, with a great metal bath by the front door in the kitchen and a chemical loo out the back. Apparently, if the baker dropped by when my mother happened to be in the bath, she simply popped the lid on and lay there quietly until he went away! My father's sister Margaret would wrangle time off from her busy job managing the medical stores and dispensing supplies for the Women's Auxiliary Airforce to visit my mother and help her out with me. There were no other children nearby, but my mother encouraged me at a young age to learn about natural history and observe all living things. I needed little urging. I always had beetles, frogs, caterpillars, and other delights tucked away somewhere about my little person, and (apparently) I was always trying to add my "wems" (worms) to the cooking. In a short while, the woods filled with rabbits, flowers, and quiet birdsong worked their charm, and I began to recover from my anxieties.

But we quickly learned that Robingate, while not directly on the Germans' flight path to London, was close enough. Antiaircraft guns were stationed close by, and the terrifying shrill sirens of air raids soon followed us to our new home. After one such week of endless raids, in which I rapidly dropped a pound of weight, my mother decided to get me away yet again. Throughout the war, we continued to move around, sometimes returning to Robingate, at other times staying with my mother's sister, Wynn, and her husband, Eric, at the big rambling vicarage in their Norfolk parish.

On many occasions, I was left at a nursery run by two marvellous women named Beryl and Gerda in South Newington. Beryl, whom I called Taya, was a wonderfully warm, buxom person who became a second mother to me during my rather frequent stays in her nursery. It was mainly in her care, away from the sirens and bombs, and with other children to play with, that my terrified existence was replaced by more normal childish pursuits.

My mother no doubt struggled with leaving her child in another's care for any length of time, but she must have been immensely relieved to have found a place where I was not perpetually afraid. I cannot imagine what it must have been like for her—and many mothers like her —with a husband serving in a dangerous war, a frightened child, and very little food.

Adding to her hardships, my mother, Joe, struggled with her health, occasionally leaving me with Taya so she could recuperate in hospital. Joe's father was a village postman and her mother the daughter of a wealthy Yorkshire tradesman. Both she and my maternal grandmother suffered from debilitating tuberculosis, which in those days had no known cure other than bracing fresh air. My grandmother was permanently confined to a shed in the garden, used her own cutlery and crockery, and never mixed with her children or society. It must have been truly dreadful for her—suffering from a painful and debilitating cough and bad chest, and being forced to live in isolation in England's damp, drizzly outdoors as a cure!

When she was young and her own condition worsened, Joe was also sometimes confined to the garden shed. Her tuberculosis was not the more common pulmonary tuberculosis. Instead, it primarily affected her left knee joint, although sometimes she also had chest problems. As a child, she was an invalid, completely bedridden for the first twelve or thirteen years of her life. But this did not cow her independent spirit and strong sense of equality. My sister, Judy, recalls a story Joe would tell about when she was first able to leave her sick chamber. Not content with the tame diversions afforded by delivering the post to the villagers with her father in the pony and trap, she persuaded her brothers to take her sailing with them even though she could not yet swim. When her brothers were beaten soundly for their poor judgement, Joe pointed out that she ought to be beaten too. And so she was, albeit with a slipper.

By the time I knew my mother, she had a permanently stiff leg because of surgery. Although this made driving almost impossible for her, it did not visibly slow her down in other ways. She was not a healthy woman, but we never thought of her as frail. Throughout my childhood, she went to London to have annual checkups at Guy's Hospital, and she would frequently pass out on these trips. It was her delicate health that prevented her from ever matriculating and ensured that her jobs were confined to tasks as an assistant, though she undoubtedly had much higher aspirations. This was probably a huge factor in why Joe pushed her own daughters' education so strongly.

When I was four, my mother heard the all-too-familiar sound of my screams. "Mummy, Mummy, there is a huge man at the door!" I announced in terror, and promptly ran away to hide. The tall stranger standing at the door in his service uniform and holding a suitcase was an unexpected and initially unwelcome intrusion for me. It was my father. The war was finally over.

Soon after his return, my father completed his training, which had been interrupted by the war, and became a Fellow of the Royal College of Surgeons. Judy arrived in due course, providing me with a companion and playmate. After my father was fully qualified, he secured a position as an orthopaedic surgeon at St. Bartholomew's Hospital in Rochester, and we moved to a great big run-down Georgian house in Kent, which we called the White House. Judy was followed by my brother, Roger, five years later.

White House was a grand place to grow up. My parents had chosen the house because of its grounds, as my father wanted to try his hand at being a smallholder. It had a large garden with orchards of cherries, apples, and plums. We also grew raspberries, strawberries, gooseberries, rhubarb, asparagus, mushrooms, and lots of other vegetables. Every week, we'd gather our produce in big baskets and send it to the Covent Garden market in London to sell. There were a series of Nissen huts left in our garden by the army, who had used the grounds during the war. These huge structures of corrugated sheet metal bent into arch-shaped enclosures were much like aircraft hangars. I am not sure what wartime purpose they originally had, perhaps dormitories for the troops, but they were ideal for cultivating mushrooms and housing pigs, geese, and chickens in peacetime. When I was eight or nine years old, I started my own microbusiness for pocket money. I incubated goose eggs and spent a fraught month nurturing the eggs until hatching day. Then I fattened the goslings and sent them to market. Along with my baby geese, I also kept a line of lambs, orphaned and passed on to me from neighbouring farmers.

BUT FOR A CHILD, the best thing at White House was undoubtedly the garden with all its glorious trees. I spent hours climbing them and have since noticed how most children love to climb—no doubt a reflection of our arboreal primate ancestry.

On one occasion, the milkman knocked on the door. "Mr. Epps, your daughter is down the road up a tree and can't come down again," he said.

My father duly appeared and informed me that since I could climb up I could most certainly climb down again. After this valuable lesson, I became very adept at this pastime, spending many hours up in the soaring trees. The practice came to an end when my mother noticed me through the window perched high up a tall sycamore tree with a live electricity wire dangling perilously close.

It hadn't taken me long to overcome my fear of the unfamiliar man who was my father. Despite being always extremely busy with work, my father was very good at making time for his children. Over the years, we spent a lot of time together on two joint hobbies—carpentry and photography. My father had studied joinery before he became a surgeon, and he was an excellent craftsman. He taught me how to do odd jobs around the house and insisted on good practice: "How many times do I tell you to hold the hammer at the *end* of the handle, that is what it is for!" And when I was sandpapering a piece of wood: "Wrap the sandpaper around a square piece of wood so that it has a flat base; you cannot sandpaper with your hand!"

We made bookshelves, a large desk, tables, and other pieces of furniture in his workshop in one of the outhouses at White House. We spent many happy hours developing photos in a small attic room we had converted into a dark room. My father also taught me how to sail and gave me my love of the sport.

My mother must have been terribly overworked trying to refurbish the huge shabby house, keep up the livestock and garden, and look after all of us while not enjoying the best of health. Along with our efforts at smallholding, my father also augmented his income with private patients. My mother recorded his dictated reports using her proficient shorthand skills then carefully typed up the patients' reports required by the insurance companies. She made it seem easy. The house was always overflowing with flowers from her garden, the smells of her delicious cooking, and the sound of classical music from a windup gramophone. She was also a talented seamstress and made most of our clothes as well as her own, even stitching elaborate suits for her visits to London. And because of her unmatched hospitality, there was almost always somebody staying with us. We all had our chores to do, but the burden on my mother was the heaviest, and more often than not, she was worn out.

Deprived of her own opportunities for further learning, my mother encouraged us to take every opportunity to learn whatever we could. Joe wanted to travel and see the world, and she did everything in her power to help her children to do so, never allowing anything, including chauvinism, to stand in the way of our interests and ambitions. She encouraged us to travel overseas, and as a teenager, I spent my summer vacation in Italy as an au pair. While at university, I looked after three children for six weeks in the United States before taking time to see some of the great American highlights. In those days, it was possible to pay ninety-nine dollars for ninety-nine days' travel anywhere in North America on a Greyhound bus, and so I did. My mother fully approved.

EVERY YEAR, we took a fortnight's family holiday in Cornwall, and my mother enjoyed some well-earned rest. We rented a small Cornish cottage called Retalick in a secluded valley not far from the sea. My parents slept in the main cottage with my brother and sister, and I slept in a little hut in the garden. I loved the solitude and the sounds of the constantly gurgling stream that ran beside the hut. It was during these happy times that I fell in love with the sea, and the many fascinating animals that lived in the rock pools provided me with hours of entertainment and kindled my interest in marine biology. We spent many hours peering into the tidal pools, identifying the myriad creatures they contained, and reveling in the natural beauty of the dramatic Cornish coast with its high cliffs, rough seas, and lonely beaches. My parents loved to surf, a passion they passed on to me, and we would ride the wonderful waves of the north Cornish coast on plywood boards my father fashioned for us.

When I was eleven, I was sent to a boarding school called Burgess Hill Parents' National Educational Union School. The PNEU were a consortium of schools with a shared philosophy, conceived by Charlotte Mason, of turning out well-rounded individuals through a strong arts-and-music focus. Unfortunately, the school fell rather short in terms of academic rigor. One evening, shortly after I completed my O-level exams, my parents sat me down beside the fire in our living room.

"What is it that you'd like to do with your life?" my father asked.

Without hesitation, I replied, "Biology."

It dawned on my mother with horror that the school only taught "general" science. This subject involved one lesson a week devoted to physics, chemistry, and biology and was totally inadequate for any student wishing to apply to university to study these subjects. Girls in those days were not expected to be scientists. Our science teacher was an uninspiring lady who frequently taught us erroneous "facts," which I occasionally corrected, but more often than not, I kept quiet. The teacher never believed that my corrections were justified.

My sister was promptly moved to a different school, and I started attending a technical college, the Medway Towns Polytechnic, to do my A-levels in zoology, physics, chemistry, and an AO-level in maths. After the sheltered and closeted life at a girls' boarding school, the freedom I enjoyed at the polytechnic was intoxicating. With the money I earned selling my lambs and working during summer holidays at a research lab run by Shell Chemicals, I had bought a rather ramshackle 1938 Morris 8 for the princely sum of twenty-five pounds. I refurbished this car myself using an old greenhouse in the garden with a sunken walkway running down the centre. It was ideal for my needs, as I could drive the little car directly into the greenhouse and use the central depression as my mechanic's pit. The hard top came off the little black car, and once I had fitted a new engine, this was the perfect set of wheels. There were hardly any girls at the polytechnic, so we had our pick of consorts. Many of the male students were a racy lot from the Royal Engineers Corps. I enjoyed this imbalance, often favouring the boys with the fastest sports cars or the best knowledge of mechanics!

As far as I can recollect, I never had a curfew, and the door was never locked. But I am sure my mother used to lie awake until I was safely home in the evenings. She had good reason. One night I had retired to my bed in the attic when my mother was roused by somebody knocking at the door. One of my boyfriends was standing blearily on the step.

"Is Meave home?" he asked, too hopefully. He had woken up to find himself in his car in a ditch and, rather bravely and honourably, had thought it prudent to check if he had already brought me home before falling off the road and knocking himself out. Although I always thought my mother was very strict, I can see now that I enjoyed a remarkably free rein.

Most of the time I was far too busy studying for my A-levels to get into too much trouble. I was determined to become a marine biologist, and I needed the grades to get into university to do so. My hard work at the polytechnic paid off, and I was admitted to my first choice, the University of Bangor in North Wales, which had a famous marine biology program. Upon arrival, I was discouraged in the most strenuous terms, from pursuing a joint degree in marine biology and zoology. The lecturers thought this combination too challenging and rigorous a course load for even the brightest students.

I refused to be deterred. In fact, their policy to discourage students from the joint program worked entirely in my favour. I was the only student foolhardy enough to disregard their advice, and therefore I was the only undergraduate in marine biology classes levelled at the graduate students. The calibre of my fellow students and the quality of teaching was far superior to that in classes geared solely to undergraduates.

I loved it. The marine biology station was situated on Anglesey, a hunk of land separated from the rest of North Wales by the Menai Strait, a small channel of water running through a steep gorge. On Anglesey's wild and desolate seashore, surf pounded against dramatic high cliffs, which rang with the cries of thousands of seabirds. Inland, the Welsh mountains offered endless hours of walks in wild country inhabited largely by sheep and the odd farmer. It was a spectacular place to live and study.

My first year, I had "digs" with a landlady who lived with her son. I shared a room with a highly religious arts student, Meghan, with whom I had little in common. She could never understand my total lack of belief in a god who had created all life and who kept a vigilant eye on each individual, saving them from the worst of life's disasters. Initially, I spent fruitless hours trying to get her to understand the exquisite logic of evolution and see my point of view. In vain, I explained some of the amazing evolutionary transitions I was learning about in my lectures, such as when fish first evolved into tetrapods that walked on the land between 380 and 365 million years ago.

"You can see it all so clearly in the fossils! It's amazing how their bodies changed over this time; the fishes' fins became legs; the tetrapods evolved necks. Evolution makes so much sense if you look at all of life from single-celled animals to aquatic vertebrates to the first amphibian that crawled onto the land."

"How could that be possible, God created everything in seven days, the Bible tells us this," Meghan inevitably replied in disbelief.

"But, Meghan, you cannot take the Bible literally, seven days must not be interpreted literally, it really means billions of years," I would respond. "It is just that whoever first made a written record of the Bible in 400 BC, or whenever it was, had no idea about evolution, so how could they write about it?"

"Because it is not true, of course!"

These discussions never led anywhere, and we eventually confined our conversations to more mundane topics.

After my first year, I moved with three other students to an apartment that overlooked the Menai Strait. Here, with my own room and able to cook my own meals, I had more freedom as I shuttled between my oceanography classes on Anglesey and my zoology classes at the main Bangor campus. I bought a scooter, which I rather reluctantly replaced my Morris 8 with to save money. This gave me ready access to explore the beauty of the Welsh countryside and the shores of Anglesey.

My zoology courses featured an abundance of both eccentric and excellent lecturers, perhaps the most colourful being a character named Mr. Jackson. Although he was hugely knowledgeable, he never had lesson plans and never knew what he would teach prior to each lecture. Instead, he brought handfuls of newspaper clippings to class and taught from the subject that happened to be on top of the pile. Inevitably, his lectures were a string of amusing and fascinating stories about natural history but had little to do with our required syllabus.

In total contrast was the man who would become my mentor. Dr. Robert McNeill Alexander was a brilliant evolutionary biologist who taught us palaeontology and evolution. He specialised in functional morphology—why the way we use parts of our bodies affects their shape. He was always bringing models to class to demonstrate a point and would sometimes illustrate the functional principles by climbing on the desk and moving his limbs. He was the only lecturer whose lessons were so well planned that he brought us a sheet of notes, which meant that we gave him our undivided attention rather than scribbling frantically.

As graduation from Bangor neared, I spent a number of weeks fruitlessly applying for jobs as a marine biologist. My interest in the subject went beyond my love of the sea. At that time, little was known about the oceans. They felt to me like the final frontier of exploration and a wonderful opportunity to do something very novel and useful with my career. I wrote to all the main marine research stations in the United States and Great Britain, but I kept running into the same roadblock at every turn. I would receive polite but short replies thanking me for my interest that went on to say, "We regret that we cannot accommodate female scientists aboard our ships."

My gender, which I had never really considered before, was a far larger obstacle than I had ever realised. Reluctant to give up my dreams but impatient to have my own job and start a career, I was becoming ever more frustrated. Eventually, I came to the realization that my chances of finding a job on a boat were terribly slim and that I would need to consider alternatives.

At Bangor, I enjoyed the friendship and attentions of a number of young men. It was one of them, a friend named John, who alerted me to an advert on the back cover of the *Times*. Given how seasick I always get and how rich my career on dry land ended up being, I suppose this twist of fate turned out to be a very positive one, although I certainly didn't see it that way at the time. The advert invited applicants for a research position in Kenya at the Tigoni Primate Research Centre and listed a telephone number. I halfheartedly picked up a handful of coins, walked down the street to the red public phone box, and dialled the number printed in the newspaper.

Louis Leakey was sitting in London waiting impatiently for the phone to ring. In his characteristically breathless voice, he immediately began peppering me with questions, hardly waiting for an answer.

"You must come to London for an interview."

"When can you be here?"

"I can explain to you then what we would like you to do."

"Have you ever been to Africa?"

"Will you be able to work in Kenya?

"When will you be free to get away?"

I didn't think he would ever stop talking — and he hadn't even told me the details of his London address. When I had grabbed a handful of coins for the phone call, I had not counted them or noted their value. I frantically fed my diminishing pile of small change into the hungry call box and desperately hoped that it would last until I had secured the essential details of when and where to meet him.

The following day, I took a train to London and arrived at the appropriate address to find Louis sitting in a heavily curtained, high-ceilinged, and overfurnished living room with Vanne Morris Goodall, Jane Goodall's mother. This was Vanne's home, and she allowed Louis to use it as a base whenever he was passing through London. I was extremely nervous, knowing of Louis's fame, but Vanne Goodall made me feel welcome and at ease. She was a friendly, easy-going lady whom I instantly liked.

The interview was conducted by Vanne with Louis keeping quiet. I later came to know that this was uncharacteristic: Louis normally talked endlessly and breathlessly about numerous subjects in the most fascinating and engaging way. But on this occasion, he interrupted only when he could no longer restrain himself, interjecting attractive and exciting titbits to encourage me. Vanne Goodall's questions, on the other hand, emphasized the problems that I would encounter on a daily basis.

"How will you manage in conditions that are far from comfortable and with little social life?" she asked.

"I am used to camping and living rough, I love to be on my own in wild places, and I am not normally terribly sociable, so I do not think any of this will be a problem," I replied confidently.

"Well, you will certainly see many new and exciting things as well as spectacular wildlife," Louis interjected delightedly, clearly pleased with my response.

As soon as I had passed that test, a barrage of other questions came from Vanne.

"Can you live on a small salary in rather primitive living conditions?"

"Can you drive a Land Rover on wet, slippery, muddy roads, and what will you do when you get stuck?"

"When your vehicle breaks down, will you have the ability to fix a mechanical problem?"

Gratefully and silently thanking my parents for their emphasis on living rough and frugally, for their never listening to any complaints, and for their encouragement of my mechanical efforts on my Morris 8, I gave answers that apparently satisfied this astute interviewer.

Within a matter of weeks, I was on a plane flying to Kenya. At the time, I never dreamt of where this would lead. I was on the brink of a brand-new career path and a completely different life from the one I had imagined for myself.

2

||||||||||||||||||||||||||||||||||||||||||||||||||||||||||||||||||||||||

A CHANGE IN TRACK

I ARRIVED EARLY IN THE MORNING. IT WAS AUGUST 7, 1965, AND I was in Africa for the first time. In those days, the international airport in Nairobi resembled a tiny provincial airstrip rather than the hub for East Africa it would grow into. With few formalities, I was soon through customs and taking in the sights and smells of a whole new continent. Everything felt different — the large horizons, the dusty air, the bright colours, and the sparkling sunshine. I was thrilled at the prospect of living in Kenya for a while, getting to know Louis better, and working at the primate research centre with live monkeys.

Louis had told me that he would meet me, and I was expecting him to arrive on time in an upmarket car. But I was wrong. Louis arrived late, which I later learned I should have expected, and in an old Morris Traveller covered in dog hair and dust. He wore baggy trousers and a short-sleeved shirt; he had dishevelled grey hair and appeared to be in a great hurry. But he had a constant endearing twinkle in his eyes. As we drove into Nairobi, I had barely any time to absorb the novelty of the passing landscape as Louis talked without cessation in a breathless, rapid voice, outlining the plans for the weekend, which was to be full of action. Louis was entertaining various VIPs, which, as I soon gathered and later informed my parents in a letter, was a catch-all title for "anyone with any influence at all in the financial world, who Louis would do his utmost to impress!" Louis had numerous projects that he was trying to fund. As well as the ongoing excavations at Olduvai Gorge, he was fundraising for Jane Goodall's studies of the chimpanzees at Gombe Stream in Tanzania; for Dian Fossey's study of the gorillas in Virunga National Park in the Democratic Republic of the Congo; for the Centre for Prehistory and Palaeontology in Nairobi; for archaeological excavations at Calico Hills, an early archaeological site in the United States; and, of course, for the Tigoni Primate Research Centre where I was destined to spend the next few years.

Louis kept his word about the busy weekend. As different VIPs came and went, I remained with Louis, who provided constant entertainment. He gave them tours of the exhibits he had built in the Nairobi National Museum (initially named the Coryndon Museum) and showed them his latest fossil discoveries, which he kept in his office there. Louis also took them to Nairobi National Park, where he kept up a constant narrative about all the animals we saw and provided insights about each. I was enthralled by both his knowledge and the fascinating information he offered.

"Look at those warthogs over there, kneeling on the ground to get their mouths closer to the grass. Why do you think they have those huge warty protuberances on either side of their heads, and why do they have their eyes so close to the top of the head?"

Then he would explain: "It makes perfect sense. They love to eat the new fresh small grass blades under the short prickly bushes that the antelopes cannot reach. The warts act like pincushions and in the dry season are full of prickles. By keeping their eyes on top of their heads, they can keep an eye out for predators while they have their heads down feeding."

We also saw many giraffes in the park, some of which had their legs splayed and were eating dirt. "Giraffes like to eat the minerals in the earth. But in order to get their heads down to the ground, they have to spread their front legs apart. But then, in order to swallow without choking, they have to lift their heads and chew the earth to soften it."

Between visitors, we repeatedly returned to Louis's home in Langata, a leafy and largely undeveloped residential area on the outskirts of Nairobi, to feed the extraordinarily large collection of family pets, including Midge, the baby rock hyrax who had to be fed every four hours. We spent the night among these pets: a beautiful genet cat, several owls, a number of dogs, many cats, and decorative fish in his aquarium.

As I was going to bed, Louis remarked, "Oh, if you hear a terrible noise in the night as though someone is being murdered, don't worry. This will be one of the hyraxes, and this is their normal call." He then added, "And make sure you leave the bathroom door open because the hyraxes use the toilet!"

True enough, I woke several hours later to the alarming and deafening screeching of a rock hyrax just outside my door! I was grateful to Louis for warning me.

The highlight of that first weekend for me was a flight to Olduvai Gorge in Tanzania to tour some of Louis's famous fossil sites. We flew in a small Cessna, and Louis pointed out landmarks as the Great Rift Valley unfolded below us in all its magnificence. First on our flight path was Lake Magadi, the second largest soda lake in the world, white from the soda ash now replacing its water. This was soon followed by the famous and utterly spectacular Lake Natron, with its vivid kaleidoscope of colours caused by algae and enhanced by numerous flamingoes adding an entrancing pink edge around the lake. To the south of Lake Natron, we flew past the flanks of Ol Doinyo Lengai, an active volcano rising steeply and dramatically from the landscape, plumes of smoke belching from its crater. Even farther south, as we neared our destination in the Serengeti plains, we saw the enormous Ngorongoro crater — a majestic circular depression thirteen miles across with a grand rim rising out of the wildlife-studded landscape. I had never before beheld such a stunning, vast, and wild landscape.

The VIP who joined us on arrival at Olduvai was carrying some huge cameras and was dressed in a black suit, bow tie, and highly polished shoes, and looked most incongruous in the rugged dusty landscape. Louis drove us at breakneck speed around the gorge, stopping at several sites before he took us to his camp for lunch, which included cold chicken and delicious bread, both of which he had cooked the previous night before we went to bed. At the camp, a simple thatched open *banda* served as the dining and work area while several uniports (metal prefabricated huts) gave some protection from lions and other predators at night.

After leaving the visitor at the airstrip in the early afternoon, Louis told me I should get some practice driving in this sort of terrain. I was delighted. It was my first time driving a Land Rover, and we were driving in the most spectacular landscape I had ever seen. I marveled at the large horizons, the abundance of wildlife as far as the eye could see, and the rich colours of the Olduvai sediments. Though Mary was away at the time, we visited several more of her sites, and Louis then showed me the dam that he had built to provide water for the Masai cattle after a precious hominin cranium had been broken into hundreds of tiny fragments by the feet of numerous cows walking into the gorge to find water.

"I really am the luckiest girl in the world," I gushed in my first letter to my parents. The amazing thing is, I have had many occasions in the subsequent decades to repeat that sentiment.

ON SUNDAY EVENING, just thirty-six hours after my arrival in Kenya —although it seemed like another lifetime—Louis drove me to the Tigoni Primate Research Centre an hour or so northwest of Nairobi. I was startled at the change in landscape that an hour's drive produced. Tigoni, at a higher elevation than the nation's capital and characterized by a series of hills and valleys, had been initially cultivated by British settlers who established tea plantations that endure today. They form a tidy mosaic-like landscape in total contrast to the wild and dry vastness of the Serengeti plains. From a distance, each tea plant, hand-plucked into a rounded stubby bush with a perfectly flat top, blends into a continuous soft carpet of the most vivid green imaginable that hugs the contours of the hills. And the temperature was considerably cooler due to the higher altitude.

After such a magical and action-packed weekend, I was a little apprehensive and unsure as to where I would be spending the next few years. However, in the face of Louis's enthusiasm and warmth, I was confident that whatever I found I could manage. We soon arrived and parked in front of a small house. A tall, blonde, rather wild-looking lady eventually came out to meet us, and she was clearly unhappy.

"Cynthia!" Louis announced, pleased with himself. "I have responded to your request for extra hands. Meet your new help!"

He stepped aside, as if pulling away a curtain to produce me. Cynthia looked me over quickly, then turned to Louis who, before she could say anything, immediately added, "Well, I'll leave you two to sort things out from here." As quickly as I could have clapped my hands, he was gone, leaving us to do just that. This was the point when I discovered that my arrival was completely unexpected and not entirely welcome. Knowing that the director of the centre would point out that she had neither the money nor the accommodation to provide for me, Louis had simply neglected to tell Cynthia Booth about me.

Cynthia turned to me. "Well, I have no idea where you will sleep, and I hope that you are able to make do with the minimum of comforts," she said, before grudgingly adding, "I suppose you had better come in."

That first evening was clearly a test. Cynthia put me in a two-bedroom flat that I was to share with a youthful male assistant, Michael Winterson—a solution that was unorthodox at the time but at least provided me with a roof and my own bed. My welcome was made clear to me in the glass of water that Cynthia solicitously brought to my room. The glass contained the largest flatworm I have ever seen. Even after we became good friends, I never asked her about this curious incident. Still, I can only imagine that it was a deliberate act to either test my mettle or intimidate me. To reach such magnificent proportions, that flatworm must have been carefully nurtured for quite some time. But

Louis had warned me in London that I would not be comfortable and that conditions would be basic at best. I was primed for far worse than a harmless dead worm in my drink.

As it turned out, Cynthia didn't have to worry too much about the cost of new help. I had done well enough in my exams to qualify for a scholarship for a PhD, and Louis let me do research toward this as part of my work, which greatly reduced the size of the salary I commanded. Cynthia was an outspoken lady who had no patience for inept employees or those who complained. She worked hard, and she expected others to do the same. She had two children, who were at boarding school much of the time, and a friendly, easygoing husband, Anthony, who doted on her and made her life livable by fixing the many daily breakdowns of equipment, cages, and cars.

My job at Tigoni soon entailed supervising the daily running of the centre with Michael. I was responsible for the welfare of all the monkeys, ensuring that they were properly fed and caged, as well as managing the other staff. Whenever a young monkey was not being properly cared for by its mother, I bottle-fed the baby myself. These infants were initially on hourly feeds and later on feeds every two or three hours, which, of course, meant night feeds as well. With more than one hundred adult monkeys and a number of babies, this took quite a lot of my time. The monkeys handled me as much as I handled them, and I was soon covered with bites and bruises. My clothes were also stamped with a very characteristic and not altogether pleasant smell that had rather a lot to do with the urine and faeces they produced. But I loved being outdoors all day and surrounded by animals.

I rapidly discovered that monkeys have characters just as we do and that some are grumpy and bad-tempered, and others loving and friendly. Louis believed that fertility depends on a good diet, so the monkeys were fed exceptionally well, with a variety of fresh vegetables and whatever fruit was available including papaya, oranges, and bananas, which we bought twice a week from the local market. The colobus monkeys, with their ruminant stomachs adapted to digest leaves, could not eat fruit so the staff would gather green creepers and leaves from nearby farms, roadsides, and woodlands. At feeding times, there were constant squabbles as to who should access the food first, and the dominant cage inhabitant always won. Between meals, the monkeys would settle down to hours of grooming with fewer bouts of squabbling. These monkeys were a daily reminder that we are all primates with very similar behavioural characteristics. It troubled me, however, to see these wild and beautiful animals in captivity. This is a conundrum I have long wrestled with—as a scientist, I am fully aware of the huge benefits that individual captive animals contribute to research questions and medical breakthroughs. But becoming familiar with the colourful and amusing individual personalities of these caged creatures contributed to the weight on my conscience.

FOR MY RESEARCH, Louis suggested that I look at the morphology of the limb bones of different types of monkeys. In characteristically theatrical fashion, he said, "Blindfolded, I can put the limb bones of different species of monkeys in my mouth and tell you what species they are." Although he never demonstrated this skill to me, I knew that he was right about how strikingly different the limb bones of different animals are. I could see this plainly when I looked at the large collection of monkey skeletons that Cynthia had painstakingly accumulated over the years from all over East Africa. I had learnt from McNeill Alexander at Bangor that different species have different bone morphology because of their different locomotion. Propelling oneself forward on four legs on the ground requires a different set of muscles than jumping and swinging through the trees, and thus, the bones to which these muscle groups attach also differ markedly.

With this invaluable collection available, I decided to study the limb bone morphology of the two subfamilies of African monkeys—the Cercopithecinae and the Colobinae. Except for vervets, patas monkeys, and baboons, which prefer the more open-country grasslands and bushlands, cercopithecines are mostly forest and woodland dwellers. In contrast, colobines are semiarboreal and spend almost all their time in the trees. I wanted to discover which features of the limb bones set these monkeys apart and what these differences implied. The work entailed taking a lot of measurements to figure out what the key differences were and what might explain them.

The problem I immediately grasped was that it is almost impossible to sort out which variations evolved in response to the animal's habitat, locomotion, and diet, and which could be instead classified as phylogenetic—inherited traits evolved in earlier lines that may or may not be useful to that species' survivorship or even have no function today. I had little inkling then how relevant this question would prove to be in my future career studying fossil monkeys and hominins.

I was fortunate that my old mentor and favourite lecturer from the University of Bangor, McNeill Alexander, agreed to supervise my PhD. With his expertise in functional morphology and evolution, he was the ideal supervisor. But because I was based overseas, I also needed local supervision, which Louis undertook, although he did a lamentably poor job of it because he simply was too busy. Cynthia's convenient collection of skeletons was carefully labelled and boxed with details of where and when each specimen was collected. She had shot most of these monkeys herself and had travelled from Zanzibar off the coast of Tanzania to the slopes of Mount Kenya to do so. Some of the specimens still needed cleaning, which was another smelly job as I had to boil the skeletons for hours to clean them of any remaining flesh.

Louis also often arranged fresh carcasses for me to dissect so I could study the muscle attachments of different species; these were given to him for his growing osteology collection at the Centre for Prehistory and Palaeontology in the Nairobi National Museum. Some were road kills, and others had been shot because the monkeys were raiding crops. These carcasses had a very limited shelf life in the heat of the Kenyan daytime, so I often worked late into the night on my dissections, which must have also added to the potent aromas that followed me around in those days. But I loved it. Soon I believed that I knew more about the differences in monkey limb bones than almost anyone else!

The days passed happily. I didn't have much time off, but whenever I did, friends showed me new and beautiful parts of Kenya. Before I knew it, it was time to return to Bangor to write up my thesis. My parents came out for a holiday before I was due to leave. I was shocked at how much my mother had aged—she clearly was not well—and I urged her to stay with me longer to recuperate. But at the end of a fantastic holiday together taking in many spectacular places in Kenya, my mother insisted on getting on the plane. She did not make it back to England alive.

The anchor of our extremely close family was gone, and we were all devastated. I returned to a sombre homecoming, and the chore of writing up my thesis was clouded and rendered more onerous by my terrible grief.

BACK AT BANGOR, I buried myself in my work, attempting to come up with a family tree of the monkeys based on the differences in skeletal morphology. I was using a technique that was very new at the time —crunching numbers using multivariate analysis on a huge computer that nearly filled a whole room and running miles and miles of ticker tape with holes punched in it. This entailed very precise programming, and the computer frequently refused to run because I merely had a comma or a full stop erroneously placed. Today, my results would no doubt look archaic and not pass muster, but in those days, it was cutting-edge research.

As I was finishing my thesis writing, word came that Cynthia had decided to give up her post at the Tigoni Primate Research Centre, leave Kenya, and travel overland to Zimbabwe with all her possessions and pets. In early 1969, PhD in hand, I returned to Kenya at Louis's behest to take over the running of the centre. I had agreed to lend Louis a hand, but this was not how I wished to spend the rest of my career. I insisted that my role there would be limited to a six-month stint while I figured out my future plans.

Not long after my arrival back at Tigoni, I received an urgent summons to Louis's son's office. Richard Leakey was running Louis's affairs for him while he was away in the United States, and he had discovered the terrible state of finances in most of his father's projects. It hadn't taken him long to figure out that the monkeys were being fed like kings, and Richard called me in to tell me that my funds were going to be severely curtailed.

Sitting behind a large tidy desk in a small bare office, he greeted me and went straight to the point. "I am sorry, but the funds for Tigoni are severely limited, and we have to find a way to spend less on the large number of caged monkeys," he began. "How do you suggest that we might do this?"

Richard was a man with a terrible reputation—irritable, impossible, irascible, and extremely arrogant are some of the epithets I'd heard with great regularity. I had approached this meeting with some trepidation and was curious to know if he deserved these awful labels. After I entered Richard's office, however, it didn't take me long to form a vastly different impression of the man. He was charismatic and charming, and we instantly liked each other.

Instead of berating me for spending too much money, he suggested ways in which I could cut costs, and we discussed the general running of the centre and how things could be made more efficient and productive. It was hard for me to imagine that this was the same man some people called impossible.

I began to see Richard quite regularly because he was publishing a paper on two important fossils—a colobine skeleton and a very early baboon skull—that he had discovered in fossil sites at Lake Baringo while leading an expedition there in 1966. Richard knew little about monkey anatomy, but by now, this was a forte of mine. I just happened to be dissecting a monkey at the time, and I invited Richard to help. Over the gory and malodourous insides of a large male colobus monkey, we shared the excitement of tracing the various muscles to their point of attachment on the long bones, and we followed nerves and blood vessels through the various fossae that enabled their passage.

"Look at this tiny thumb, Richard, it is nothing like the thumb of the baboon that I was dissecting the other day," I remarked.

"Yes, all modern African colobines have small thumbs," Richard observed. "They don't need long thumbs to grab the leaves they eat, but baboons and guenons need a very precise grip to pick up the tiny seeds and fruits that they relish."

"True, and I am sure that thumbs really get in the way when you are swinging around in the trees the way the colobus monkeys do!" I laughed.

This collaboration soon led to another, and Richard invited me to join his expedition on the eastern shores of Lake Rudolf (now Turkana) to study the fossil monkeys he was finding there. Naturally, I thought this was a grand idea.

IN JULY 1969, Richard drove me to Koobi Fora, a spectacular sandy spit extending into the lake where he had based his camp. We drove nonstop for hours in a bouncy old Ford Bronco with a fully laden trailer, leaving Nairobi sometime before dawn and arriving at one in the morning the following day. It was too dark to see very much, but I could just make out a line of tents along the shore. Richard showed me a small tent near the end of the line next to his larger one, and I gratefully jumped into the metal-frame bed inside, made up for me with sheets and a blanket. Before dropping off to a very deep and relaxed sleep, I briefly lay awake listening to the sound of lapping water on the lakeshore just outside my tent.

I woke early, excited to see my surroundings and curious about the strange barks I had heard in the night. There was a large herd of zebra near my tent, which explained the unusual noises, and there were other herds of tiang and zebra grazing peacefully along the lakeshore in the early morning light. The camp was spread along the leeward beach, the small waves almost reaching the entrance to each tent. The horizon stretched for miles in all directions. The lake was a beautiful green-blue colour, which led to its informal name of the Jade Sea, and the hills on the far shore were just visible. I took a deep breath—it was magical.

I made my way to the large mess tent for some tea, and I found it abuzz with excitement. The previous day, Kay Behrensmeyer, a Harvard University geology student who had joined the field expedition that year, had discovered some stone flakes eroding from a tuffaceous (volcanic ash) horizon about a two-hour drive from Koobi Fora. Based on the features of the fossils in the same layers, she believed these to be some of the earliest stone tools yet known. In spite of our late arrival the previous night, we immediately jumped back in the Bronco to check out Kay's discovery.

We drove with Kay and some of the field crew to a spot within walking distance of her site. Although we had driven this same road the night before, it had not been possible to see anything in the dark. Now I found the sandy ground covered by small bushes and shrubs, almost all armed with prickles. Acacia trees occasionally graced the landscape with their elegant slender trunks and tiny leaves. I was amazed by the abundance of wildlife—oryx, gazelle, Grévy's zebra, Burchell's zebra, ostrich, giraffe, dik-dik, hares, and exquisite gerenuks with delicate long necks and long limbs.

After parking the car, Richard led us at a fast pace to the site. I wished we could go more slowly; I wanted to stop and look at the many intriguing rocks everywhere and check out those that I thought might be fossils. I was enthralled by my surroundings. Here, in contrast to the steep slopes of Olduvai Gorge, relatively flat badlands of sedimentary and fossiliferous exposures stretched in all directions.

After about half an hour, we came to a small hill with a visible white tuffaceous bed halfway up.

"Look at these," Kay announced pointing to some smallish grey flakes lying on the surface of this tuff and on the slope just below it.

"Surely these are tools. They are clearly flaked, and there are no rocks of this type in these sediments. They must have been carried here from elsewhere."

"Yes, indeed," Richard said, smiling broadly. "These surely are stone tools. Congratulations, Kay! A wonderful, wonderful find!"

This tuff later came to be called the KBS tuff after Kay Behrensmeyer, and the tool site was named the KBS site. Having seen what Kay had found, Richard immediately realised that the previous year he too had seen similar flakes eroding out of this same tuffaceous horizon not far away.

"Just wait here, I have to go to check something," he remarked, before disappearing into the surrounding hills, mounds, and gullies made of exposed ancient sediments, which we call "exposures." After walking for about half an hour, he found the spot from memory. Just as he had remembered, there were similar tools eroding out of the same tuffaceous bed. It was another tool site of the same age.

Later that month, the field crew began a small excavation at Kay's site to try to locate flakes in situ, as this would prove that the tools actually did derive from the tuff itself. This test-trench excavation was successful, and after about ten days, several flakes had been recovered in the tuffaceous horizon. This was a tremendous bonus for the expedition. The ability to make and use stone tools is one of the major milestones in our path to becoming human. Not only would this momentous discovery help our research, but this was the sort of find that would surely help persuade the National Geographic Society to extend their support of our fieldwork in future years.

My own first intoxicating taste of discovery came the day after we had been sitting around the cooking fire arguing with Nzube about Neil Armstrong. Little did we know that our own exploration would take its own giant leap forward. After a simple supper of rice, baked beans, and corned beef cooked on a small fire, we had spent the night on tarpaulins under the most magnificent carpet of stars I had ever seen. We woke early, and after a quick cup of tea, we split up, with

Kamoya and Nzube taking one area of exposures and Richard and I searching another.

"If you find anything good, don't pick it up but come and find us," Richard called to them as they left. "And we will do the same."

Our plan was to find out how many fossils there were in this area and if it would be worth returning to spend more time here with a larger team. So we began searching the sediments carefully, peering at the ground and checking each and every fossil to be sure there were no hominins or other rare species. To my untrained eye, the ground seemed to be covered by a mass of rocks, pebbles, and stones of assorted shapes, colours, and sizes. I was quite put out and nonplussed when Richard immediately bent over and exclaimed, "Look, lots of fossils!" There was clearly more to this than I had anticipated. "Try focusing your search on the sides of erosion gullies where new bones are being exposed each time it rains," he advised. But I still had little success. He, on the other hand, gleefully picked up and identified fragment after fragment of bone. "What am I doing wrong?" I finally asked with a note of desperation and frustration in my voice. "Well, it should be easy for you since you have spent so long looking at all those monkey bones you like dissecting so much!" he said unsympathetically. "What do you mean?" I retorted, starting to feel irritated with both of us. Eventually, Richard was able to explain to me that I needed a search image in my mind of the shape and size of the various parts of bones of different animal species. Without this mental template, it is impossible to pick out a partial bone from the mass of pebbles and stones that it is hidden in. The larger your reference library of search images, the better you are at finding fossils. Armed with this valuable insight, I was soon exclaiming excitedly over my own discoveries and entering into a spirit of competition about it all.

It was a typical scorching morning, with not much wind. We had gone by camel to explore some exposures far away from camp. By eleven a.m., the temperature had easily climbed to around thirty-five degrees Celsius (the high nineties in Fahrenheit). Richard's back was feeling the effects of a long camel ride, and I was thirsty and more than ready for a break. So we began heading back to the camels and were wandering down the sandy bed of a small dry stream when we came upon a sturdy, rounded object lying on the ground ahead of us. Staring at us from the sand was a perfect skull. We sat down and stared back.

"Austr . . . Austr . . . Austr," Richard gasped, looking very much as though he had seen a ghost. This was one of the only times in our entire life together that I have heard Richard fail to complete a word or a sentence coherently. The last rain must have eroded the stream bank enough to free the skull, and it had rolled into the riverbed. One more heavy downpour, and it would have been washed down the river and lost forever. It was one of those unforgettable moments when time stands still and you think you must be dreaming. Concerned that we might fail to find this site again, we erected a pile of stones on top of each of the low hills in the proximity of the sandy streambed before going off in search of our companions. When Kamoya and Nzube saw this spectacular fossil skull, they were as delighted and astounded as we were. They grabbed Richard, picked him up, and carried him on their shoulders shouting excitedly. Never again have I experienced quite such a thrilling and perfect discovery.

Our camel trip was over barely before it had begun. We simply could not risk taking such a valuable find onwards with us on our expedition, for we were heading into country where dangerous, heavily armed *shifta* bandits rustle cattle. Instead, we turned around and rode triumphantly back to our base camp at Koobi Fora. Richard shared his mount with what was probably the earliest hominin to ever ride a camel.

This skull was the 406th specimen from East Rudolf that we catalogued in the Nairobi National Museum collections, so it is, quite unimaginatively, called KNM-ER 406. It was the second-known complete skull of its species, *Paranthropus boisei,* and it was found almost exactly ten years to the day after Richard's mother, Mary, found the first one at Olduvai George—the skull nicknamed Dear Boy and otherwise known as Zinjanthropus. We would have to find many more fossils over a span of decades before we could begin to properly comprehend how these two iconic specimens fit into the bigger picture of evolution.

At Koobi Fora, I soon fell into the camp routine. The days went by rhythmically, each morning bringing a sense of purpose and anticipation. Would we find something important? What would we learn? Every day brought exciting discoveries, logistical obstacles, broken vehicles, or general mishaps. We would wake early, usually at five thirty a.m., just as the sky to the east began to lighten; emerge from our tents for a quick wash; and sometimes take a dip in the lake. After a quick cup of tea, we were off to the exposures to search for fossils just before the sun crept over the horizon and the scorching heat began.

Although the idea of prospecting from camels sounds wildly romantic and exciting, the reality, apart from our astounding and miraculous discovery of 406, had proved rather more challenging and impractical. The camels were the most stubborn and wayward creatures I had ever encountered, and they walked so slowly that the entire endeavor was hugely time-consuming. Riding them is uncomfortable to put it mildly, so we usually preferred to walk beside them. At night, due to the large lion population, the camels had to be hobbled in a tight circle. We formed a surrounding protective ring around the camels, feeding a campfire through the night as a lion deterrent. Glad to have experienced a once-in-a-lifetime camel safari, I was relieved when we quietly returned to a motor-propelled conveyance for future forays into the exposures. This vehicle was a grey, battered, stripped-down Land Rover with no roof that we called Kilaloma. It could defy the most formidable rocks and slopes under the expert driving of both Kamoya and Richard, and it didn't need night protection from lions or cajoling to go in a particular direction or at a particular speed.

The drive out to the sites always gave us some unexpected pleasure as we would invariably see spectacular wildlife—herds of oryx; the spiral-horned, very shy kudu; and skittish gerenuk. If we were lucky, we would see some of the large carnivores, most often hyaena and lion but occasionally wild dog, leopard, or cheetah. We carried very little with us in the field—just a small bag containing a dental pick, some Bedacryl (fossil preservative), thinner, glue, and a paintbrush. Richard taught me how to use these staple items in a palaeontologist's kit. Many fossils are not lying fully exposed or ready to pick up, and they are rather fragile. We would meticulously pick away at the ground they were embedded in with retired dental picks that had been kindly donated by Nairobi dentists. After gently brushing away the loosened soil, we would paint the Bedacryl onto the exposed part of the fossil until it could be removed and carefully packed into reams of loo paper so it could withstand the bumpy journey back to Koobi Fora. Back at camp, I would often get to spend time reconstructing these tantalizing puzzles of broken bones into more complete specimens. My search image, initially restricted by my focus on primate anatomy, rapidly expanded as the richness of the Koobi Fora sediments yielded species after species in the bountiful fossils we were finding and retrieving.

If we found something that needed time and patience to excavate, we would generally return later with additional tools. Very fragile specimens needed to be encased in plaster of paris. "The sacking has to be wet before you start or it doesn't adhere properly," Richard explained. "And if the fossil is heavy, you need to use sticks to build in supportive braces to protect the fossil." The plaster of paris dried quickly, and I was soon adept at judging the correct amount and consistency to help prepare these packages for their journey back to Koobi Fora and then to Nairobi.

We carried water in the vehicles but not in our field bags, so by the end of the morning, we were always ready to return to the vehicle for a drink and some shade. In those days, no one had made the connection between sun and skin cancer; we wore the minimum of clothing and used no sunscreen, and I never wore a hat. Lunch was sparse, usually a few slivers of dried meat from a tiang that Richard had reluctantly shot and dried in the sun earlier. We often ate under the shade of an acacia tree on the bank of a dry sand river, or if we returned to the base camp by the lake, we enjoyed a more gourmet lunch depending on the camp's supplies. We discussed the morning's discoveries over this repast, and the field crew would always have something to show us in the afternoons. Although temperatures were normally in the high thirties or low forties (Celsius), we became used to the heat, and for much of the morning, there was usually a welcome cooling wind. After lunch, we would gratefully rest for an hour or two until the heat began to lessen.

We always looked forward to the late afternoons and evenings back in camp when we could take a refreshing swim and enjoy the cool breeze blowing off the lake.

Over dinner, we would chat about the day's discoveries, often enjoying spirited discussions about what they meant and what questions they raised. The evenings were always full. The fossils that we recovered each day needed additional work to remove loose matrix, reassemble broken fragments, and apply preservative to harden the bone. We wrote notes about the day's activities and finds, and updated our field catalogues. The fossils were then carefully packed for their return over bad and bumpy roads to Nairobi. We did much of this work each evening by the light of a kerosene pressure lamp on a large table in a work tent fanned by the cool wind off the lake. This hurricane lamp emitted a loud hissing sound due to the pressure and attracted an inordinate number of insects considering our arid surroundings. It was perpetually surrounded by a cloud of lake midges and other small flying insects, all intent on death-defying swoops towards the mesmerizing light source.

Although I had been at Koobi Fora for only a matter of weeks, I was lost to any other life. I had never experienced anything quite so exhilarating, and I was completely hooked on the excitement and allure that this profession might offer. Koobi Fora was remote and rich with an abundance of spectacular wildlife. Great herds of oryx, tiang, and Grévy's zebra as well as jackals, lions, hippos, crocodiles, and glorious clouds of pelicans and flamingoes were just some of the abundant game that frequented the Koobi Fora spit. It felt as though we were the only people around for miles. I loved the remote setting, the huge landscapes, the quiet, and the knowledge that I had to always be on my toes to avoid hidden dangers. And as I got to know Richard, I knew I was falling in love with him. But Richard was married, and his wife was expecting a child. I was appalled at the idea of an entanglement with another woman's husband.

Upon our return to Nairobi in August 1969, Richard and I continued to collaborate on the fossils. I had stopped working for Louis at the primate centre, and my days were spent at the museum poring over the bones and trying to make sense of what they all meant. I moved into a tiny cottage in a leafy lane in Karen, a quiet suburb of Nairobi. My landlady was rather a dragon, and I am sure that she disapproved of Richard's clandestine visits to an unmarried young lady!

Despite all the young men who had danced attendance on me in England, I was actually rather a prude. Apart from one serious boyfriend, I had kept them firmly at an arm's length and often wanted less from them than they wanted from me. I had to keep Richard at the longest length of all. Still, there was no denying the strong attraction that we felt—or that this added to the excitement and magic of those early explorations. None of my boyfriends could hold a candle to Richard—or to his skill and speed behind the wheel. And England's orderly countryside could not compare to the enthralling scenery that he drove me through or the tantalizing life that a relationship with him offered. Richard's marriage was not a happy one and was not destined to last, but it cast a dark cloud on what would have otherwise been a fairy-tale story until his divorce was settled.

3

RACING AGAINST THE CLOCK

ONE DAY LATE IN OCTOBER 1970, RICHARD AND I WENT QUIETLY down to the registrar's office and were married without fanfare during our lunch hour, with Kamoya as our witness. Our marriage was remarkable for the lack of planning or ceremony. I was sharing an office with a visiting researcher from Germany, and when I came back after lunch a little late and looking happy, she asked me conversationally where I had been. "Oh, I went to get married!" I ventured nonchalantly to her great surprise. My mother-in-law, Mary, was absolutely furious with us for not inviting her as she had specifically said that she wanted to be there when we did this. But in contrast to my mother's great attention to celebrating anniversaries and special occasions, this total lack of ceremony would be a long-standing feature of my new life with Richard. In any case, I have always felt that the richness of our experiences together far outweighs any fleeting memories that a more elaborate wedding ceremony would have offered.

Thus began a golden decade. At Koobi Fora, wonders were pouring out of the ground on an almost weekly basis. Although we still knew almost nothing about what these fossils might mean or how old they were or where they fit in the family tree, we knew we were part of something truly remarkable. We tried to get up to the lake as often as we could and thought nothing of a quick weekend trip. We left the office at lunchtime on Friday and drove through the night at a breakneck pace, usually in Richard's comfortable Volvo sedan. The road diverted through Uganda, and we passed border-post controllers who were invariably drunk. We arrived on the west side of the lake at dawn. Sometimes we then had to wait until later in the day for the wind to die down before attempting to cross by boat from Ferguson's Gulf to Koobi Fora on the opposite shore. These enforced delays cemented our lifelong friendship with Eduard and Laurette Dandrieux, a young couple who ran the lodge at Ferguson's Gulf.

In those days, the road was corrugated sand and gravel rather than the pot-holed tarmac of today. You certainly couldn't go that fast now, and it is doubtful that a Volvo sedan could make it at all. Nevertheless, to get there in the shortest amount of time, Richard drove like the clappers. One time, he took the road a little too fast and completed a 360-degree turn before we carried on as though nothing had happened — even though this occurred in the Marich Pass, a steep road of hairpin bends that takes you down the Cherangani Hills to the plains of the rift valley below. After one night at Koobi Fora, we set off back across the lake on Sunday afternoon and drove through the night to be back in the office bright-eyed and bushy-tailed early on Monday morning. It was crazy — and wonderful. Those compressed weekends left one as refreshed as several weeks' holiday anywhere else.

Like our rushed weekends, we did everything in a hurry. For Richard had a terrible secret. Few people knew that he had been given ten years at the most to live because he was suffering from debilitating kidney failure. At the time that he took over the museum in 1968, his kidneys were seriously compromised by an autoimmune reaction to a throat infection. I was in complete denial. I don't think I ever truly believed that he would die. Nevertheless, the knowledge of his impending illness certainly added great intensity to our life together. We lived richly and fully, savouring every minute.

For his part, Richard was determined to cram a whole lifetime of accomplishments into that decade. He wanted to build the still small National Museum of Kenya into a truly Kenyan enterprise, with museums around the country that reflected the rich heritage and diversity the country had to offer. And so he did. When he took over running the museum, it boasted twenty-three mostly European employees and a few small dilapidated buildings housing the main museum along with his mother Mary's laboratory, some very basic offices for an accountant and a secretary, and limited storage space for the natural history collections. Richard transformed it into a country-wide regional network of museums staffed largely by indigenous Kenyans that reflected many aspects of Kenya's culture. He significantly upgraded the research facilities too, with improved offices and laboratories for entomology, mammalogy, archaeology, palaeontology, osteology, and geology. Richard also built a bomb-proof, high-security hominin vault to safeguard the priceless fossils that were pouring in from Turkana and other parts of Kenya.

In March 1972, we had our first daughter, Louise. According to Mr. S. J. Vyas, the museum accountant who happened to be a Brahmin priest, Louise was born in a very lucky time of the stars and was destined to bring great happiness to her parents in their field of life. And so she would! She was followed in 1974 by another daughter, Samira. Because I had no wish to miss the excitement of fieldwork, both children were hauled off to Turkana within weeks of their birth.

The usual adventures of new motherhood were accentuated in the harsh and remote setting of Koobi Fora. My milk dried up instantly in the heat, and dehydration in the children was a major worry. My one and only effort at shared parental responsibility of a young infant nearly ended in disaster. I asked Richard to bottle-feed a very young Louise while I took some visiting friends to see the fossil exposures. Impatient with this as he was with everything else, Richard soon grew bored with the leisurely pace of Louise's sucking and decided to enlarge the nipple with his penknife to increase the flow. She nearly choked, and I came home to an enraged child and unapologetic father!

Richard's more helpful efforts were of a practical nature. He added chicken wire to the windows of our new stone *banda* to stop the lions and hyaenas from jumping in to seize one of the little ones, who slept slung in a hanging cot to keep cool and safe from the numerous snakes and scorpions that came out at night. Alas, we did not consider the perils from the roof. One night, we were urgently summoned from the dinner table by the kindly childminder we had left with them. He had heard a telltale slapping thump, and his torchlight revealed a large angry spitting cobra that had narrowly missed falling directly into the cot with the sleeping Samira.

When the girls were young, we took them almost everywhere, and they spent much of their early childhood in a very unconventional manner, usually sitting in a basin of water keeping cool at Koobi Fora or trying to keep up with the furiously fast pace Richard set in the fossil exposures while they looked for their own bones, with little Samira making a herculean effort not to get left too far behind. I sometimes had to travel outside Kenya, and I would leave the children with my sister. Judy had moved to Kenya after completing her master's degree in potto behaviour at Makerere University in Uganda. The atrocities of Idi Amin's rule had forced her to leave, but I was fortunate enough to share the new experiences of motherhood with my sister who was also newly married—to a colleague of ours, John Harris, who specialised in the evolution of bovids. Samira would give me an extreme snubbing whenever I left her, making it quite clear how she felt about being left behind. But in spite of the cold shoulder I had to endure, she loved Judy and enjoyed staying with her. They forged an exceptionally close bond that has endured to this day.

We took holidays on Kenya's beautiful north coast on the historic island of Lamu, where we first rented and later bought a property. During these brief interludes at the seashore, I was able to indulge my love of all things marine, poking about in tidal pools and spending hours with the children snorkeling on pristine coral reefs and, as they grew older, peering at plankton under the microscope. When the children were old enough, we took them with us on our small sailing yacht as we explored the archipelago around Lamu. For our fieldwork, we continued to explore the rich sediments around Koobi Fora, which we extended with surveys of as many of the other exposures in the Omo-Turkana Basin as possible. Throughout the 1970s, the family of hominin finds from Koobi Fora continued to steadily grow.

I was in Nairobi with a rather unwell three-month-old baby Louise and trying to prevent dehydration by feeding her five ounces of water with a teaspoon of glucose and a pinch of salt every two hours when one of our most momentous and intriguing discoveries occurred. Poor little Louise got immediately dragged off back to Koobi Fora, glucose and salt in hand, as soon as the weather cleared enough for us to fly out beneath the low-lying clouds and over the edge of the Rift Valley escarpment. Kamoya and his team had found several skull fragments, which Richard impatiently attempted to reconstruct while I hovered over him, itching to get my hands on them. Over the next few weeks, more and more fragments were gradually recovered.

Much to my relief, Richard gave up before long and handed the puzzle over to me. I have wonderful memories of those days spent in the shady verandah of our *banda* at Koobi Fora. A young hippo would come every day to play with the drum attached to our boat anchor, behaving more like a seal than a hippo. And I had the indescribable satisfaction of watching the hundreds of pieces of broken bone come together. Before long, I could discern a curiously large-brained, flat-faced, and surprisingly humanlike skull, which came to be known by its museum accession number, KNM-ER 1470. It looked entirely different from the sturdy, small-brained skull that Richard and I had found in the riverbed on our camel safari. Because of its small brain and delicate features, we grouped it together with some similar fossils from Olduvai, which had been named *Homo habilis*.

Almost exactly one year later, Kamoya noticed some teeth eroding from the rocky ground of another site near Koobi Fora, which we reassembled into another stunning skull, KNM-ER 1813, that was remarkably different from 1470. Did this curious creature represent another species yet unknown? It was immediately apparent to me that all the fossils that had then been grouped into a single early hominin named *Homo habilis* would need re-examination because some of them more obviously resembled one or the other of these two iconic new skulls. Little did we guess then how many decades would pass before any light would be shed on the conundrum into our past that these two fossils presented.

Eleven years elapsed before Richard's kidneys finally gave up completely, and in 1979, it became clear even to Richard that he simply could not go on without treatment. To live, he would have to either remain on dialysis for the rest of his life or receive a kidney transplant. Those were the early days of transplant success, so the outcome was far from certain. But a life in Kenya as a fossil hunter was not compatible with being tied to the dictates of dialysis. Kenya lacked the facilities, even in Nairobi, to support this choice, and Richard had no intention of leaving Kenya permanently. He opted for a transplant, and all his brothers came forward as possible donors, and his younger brother proved a perfect match. But Philip was standing to be a member of parliament in Nairobi, and the dates of the election kept being postponed. Several agonizing months passed before Philip could come to England for the operation. We moved into a tiny flat near Victoria Station in London, and once I had found a nearby school for the girls, Alan Walker, a close friend and colleague, flew the girls over to join us.

Alan was an anatomist by training and was working at the University of Nairobi as a lecturer. His extensive knowledge of anatomy and love of fossils had proven invaluable to Richard on many occasions, and they made a complementary team. At times crusty and argumentative with adults, Alan was exceptional with the children. He would handwrite and illustrate amusing stories about animals to distract the kids from the difficult time they were having. Our friendship developed into a strong professional relationship, and he collaborated on many of our most significant fossil discoveries.

As we waited, Richard became progressively sicker. There were not enough dialysis machines for the number of kidney patients, so it was possible to receive the treatment only when you were on your last legs. It was simply dreadful to watch him suffer. He was constantly so cold that in the height of summer he would huddle on a park bench trying to catch the sun's rays wrapped in a thick winter coat, scarf, and hat, and attracting many curious glances from passersby. He spent much of his time listlessly lying on the couch while he waited. When Richard was finally pronounced ill enough for dialysis, the transformation was re-markable. Life and energy seemed to seep back into his veins. He spent the hours he was harnessed to the machine writing his autobiography, which I then painstakingly transcribed using an ancient typewriter with a ribbon so well used that it barely left an impression on the page. I have no idea why it never occurred to me to buy a new ribbon — but such simple expediencies were clearly beyond me as I tried to juggle two young children and the need to be almost permanently at Richard's bedside.

Although only five and seven, the children must have perceived the gravity of the situation because they were incredibly well-behaved and considerate. But the strain showed on Samira, who lost almost all her hair from anxiety during this terrible time. She would come downstairs in the morning like a rather scrawny and heart-wrenching scarecrow, great balls of hair sticking to her nightie. My father and his second wife, Elizabeth, looked after the children during Richard's most critical times, and many other friends helped in every way they could. The transplant went ahead at last in November 1979. All seemed well, and we began to make plans to enjoy Christmas with family members in England and go home shortly after. But just three weeks after surgery, Richard's body began to reject the new organ, and his kidney function stopped completely. He was administered a massive dose of immuno-suppressive drugs to halt the rejection, which fortunately worked. But then he became still more gravely ill. With no immunity to infection, his beleaguered body succumbed to pleurisy. It really seemed that all was lost. As he hovered between life and death, I sat by his bedside through the night, talking to him incessantly and willing him to live. I still don't know how Richard pulled through this. This was one of the darkest moments of our lives, and I have blocked it out so successfully that I can remember very little else about it at all.

EARLY IN 1980, we resumed our lives in Kenya, and the nightmare of the previous year slowly receded. Just before his kidneys gave out, Richard had been planning a comprehensive documentary film on the many facets of human evolution. Filming began on this stalled documentary, *The Making of Mankind*, which meant considerable international travel for Richard. When I was able to join him, our times together only underscored our efforts in the field. In China, we visited the site famous for the early discoveries of Peking Man; in France, we filmed the rock art in the Dordogne. At around this time, I was also helping Mary with her book on the emotive Tanzanian rock art, *Africa's Vanishing Art*.

Over three months in 1951, Mary, with the help of Louis and an Italian assistant, Giuseppe Della Giustina, had painstakingly traced these evocative images from the rock faces at 186 sites in central Tanzania. The original reproductions had been stored away for thirty years until I came across them while helping Mary clear her office. I was astonished at the detail in these paintings. In 43 paintings, 1,600 figures were recorded. I was enthralled by these lifelike images and the scenes depicted in this art, and I persuaded Mary that they needed to be shared with a wider audience. She agreed to put together a book but only if I helped. I questioned Mary about her interpretations of their meanings, and she would always have an answer to my many questions.

"Mary, look how they have picked out and exaggerated the essential characters that make it easy to identify the species being painted —these snakes have been given so many coils, and the antelope horns have an exaggerated number of spirals, the eland have exaggerated dewlaps and the cats long thin tails—but whatever do you think these people are doing?"

"They seem to dancing in a river to celebrate some event," Mary responded. "One of my favourite figures is this pipe player with music falling from his pipe."

I pulled out another painting. "Mary—in this painting, see how these figures are excitedly dancing and cavorting around an elephant that they appear to have caught in a trap."

WE WERE BOTH INTRIGUED by the scenes depicted and the lifelike images projected onto the uneven rock faces. They were accurately drawn and gave insights into the lives of these ancient people. The poignant images of daily life—hunting, playing music, seeing herds of game—evoke the perennial questions motivating our search: what connects, and sets apart, modern humans from our ancient ancestors? This time with Mary sharing our interpretations of images' possible meaning was profoundly rewarding both professionally and from a personal standpoint as I got to appreciate my mother-in-law for all the facets of her complex and special character.

After the horrors of the past year, we resumed the threads of our lives at a slightly more measured pace than we had lived in the 1970s. Although very full, these years were uneventful from a personal perspective—no births, no deaths, and no life-threatening illnesses. The dictates of a school schedule meant that I spent less time in the field, and as I took over as head of palaeontology at the museum in 1982, I had an increasingly large load of office work. Spending more time in Nairobi meant that I was also more directly involved in writing the scientific papers on the fossils, often collaborating with Alan Walker. But Richard, with his new lease on life, took the fieldwork farther afield in terms of the geological age span and in geographical distance from our Koobi Fora base camp. As Richard and Kamoya explored far and wide throughout the 1980s, they discovered many new sites on both the east and west sides of the lake. Unlike the sites we had concentrated on in the 1970s, which were all between two and one million years old, these new sites ranged in age from thirty million years to one million years old. This period spans almost the entire evolution of apes and hominins.

Our exploration of the sites on the west side of the lake began in 1981. Because these sites cover a much longer geological time frame, they were key to amass, piece by piece, an unprecedented knowledge of the evolutionary history of the vast Turkana Basin. As expected, our fossil collection of human ancestors steadily grew.

In 1983, Richard and I visited an eighteen-million-year-old site called Buluk and had a truly memorable weekend. On the first day, we had gone for a walk across the exposures. It was very hot, and Richard was in the lead, walking extremely fast as usual. Little Samira simply could not keep up, so the two of us were trailing the others. It turned out to be the very best place to be. Everybody else walked straight past the discovery that was awaiting us — beautifully preserved parts of both the upper and lower jaws of an ancient ape, which must have tumbled out of the bank above the path during the last heavy rains. I knew at once that I had never seen anything that looked quite like it. Sure enough, it was completely new to science! Richard and the others retraced their steps, thrilled about the discovery but privately chagrined to have missed such an important find. We collected the precious fossil, wrapping it in loo paper and recording its exact location on the aerial photo with a pinprick. But no sooner had we set off again, with Samira and me still lagging far behind, than we came upon something else that everybody had narrowly missed stepping on. But this was alive and much deadlier. It was an agitated carpet viper hissing and writhing in the sand and ready to strike!

The period before the origins of our hominin lineage is known as the Miocene (lasting between twenty-three and five million years ago), a time when many different apes existed in both Africa and Eurasia. Most of the ancient African ape fossils known in 1983 originated from two islands on Lake Victoria, Rusinga and Mfangano, which Louis and Mary Leakey had explored in the 1940s. There are so many apes known

from these two rich Early Miocene sites that we had not been expecting a new ancient ape to be found at Turkana. But the ape that Samira
and I stumbled upon looked so different that we were compelled to
search in other sites of this age in Turkana as well.

Richard asked Kamoya and the crew to stop off at site on the western shore called Kalodirr on their return overland to Nairobi, a site
Richard had noticed some time before from the air. To his astonishment, Kamoya turned up in his office some days later grinning broadly
and carrying a precious bundle. During his lightening visit to Kalodirr,
he had found a delightful little skull completely new to science.

We planned a longer visit to Kalodirr the following January. For
once, we had funds to do a second field season. A certain Mr. Brownlee
had been persistently trying to see Richard at the museum on a day
when Richard had left clear instructions that he was not to be disturbed.
But Mr. Brownlee was not to be deterred. In contrast to the usual run
of visitors who urgently requested some form of assistance, Mr. Brownlee wanted to give us a very generous gift of one hundred thousand
dollars for our research. Mr. Brownlee had made his money inventing
medical equipment, but his hobby was studying ants. He came to Kalodirr to visit for a few days, and I remember spending an extraordinary
hour with him one very hot lunchtime while we waited for Richard and
the crew to return. Mr. Brownlee wanted to persuade me that different
ants taste different and that this was a sure-fire method of identifica-
tion. I was happy to oblige this kind benefactor, so we sampled ants. A
most unusual hors d'oeuvre!

This second visit to Kalodirr produced many new species of animals. There were several new examples of the ape we had found at Buluk, which we named *Afropithecus turkanensis*. We gave the skull that
Kamoya brought to Nairobi the name *Turkanapithecus*, and a third
brand-new fossil ape also cropped up. We called this tiny monkey-sized
creature *Simiolus enjiessi* as a nod to the National Geographic Society,
which had funded so much of our research over the years. There was
an even greater diversity of early apes than we had imagined. One ape

would eventually split from the others to give rise to the human lineage — so all these different early examples are of considerable interest. But the subsequent period of prehistory, called the Late Miocene, is poorly recorded in the sediments of the Turkana Basin. We simply don't know what happened next or where these early apes led. Indeed, we know very little about any of the apes that evolved from these ancient creatures, let alone the enigmatic species that led to humans.

During the school holidays in the summer of 1984, Kamoya and his team completed their survey of the Miocene sediments and moved on to search younger exposures to the north of Lomekwi on the west side of the lake. I was at Koobi Fora with my girls for their first proper excavation — an enormous giant tortoise that must have died when it somehow landed upside down and was unable to right itself. This spectacular specimen remains on exhibit to this day and is one of the most impressive field exhibits we have in Kenya, both because of its size (the carapace is nearly two metres) and completeness. The girls could sit quite comfortably inside the inverted shell as they worked and had ample room to spare. After work, we'd all cool down in the lake and enjoy Koobi Fora's majestic beauty.

All this came to an abrupt halt one Sunday when Kamoya took a short Sunday stroll from the camp across the lake and stumbled upon an unpromising matchbox-sized piece of skull. It soon became clear that the little skull fragment was just the beginning of a miraculous and groundbreaking discovery as fragment after fragment followed that first piece out of the rocky ground.

Once the enormity of what was happening became apparent, Richard ferried us over the lake to join in the excitement on the west side. We flew into a short airstrip close to camp that gave the uncomfortable sensation of landing on a very narrow aircraft carrier because it was built on a skinny plateau that dropped steeply off on all sides. The site was on the bank of a tributary running into the Nariokotome sand river and was a short walk from a lovely shady camp set among the acacia and doum palms fringing the river.

"Come and see the site!" Richard said enthusiastically before my feet had even touched the ground. The field crew, under his supervision, had begun the excavation.

"What are you going to do about that tree?" I immediately asked, pointing to a small toothbrush tree right where the skull fragments were being retrieved.

"The tree stays. Removing it might damage anything in the ground that we can't see!"

"Humph. Or you might not see the pieces you need to remove without addressing the tree!" I retorted.

I was soon caught up in the thrill of what was unfolding. Back in camp, under an improvised canvas laboratory tent, there were already a number of wooden trays with numerous fossils in them. Not all of these were hominid as we collected every bone fragment we could find, and these would later help us understand the setting and circumstances at the time the individual died. The thrill of finding a hominin fossil is hard to describe. But to witness an outpouring of bones completely new to science and in immaculate condition is something quite extraordinary. Camp was buzzing with excitement and an air of anticipation as to what might follow.

Back in camp, Alan Walker and I gradually reassembled a complete skull with a face. Richard oversaw the excavation work at the site, and he and the field crew uncovered more skull fragments, the mandible, and then, to our astonishment, the first bones of the skeleton. Here, for the first time, were well-preserved elements of a *Homo erectus* skeleton associated with a cranium and mandible.

As the scale of the discovery became more obvious, Richard invited his mother, the matriarch of proper excavation techniques, to observe our efforts. Far from impressed, she soon was criticizing him and Kamoya for their sloppy work. "Do you think you are still digging potatoes?" she chastised Kamoya with her fiercest glare. Kamoya and Nzube grinned ruefully, having cut their teeth on Mary's excavations in Olduvai and endured many similar tongue-lashings in the past. Measuring tape and string in hand, Mary soon imposed her meticulous standards and grid method for plotting the precise position of each bone fragment. Stakes were placed precisely in the ground to form a level grid, and as the excavation got deeper, the elevation of each grid was plotted so that the exact layer the bones are eroding out of could be recorded. This method is still our standard practice today, and her input would later form a key contribution for piecing together the environment and circumstances when the fossil individual died.

We employed a number of local boys on their school holidays to help with the multitude of camp chores entailed in running such a big excavation. Louise and Samira spent long hours with the Turkana children, all enthusiastic workers enjoying the unprecedented and plentiful meals along with their salaries. These children were also hungry for any opportunities to learn, their school having only minimum facilities. They eagerly pored over the girls' old schoolbooks, National Geographic children's magazines, and anything my girls could find for them to read. Louise and Samira taught them card games and in turn learned traditional Turkana games. The children remember these field seasons as being far more fun than usual—instead of being confined in silence to our private *banda* at Koobi Fora, out of sight and earshot of the "very important scientists" with whom they were only rarely allowed to mix, they suddenly had their own companions.

In 1985, when Louise was twelve, she was given greater responsibility for maintaining the camp drinking-water supply by being allowed to drive all the kids to the nearby waterhole. Each day, it took them precisely four hours to fill up the drums of drinking water using an old Kimbo tin (a type of vegetable shortening we used for cooking) to scoop the clear water out of the hole dug into the sand river. One tin at a time, Samira, who was the smallest, would patiently fill a bucket that was then passed up the chain of children to the biggest and strongest of them, Christopher, who was manning the drum (Christopher would later become camp cook when he left school). The whole operation was completed with a great deal of noise, singing, and storytelling, especially during the frequent breaks they were obliged to take while they waited for the depleted water in the waterhole to slowly rise again. Louise was barely able to see over the steering wheel, but she quickly learned to drive in difficult terrain. The children were proud to at last be given a modicum of responsibility and play a role in helping facilitate the historic find that was unfolding in camp.

Meanwhile, the size of the excavation continued to grow as tons of rock were gently removed piece by piece from the site. All the excavated soil and pebbles from the layers containing bone fragments had to be sieved and picked over to make sure nothing was overlooked. The soil was placed in a wooden frame with handles and a wire mesh bottom. Two people vigorously shook this sieve to remove the dust and smallest particles, and the rest was looked over meticulously before being discarded. This process is accomplished much more effectively when the material is wet because the fossils stand out much more distinctively. With Samira and the Turkana schoolboys as helpers, Louise drove the pickup truck to the lakeshore and dropped off sacks of sediment for the sievers to pick through. After trying their hand with the sieve and taking an obligatory and wonderfully refreshing dip in the lake, the children then returned with the washing water for camp.

Over the course of five years, more and more bones were exposed —arm bones, leg bones, ribs, vertebrae, and a pelvis. But the hands and feet remained hidden. Each year, during the last week of fieldwork, a bone would turn up in the far back corner suggesting that perhaps these precious missing elements were just under the next part of the bank. The excavation was then extended into a high bank, which meant that the tons of soil covering the horizon in which the bones were buried had to be removed.

At the end of the 1987 season, we accepted defeat and gave up the hunt for the elusive hand and foot bones. The cost of continuing the enormous excavation outweighed the remote chance that we might ever find them. But even without these missing appendages, we had an amazingly complete skeleton, skull, and mandible of a single individual. This fossil is a unique discovery—the skeleton of a child, seven to eleven years old, who lived around 1.5 million years ago. It is known all over the world as the Nariokotome Boy, or the Turkana Boy, and is celebrated today by a monument and bronze replica at the site where it was found. The completeness of the boy's skeleton would permit us to make many deductions hitherto impossible from single limb bones or unassociated skulls. The wealth of information contained in these bones would take many years and many different specialists to uncover, but the search for our last ancestor had just taken a quantum leap forward.

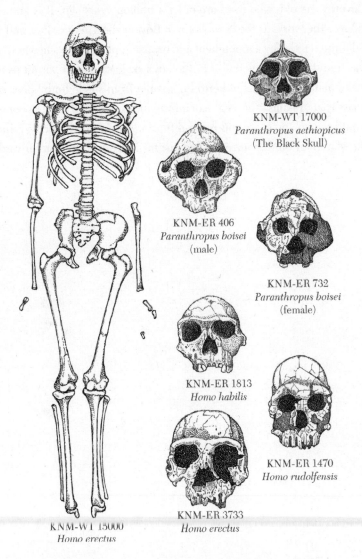

KNM-WT 17000
Paranthropus aethiopicus
(The Black Skull)

KNM-ER 406
Paranthropus boisei
(male)

KNM-ER 732
Paranthropus boisei
(female)

KNM-ER 1813
Homo habilis

KNM-ER 1470
Homo rudolfensis

KNM-ER 3733
Homo erectus

KNM-WT 15000
Homo erectus

Some of the most iconic finds we made in the 1970s and 1980s
(not drawn to scale).

4

CHANGING OF THE GUARD

LATE ON A THURSDAY AFTERNOON IN APRIL 1989 SHORTLY AF-
ter the one o'clock news bulletin on the state radio (which neither of
us was in the habit of listening to), Richard received an extraordinary
phone call from a colleague.

"Congratulations!" began the colleague. "Thank you. Whatever
for?" asked a bewildered Richard. "You must be joking!" replied the
now equally bewildered colleague. "No. What have I done?" queried
Richard bemusedly. Thus Richard came to know that he had been
handpicked by President Daniel arap Moi to head the body oversee-
ing Kenya's national parks and reserves, the Wildlife Conservation and
Management Department, effective immediately. I didn't believe it either when he called me shortly afterwards to share the news.

Richard's disbelief was not only caused by this unorthodox method
of learning about his new job. He had been engaged in a ferocious public spat with high-ranking government officials during the preceding
weeks, so the announcement was a bold and unexpected step by President Moi. At the time, elephants were being slaughtered in increasing
and alarmingly high numbers, yet the government had been categori-
cally denying that poaching was a serious issue. Never one to hide his
head in the sand, Richard had launched a scathing attack on the minister responsible for wildlife and tourism, George Muhoho, going so far
as to allege a high-level government cover-up of the scale of poaching.

This must be seen in its proper context: in 1989 Kenya, multiparty politics had not yet made its debut. Put plainly, a civil servant did not under any circumstance openly and publicly challenge the government, which was synonymous with the party line. With the whiff of corruption added to this scandalous departure from convention, the media had pounced on the story. As a consequence, the decimation of Kenya's elephant population was hitting the headlines daily, and gruesome images of the great hulking carcasses, their tusks violently hacked away, were regularly splattered across the front pages. The latest salvo from Right Honourable Minister Muhoho had just dismissed Richard's "cheeky white mentality," so the last thing we had been expecting was for Richard to be the very one tasked by the government to sort out the problem. But strange and unexpected things happen in life.

That afternoon, we hurtled home from work in complete silence, each of us deeply engrossed in our own thoughts. Although driving slightly faster than his usual high speed, Richard was nevertheless dodging the potholes in the narrow road with characteristically surgical precision. Our drives were usually filled with careless talk about goings-on at the office, news of colleagues abroad, and the endless perplexing questions about particular fossils that make our shared profession at once all-consuming yet sometimes highly contentious and frustrating. But Richard was a million miles away from humdrum museum politics and fossil bones, and the silence in the car was pregnant with possibilities.

Ever the strategist and never one to turn down a seemingly impossible challenge, Richard was already mapping out his first one hundred days in office and beyond. My thoughts were of a different bent — how this appointment would fundamentally turn our lives upside down. Twenty years had passed since I had first set foot in Kenya in response to Louis Leakey's ad in the *Times* — and nearly twenty years of a deeply fulfilling and successful partnership that had blurred the lines of work and play, home and office, husband and colleague so absolutely that the prospect of disentangling this perfectly meshed web that constituted our life together was quite simply impossible for me to imagine.

But the facts were unavoidable—Richard's new appointment as director of wildlife meant that he could no longer coordinate the palaeontological fieldwork at Turkana. He would now have other interests, another life, and even further constraints on his time. With a deep pang of regret, I realised we were unlikely to ever again share those indescribable magical moments that come from working so closely together in the field with a shared passion.

My thoughts then turned to my mother-in-law and how Mary's life had suddenly changed with her discovery of the australopithecine skull she called Dear Boy. With the increased funds and publicity resulting from this find, Louis no longer had the time to spend with her at Olduvai. Those halcyon years they had shared together were abruptly over. This may well have been on Mary's mind too, for when Richard broke the news to her later that evening, she uttered a few choice and succinct words, then retreated behind a wall of stony disapprobation that lasted for many months. Mary, quite presciently, was deeply worried about what the appointment might mean not just for our family life but also for Richard's safety and welfare. Later that evening, I called our teenaged children to tell them the news. At boarding school, Louise and Samira had been entirely shielded from the daily bombardment of press articles and had little inkling of how significant this would be to their lives. "Okay," they said—and that was that.

Mary's icy disapproval was not an option for me: Richard's excitement was palpable and contagious. Under his directorship, Kenya's national network of museums had burgeoned and had been running well for a number of years. He yearned for a new challenge. And I immediately recognised that with Richard's appointment new doors might open for me too. Someone would have to take over the leadership of the field research, and I was in a prime position to do so. The only thing holding me back was an awareness of how big the shoes I needed to fill were—Louis, Mary, and Richard had built a formidable legacy. I couldn't help but wonder if I was up to the challenge.

Yet, having digested the enormity of Richard's career change, I couldn't deny that there was a satisfying symmetry about the timing of all of this. Richard, a big-picture person, was most at home performing what other people would view as impossible tasks—finding Koobi Fora in the first place, exploring the gigantic spread of the Turkana Basin, raising the money and public profile we needed, and now, of course, turning Kenya's severe poaching crisis around. In our partnership, I always tended to the minutiae—the details of parenting, studying the fossils, and penning the research papers with our colleagues.

Under Richard's leadership, our work over the previous two decades had given us a good picture of the history of the Omo-Turkana Basin as a whole. Apart from the brief forays into the Late Miocene deposits during the early 1980s, we had concentrated most of our efforts on sites from the Late Pliocene and Early Pleistocene between two and one million years ago. This is relatively recent—a time when the human lineage had already long since differentiated from the ape lineage. But we had also identified a number of potentially important older sites and were at last beginning to connect the dots as to how disparate sites scattered around the modern-day lake relate to each other and reveal slices of time cut from a single geological sequence.

So perhaps the time was now right to change lenses, to zero in on a more micro level. At least another lifetime of exploration and discovery awaited me in near-virgin territory, palaeontologically speaking. I knew that each site might hold the secrets of different twists and turns in the story of our past. Wistfully, I acknowledged that I simply wouldn't have the time to do it all. But where should I begin? In my mind's eye, I reviewed what we then knew about the Turkana Basin and its geology, and thought that what I really wanted to explore was the earliest phases of human evolution—look for sites that predated all those we had been working in for the previous twenty years. But did those sites exist in Turkana?

· · ·

THE OMO RIVER has drained the Ethiopian Highlands for millions of years, flowing southwards and leaving in its wake a thick floodplain of sediments rich with fossils. The secret to the Omo Valley's perfection as a fossil site is the combination of these ideal conditions for fossilisation with repeated tectonic activity due to its location in East Africa's Great Rift Valley. As a result, faulting cracks, uplifts, and tilts the beds of sediments so the older layers buried far beneath the present-day surface are exposed again. And periodic volcanic eruptions leave convenient layers of ash called tuffs in between the fossil layers, which can be easily and accurately dated.

In the past, the Turkana Basin didn't exist. Some 4.2 million years ago, there was a great deal of volcanic activity and upheaval in the area, and we believe that this is what reshaped the landscape. As the Great Rift Valley was expanding, the earth's crust became dangerously thin, causing it to weaken and sag, and this led to the formation of a depression. In the weakened area, a narrow zone about twenty-five to thirty kilometres wide, there was a great deal of tectonic activity. New faults appeared, and huge outpourings of magma led to the formation of sheets of basalt that can be seen today all over the basin.

The Omo River flowed into this new basin, and over thousands of years, sediments gradually filled the basin until the inflowing water overflowed the southern rim and the river again continued its path to the Indian Ocean. New subsidence and tectonic activity in the ever-active Great Rift Valley then deepened the basin further, and the process of filling with water and silt began anew. Thus, due to a combination of climatic fluctuations and tectonic activity throughout the last 4.2 million years, a river system has dominated the Turkana Basin while being superseded for short intervals by a lake collecting in the depression. This process is ongoing to this day.

Looking at the lake today, it is hard to conceive that it is an occasional presence, not a permanent feature. This succession of ancient rivers and lakes through time is the reason the Turkana Basin provides a lifetime of places to look for fossils—for all along the migrating lakeshore and river floodplains, animals were dying, and their carcasses were becoming fossilised and buried, with new tectonics later exposing them for us to find. These sediments are called the Omo Group deposits. They lie on top of sedimentary deposits that were laid down before the Turkana Basin was formed. These older sediments extend back for a staggering amount of time and include a few that are 165 million years old, where dinosaurs have been found.

Along the Omo River, the sediments in the Omo Valley are conveniently stacked up, and as a result, the fossil deposits they contain are easy to correlate in terms of their relative ages. Back in 1967, these immensely rich Omo sediments were thought to be very localised. The maps marked much of the rest of the Turkana Basin as "lava" barren of fossils and therefore of little interest to palaeontologists. It was Richard who first noticed that the maps of the Turkana Basin were wrong. Much of the vast depression swathing 146,000 square kilometres is indeed devoid of fossils, but within this expanse, there are a number of pockets of sediments that are very rich in fossils. However, unlike in the Omo Valley, the sediments are not conveniently stacked up like a sandwiches on a tray. They are often many hundreds of miles apart. Compounding the challenges—although there are numerous ash layers from countless volcanic eruptions across the Turkana Basin—the volcanic horizons, so easy to spot in the Omo Valley, are often much harder to find and are not always present. And the same ash layer derived from the same volcanic eruption might look very different in two different locations, which makes it impossible to discern its identity through field observation.

While you have a whole proverbial elephant to look at in the Omo Valley, you might have only a trunk, a tail, or perhaps a toe in the rest of the basin. Putting the whole picture together and tying it back to the well-dated sequence in the Omo Valley proved to be almost impossible. Even when we did have a fossiliferous layer of sediments sandwiched between datable layers of volcanic ash, we could never be sure which of these tuffs derived from the same eruption, so we still could not correlate them with any certainty.

All we had to rely on for years were a series of maps that had been drawn in the 1970s when a professor from Iowa State University named Carl Vondra and his students came to Koobi Fora. Their interest was lithology—the waxing and waning of the lakeshore and river floodplains over the millennia—and they drew up a series of geological maps that we still use to this day. The shortcoming of these maps was that there was no way to tie them together securely and know if the lakeshore present in one area was the same age as that in another. Iowa State did not have a dating lab, so the maps had no dates associated with them. The only way we could guess at the correlation was to look at certain fossils that are really good age indicators because they change distinctively over time. The best of these are pigs, whose teeth have distinctly different characteristics according to the fossil's age. By comparing the Turkana fossils to similar specimens from well-dated sites, it was possible to narrow down the likely age of our fossil sites. But the pigs provided only a rough date range.

Our current accurate understanding of the different ages of the sites around the Turkana Basin and our ability to relate them to one another is largely thanks to the tireless work of Frank Brown. Frank was a remarkable man—and a brilliant geologist to boot. Frank began his lifetime association with the Turkana Basin in 1967 as a young PhD student participating in the International Omo Research Expedition in Ethiopia, a joint affair between Kenyan, French, and American scientists led by Richard, Yves Coppens, and Clark Howell. When he came to work with us at Koobi Fora in the early 1970s, Frank already had a far greater understanding of the sequence of volcanic horizons than anyone else.

Frank's genius was to develop a technique to analyse the ash layers for their chemical signature. It turns out that every single volcanic eruption, including different eruptions from the same volcano, has its own chemical signature, a unique composition of elements in the erupting magma and ash. This "fingerprinting" method finally gave us what we had long been looking for: a way to tie all the disparate groups of sediments in the basin to one another and the stacked sequence in the Omo Valley. Over decades of collecting ash samples from all over the Turkana Basin, Frank built a single picture of the area's geology across a remarkably wide breadth of time.

Right around the time that I took over from Richard, one of Frank's students at the University of Utah, Craig Feibel, also began to collaborate with us. He complemented Frank's work by putting together a geological history of how the palaeo-environments changed in the last few million years. The sedimentary deposits laid down by river and lake systems carry a history of the particular environment that existed at the time. By being able to read these rocks, it is possible to reconstruct the habitat to a remarkable degree. Between these two great geologists, and others along the way, a relatively complete history of the changing habitat and climate of the Turkana Basin over four million years became possible decades after we found the first hominin skull in the riverbed at Koobi Fora.

"I'VE DECIDED WHERE I want to work," I told Richard not long after his new job began. Richard, up to his neck in seemingly insurmountable problems, had very little time or inclination to think about fieldwork. Nevertheless, I was keen to bounce my ideas off him and tap into his experience leading a field exploration.

"I want to focus on the time when humans split off from the other apes."

"Nobody has ever found anything at all at that age in Turkana. You probably need to do what I had already planned."

"That's exactly why I want to look!" I retorted.

Given the enormous and intractable poaching and mismanagement problems he was facing, Richard simply couldn't generate any enthusiasm for my plans and was totally unable to share my excitement with me. Reluctantly, I realised that I was on my own in terms of making decisions and winning over the confidence of my inherited field crew.

I knew that the initial split between our ancestors, who walked bipedally, and those of the chimpanzee took place between six and five million years ago. This had been established back in the late sixties by Alan Wilson and Vince Sarich, who had literally counted the differences in the molecular composition of human and chimpanzee proteins to arrive at this relatively youthful time span. Proteins consist of long strings of amino acids, only some of which are vital to a protein's structure and function. So it follows that mutations that change the structure of the nonessential amino acids will not affect the way a protein acts in the body and will therefore not be eliminated through natural selection. Because these mutations occur at fairly regular intervals, they are molecular equivalents of units of time. Wilson and Sarich thus arrived at an approximate measure of the amount of time that had passed since we set off on distinct evolutionary paths away from a common ancestor. Since this pioneering work, molecular clocks have become extremely sophisticated, and instead of proteins, scientists can work directly on the differences in our DNA sequence — our genome and even the ancient genomes of Neanderthals. But the main insight of Wilson and Sarich has not really changed: humans and African apes have been evolving along separate lineages for some six million years.

Richard was correct. In the two decades that we had been working at Turkana, neither our team nor anyone else's had found any fossils to document this momentous evolutionary leap forward — the first apes to walk on two legs — or put forward the evidence for a sensible explanation as to why it happened when it did. Something monumental must have prompted this profoundly new evolutionary response in the human lineage that conferred an extraordinary advantage over previous adaptations. What this was and why it happened seemed to me to be the key to our subsequent evolution into the unique species that we are today.

I was more determined than ever to find that crucial evidence. Even more, I longed to crack the conundrum as to why it occurred. Charting the course of my research, I decided to hone in on that critical split between ape and human and work my way forward. From this great question, I knew that others would inevitably follow in a logical progression of currently unanswerable, deeply contentious questions right up to the birth of modern *Homo sapiens*. If I could answer any of these questions, then the answers would raise a slew of new questions of their own.

For that first season, my goal was to gain my team's trust, cut my teeth as a leader, and identify which fossil site was most likely to yield the answers I was seeking. I knew I would need to prove my mettle fast. In this field, no fossils means no funding. I needed some important discoveries under my own belt to ensure that the project's finances didn't simply fizzle out under my leadership. Fortunately for me, I had been granted a vital window of opportunity: Richard had recently raised enough funds for five years of fieldwork from the NOLS Outdoor Leadership Foundation and Amoco, which was exploring the Turkana area for oil. Five years on the job is a comfortable probation period—but fossil hunting is a treacherous, fickle business that relies heavily on luck.

I decided to begin on a set of exposures in West Turkana just south of the large seasonal Turkwell River. Since this was new territory, we had no aerial photographs; and procuring such photos entails a plane flying methodically back and forth to capture a grid of continuous images of the terrain below. Richard took our first set of aerial photos of East Turkana in 1970 on a shoestring budget. This was an uncomfortable and sweaty task for him as he perched with his camera above a small aperture that had been fashioned in the belly of the plane while Keith Mousley, a skilled bush pilot, flew back and forth, and kept the altitude and direction as steady as possible. At our first Koobi Fora camps, we all pored over the photos and patched them together to form a continuous area with only a few missing snippets of land. There are companies with specially adapted planes that do this for a substantial fee, and once we had the funds, we redid the East Turkana images professionally. But for the southern sites on the west side, I first needed to be sure that it was a justified expense.

Accordingly, I planned a series of quick survey visits to figure out the potential of the sites I was thinking of exploring. We would be based at South Turkwell, a site that we estimated could be 3.5 million years old. The steep and colourful cliffs bordering this stretch of the river make the forty-square-kilometre site a particularly stunning one and provide good relief on many outcrops—perfect for finding fossils. From Turkwell, short excursions would be possible to other promising sites. These included Loperot (an Early Miocene site where we hoped to find apes that lived some eighteen million years ago), Napudet (an extension of the South Turkwell exposures), Eshoa Kakurongori (a site that, like South Turkwell, might also give us early hominins), and Kanapoi (a slightly older site far to the south). I also planned to visit the site that most captured my imagination, Lothagam—an island of exposures that rises dramatically out of miles of low-lying, windswept desert plains between the Kerio and Turkwell rivers.

My first field season as the boss was a success. Everywhere we went, we found significant fossils, which encouraged us to return for longer, more concentrated surveys. At the end of the field season, I collected my courage, committed the funds, and arranged to have aerial photographs taken. Then I set my sight on Lothagam as the first place to look for our enigmatic early ancestors who first stepped out of the trees.

PART II

5

WATER, WATER EVERYWHERE

It was 1989, and I was embarking on my first exploration of the early exposures at Lothagam, determined to find evidence of those early bipedal ancestors. The site is stunning—a dark imposing mountain protruding in isolation from the surrounding plains and split north to south along the middle by a tectonic fault. Since it is one of only a handful of sites known in Africa that encompasses the critical age of around five million years ago when apes and humans split, Lothagam seemed an obvious place to start. An American team from Harvard University led by Professor Bryan Patterson had surveyed Lothagam in the 1960s. Among the many fossils recovered by Patterson's team is the tantalising mandible of an early human ancestor—proof that they were indeed there. But would there be anything left for us to find?

Lothagam was an even bigger gamble than usual because it is notoriously hard to find fossils there. Erosion at Lothagam happens at a glacial pace because the sandstone is so hard and there is little vegetation to pry open new cracks for the scant rainwater to worry its way through. But Patterson hadn't had the benefit of my newly inherited and highly skilled team of fossil hunters.

At midday on our very first day, not yet acclimatised to the ferocity of Turkana's soaring temperatures, I had been cheating. Limp from the fierce intensity of the noon sun, I had been hugging the banks of the Nawata River, a dry sand river that flows through Lothagam only during occasional rains, hoping to avail myself of the occasional shade offered from its steep banks as I made my way back to the others for lunch. Centuries of weathering have carved the deep red rocks into extraordinary shapes and deep gorges. While scenically arresting, they also serve like giant thermal sponges that radiate absorbed heat and block out any whisper of a cooling breeze.

But Lady Luck must have been with me. As I lingered at a shady bend in the river, I spotted a row of fossils peeping out of the sandstone bank at eye level. At my feet, a number of bones were collected in a small sandy depression in the riverbed. Looking more closely, I saw that the bones were unmistakably the fossils of an ancient carnivore. Pleasure, relief, and anticipation washed over me in an exuberant wave —the entire skeleton could well be concealed in the cliff face! Carnivores, being predators at the top of the food chain, are even scarcer than most other fossils, and so are all the more valuable. No doubt about it, this was a significant discovery, and it would earn me the merit badge I sorely needed to seal the respect of the crew in my first field season in charge of the expedition.

My carnivore gave me the final nudge I needed to commit to returning to Lothagam for a full season in 1990. First, though, we had to retrace our steps to some of the other sites we had surveyed in 1989, to collect important specimens and plot their locations on our new aerial photographs and begin to build a picture of that unchartered period when early bipedal hominins roamed in Turkana. We therefore only arrived in Lothagam in early August 1990 to begin what would be the first of five seasons of excavations and surveys. The work at Lothagam was remarkable both for the extraordinary challenges and the unimaginably rich rewards.

We made our camp on land belonging to Mr. Ekuwom, an older man with enormous charisma and an air of great wisdom. He had several wives, many children, and even more grandchildren. Kamoya explained that he was an elder who served as a sort of counsellor or advisor in the community and had become very rich because people seeking his advice paid for the privilege in goats and camels. This is what enabled him to have such a big family—the high bride price for Turkana wives (paid in the form of livestock) prohibits most men from taking more than one wife even though this cultural practice is considered a sign of high status among the Turkana to this day.

We first camped quite close to the northern end of the site in a small dry sand river with very little shade, lots of scorpions, and an extraordinary number of flies. When we finished prospecting the northern end, we moved our camp farther south. The new camp offered more shade and better protection from the wind, but it still served up the same quantities of dust and prodigious numbers of flies. At least one of these tenacious creatures inevitably found its way into my mouth and down my throat every day. The ghastly gagging sensation did not improve with repetition!

At camp, when we were not besieged by flies and heat, it was because of the strong easterly winds that blow off the lake and whip up great volumes of gritty, sandy dust that gets into everything. We spent half our time wishing the hot gale would cease. Then, when it occasionally did, the hordes of flies immediately made us wish it back again.

For a long progression of afternoons, I worked side by side in camp with Kathy Stewart, an easygoing Canadian expert in African fossil fish and a veteran of fieldwork. Together with our field assistant Samuel Ngui, Kathy systematically sampled the Lothagam fish over the course of three field seasons and unlocked vital clues about the ancient ecology from their minute and fragile remains. The two of us were never able to decide which was worse: the wind that threatened to blow our tiny bone fragments away or the mass of buzzing, tickling, pesky flies that made concentrating on the bones extremely taxing.

Richard had always dealt so competently with the logistical side of things that I had been able to focus single-mindedly on research. I am happiest with my eyes glued firmly to my particular obsession: fossils. When I was confronted with a mass of logistical and other practical obstacles, I initially assumed that I was merely experiencing the full brunt of responsibility for the expedition's safety and success for the first time. Even with a stalwart Kamoya soldiering by my side, it was an onerous burden. But years of subsequent fieldwork showed me that Lothagam regularly served up more testing trials than usual.

Bouts of sickness including several serious cases of malaria as well as various eye infections due to the dust and flies affected everyone in camp. These health issues were compounded by a prolonged period of drought that had drastically lowered the water table in the seasonal sand rivers. The water in a nearby waterhole was consequently brackish and brought on such a severe spell of diarrhoea and vomiting that we were all incapacitated. The only solution was to transport our water for drinking, washing, and cooking from Lodwar, some two hours' journey each way. With some twenty to twenty-five people in camp, we were consuming two and a half drums, or five hundred litres, each day despite the strictest economy. Although this constituted a considerable additional expense in terms of fuel, wear and tear on our vehicles, and time, there was no alternative if we were to stay healthy.

We even had an anthrax scare after many camels began suddenly dropping dead of disease. In Africa, anthrax does not arrive in a potent, concentrated form prepackaged in anonymous envelopes delivered maliciously through the mail. Instead, it lies dormant for years on the plains in the dried carcasses of the wildlife and livestock that died in a previous outbreak. It takes a sudden rain to bring the highly virulent anthrax spores to life. We bought our meat "on the hoof," negotiating with local tribesmen who sold us one or two sheep or goats at a time, which we slaughtered and butchered in camp. We therefore worried about what to do in the face of anthrax, but the veterinary officers at Lodwar assured us that goats were not susceptible to the disease. I had not known this before, so I later checked its veracity. Actually, goats can get anthrax, but they are less susceptible than other livestock. The Lodwar officer had been a bit too approximate for my comfort—and we were fortunate that no one caught this serious, occasionally fatal disease.

Without Richard's regular plane runs with fresh produce from Nairobi, our diet consisted largely of durable canned and dried goods, rice, maize meal, and great quantities of cabbage, potato, and onion. Laid out on tarpaulins in our shaded store tent, these humble vegetables kept for weeks. But camped so far from the lake with its bountiful supply of fish, we needed fresh meat for the variety and protein it provided our diet.

In addition to the responsibility I felt for keeping my team healthy in the face of these considerable risks, I was presented with a greater quota of vehicle breakdowns than usual. Communicating with Nairobi also presented challenges for me. Some mornings, we spent interminably long periods of time impatiently awaiting our turn to talk on the shared radiophone network used by missions, hospitals, and safari operators across the remote parts of Kenya. After the sun and the temperature soared high, Kamoya and I were at last granted our share of air time. Our patience was all too often rewarded with a message from the operator: "Sorry, Richard says he has had to go to an important meeting!" Sometimes, I reflected grouchily that I should adopt Richard's efficient policy of never allowing himself to be kept waiting for even a minute. But then we'd probably never communicate at all. I sorely felt his absence as well as the loss of the many practical advantages conferred by having a plane and pilot on the expedition.

Whenever we moved to a new site, one of the first tasks was to locate a suitable flat area and build an airstrip. This is usually hard physical labour involving every pair of hands and at least one hot day of hacking at thorny scrub, shifting large quantities of rocks, and a fair amount of digging. The objective is to level the many bumps and dips on the surface of the strip so the airplane does not land nose-first or belly-up in the dirt with a bent propeller. But the strip at Lothagam was one of the easiest and longest we had ever made. All we had to do was move a few large stones from the flat sandy plain and lay out markers of piled stones splashed generously with white plaster along the length of the strip so pilots could locate it from the air easily. Once the strip was prepared, Richard dropped in occasionally en route to some of the northern parks. The whole crew welcomed his visits, which were invariably accompanied by cartons and trays of delicious fresh fruit and vegetables as well as the inevitable and unequivocal advice on how to improve some aspect of our camp arrangements.

I treasured these weekends stolen out of his busy new life although they gave me a nasty inkling of the enormous stress he was under. Richard had successfully lobbied to have the Wildlife Conservation and Management Department moved out of the direct control of the Ministry of Tourism and stand as a separate entity, the parastatal Kenya Wildlife Service. This reorganisation had been conceived before Richard took over, but there was a lot of resistance to the idea of an armed service with a high degree of autonomy from the army and police, so the legislation had been held up in parliament for some time. But Richard took the view that without an effective ranger force the war on poaching could not be fought or won.

Richard flatly refused to talk about his many problems, and his reticence considerably stretched the new physical distance between us. From what little he did say, I gleaned that there was an all-out war to contain the poaching situation. There were successes, which he mentioned with a hard glint of satisfaction in his voice. Elephant and rhino killings were on a steady decline, and staff morale was up. A new fleet of vehicles was smartly painted in a distinctive olive colour now called "KWS green," and buried caches of ivory were being recovered regularly through improved intelligence. But there were setbacks too. Rangers were being shot dead in vicious combat, which Richard found enormously distressing. I also had the feeling that Richard was stepping on rather too many political toes for comfort. He needed an armed bodyguard and escort car, and at our home on the edge of the rift valley, elite, trusted KWS rangers prowled the premises twenty-four hours a day to keep him safe.

He told me about the new bright-red Soviet-style hotline phone that sat by his bedside and connected him to State House and a small circle of high officials. There seemed to be a degree of revolving membership involved in the red-phone list, depending on who was currying favour at any one time. And sure enough, Richard's own red-phone status would later prove to be a very accurate barometer of his political fortunes. The red phone didn't get nearly as much use as another new piece of equipment, a multifrequency VHF radio that linked Richard to the KWS headquarters and the control rooms of all the national parks around the country so he could be in constant touch with his wardens and rangers.

Our regular telephone had begun making a whole new repertoire of clicks and crackles. When this white noise got overly loud, Richard had conversations with the men who were wiretapping it. "Wouldn't it be better for all of us if I can hear enough to actually use the phone to talk with people other than you?" he asked them, and they proved surprisingly adept at prodding their peers to fix the problem. We enjoyed uninterrupted service for the first time in years. This seemed truly bizarre and ought to have rendered the phone tap pointless. But Richard judiciously fed these intelligence men scraps of information and misinformation, and discovered that phone tappers have their uses to both the wiretapper and the wiretapped!

I was being fed only fractions of facts too, no doubt to shield me from the grim, grizzly realities. I hated it. Feeling constantly torn between wanting to be in the field doing the work I love and wanting to be in Nairobi with my embattled husband was a new and unappealing aspect of fieldwork. What I didn't know was that this new tension in my life would only get worse in the years to come as Richard's career diverged further away from mine into the realm of politics and public service.

In spite of these testing times in our personal lives—along with the array of challenges thrown at me by Lothagam—the rewards of our work on early hominin sites were richer than I could have dreamed. The fossils were exceptionally difficult to find, but committed and patient searching was rewarded by beautiful well-preserved specimens.

Patterson's team cherry-picked the best of what they did find—but they missed a lot too. There were many thrilling days with discoveries of new and strange species and many others that had never before been found in the Turkana Basin. Some of the specimens were so complete that we could reconstruct the whole animal; others were less so, but they yielded exciting new details of the skull shape and dentition of numerous animals that shared the land with our remote bipedal ancestors.

As we became acquainted with the site and started to find fossils, we began plotting each specimen onto our newly acquired aerial photographs so we could return to collect them. I had arranged to have enlarged prints made of the two photographs that covered Lothagam, and these gave us surprising detail. We could distinguish individual trees and bushes, small stream channels, and landmark rocks. Even so, our fieldwork at Lothagam followed years of comfortable familiarity with the Koobi Fora aerial photographs and the years working at Nariokotome in the ever-expanding earth pit where we excavated the spectacular skeleton of the Turkana Boy. So perhaps it is not surprising that I initially had trouble orienting myself in the sixty square kilometres of the Lothagam site. I lack Richard's uncanny ability to always know exactly where he is; I couldn't effortlessly separate this particular patch of streambed, sediments, and thistle tussocks from the surrounding vast stretches of the bald terrain. So Kamoya and I spent many a long hot morning clambering to the top of the nearest vantage point to figure out where we were. Sometimes we had to repeat the exercise the next day after getting something wrong. This task became much easier as we grew more familiar with the landmarks, the geology, and the photos.

We also started to go through the records from Patterson's expedition to try to locate the exact position of his specimens on our photographs using his field-log descriptions. Patterson's team was not meticulous in their collection methods, to put it politely. They did occasionally use an aerial photograph with a grid measured on it, but there was no scale provided anywhere. I never did discover which photograph or what scale they used even though I devoted a lot of time and effort to trying. To locate the finds, we resorted to following the eyeball descriptions in their field log, which often proved to be a wild-goose chase. For instance, specimen 328–67K (a hippo tibia that was the 328th specimen to be collected in 1967) is described in their log as being retrieved from "TDS 100 ft N of 316." TDS is their shorthand for an isolated, curiously shaped rock that vaguely resembles a toadstool and forms a distinctive landmark. This entry put us in the vicinity of the so-called toadstool and roughly one hundred feet north of specimen 316–67K. The entry for 316–67K similarly reads "TDS 100 yards SE of 027." No closer to establishing where the original hippo tibia was found, we were then directed back to "500 ft SSE of TDS" in the log entry for 027–67K. And so we were referred seemingly ad infinitum to yet another specimen an approximate number of yards or feet away.

We managed to start disentangling this mess from entries where they helpfully mentioned some features that we could still recognise at the site, such as the so-called toadstool. But to make matters worse, it was clear that they had made mistakes—the catalogue occasionally describes something as being to the west of a particular feature when it could only have been to the east. This task was tedious and confusing —but important. We needed to pinpoint exactly which horizon Patterson's team collected the fossils from. A fossil without provenance—an associated point on a map with accurate geological information about its age—has no value for research. A few of Patterson's best specimens were marked directly onto a geological map that Kay Behrensmeyer had made, and we managed to decipher where some of the others were located based on the log descriptions, but for the majority, we never managed to figure out locations conclusively.

This loss was felt most keenly for the elephant data, which were studied by a palaeontologist on Patterson's team named Vince Maglio. Elephants are huge and robust, so they tend to erode out slowly and stay on the surface for years. Vince found and collected almost all of the best elephant specimens known from Lothagam—but for many of these, we simply do not have provenance data. We therefore had to make do without many of the most complete specimens for our study. The elephant story at Lothagam is one that would turn out to be one of the most interesting, and these beautiful but worthless specimens will always be the source of great frustration and regret for me.

One memorable day quite late in the season, Kamoya and I were putting the final few specimens on our photographs in the heat of high noon. We were in the Kaiyamung, a small area of fossil-rich sediments in the western reaches of the site, and we were using several of the small-scale photos since the enlarged ones did not extend this far. We had the hang of this at last. All the photos in the series were painstakingly marked with tiny pinpricks, each one representing a fossil whose accession number had been inked on the back of the photo.

A particularly strong and sudden gust of wind unexpectedly snatched the photographs out of my hands, and they took off in an up-current, over the next hill. We tore after them, gathering them up as we went, but the last photograph continued to elude us, snatched out of reach by the capricious wind each time that one of us bent to grasp it. We scrambled over rocky mound after rocky mound until we could grab the renegade photo. I looked at a flushed and dishevelled Kamoya—usually slow and deliberate as he meticulously scours every inch of a surface for bones. We simultaneously exclaimed with new-found respect and admiration: "I didn't know you could be so tough!" We collapsed in the shade, exhausted and laughing with relief that our precious provenances were safe. Back in camp, I carefully transferred all our specimen numbers onto tracing paper so I had some sort of second record should another dust devil or unanticipated misadventure catch us unawares.

This was not long before pricking a hole in a photograph as the sole record of where a fossil was collected would become quaintly archaic. We still use photographs to this day—but all of our provenance data are also now recorded with GPS coordinates and backed up on computer. It was while we were at Lothagam that I first came across global positioning. Richard flew in for a weekend with his chief pilot, Phil Matthews, who proudly demonstrated his new Trimble device—one of the earliest commercially available GPS models, which was huge in comparison to the pocket pieces that would follow. Coming hot on the heels of our epic chase up and down Lothagam's rocky peaks, a large portion of that day's entry in the field diary was devoted to extolling the virtues of this wonderful new machine and what a difference it would make if only we had one. But this was all still in the future.

One of the main intellectual challenges for me when we first arrived at Lothagam was to make sense of the many huge geological faults. These are, of course, one of Lothagam's greatest assets since the faulting is what has exposed such a long time sequence in a relatively small area. However, the faults do make it harder to interpret what's going on, especially for the uninitiated. "How can you tell which is up and which is down?" I wrote despairingly in my field diary after Kenyan geologist Patrick N'gang'a, Kay Behrensmeyer, and I had walked the entire length of the site for three hot, exhausting days. A case of too many cooks, perhaps, for they vehemently disagreed about whether we were picking up the main marker bed or something else as we followed it through the southern exposures. As for me, I no longer knew what to believe.

Lothagam's geology was first described by Bryan Patterson and William Sill during the 1967 American expedition. While still a student at Harvard, Kay did some further studies in 1968 when she was considering doing her doctoral thesis on Lothagam's geology. Kay had gone on to do her PhD at Koobi Fora, studying the processes of fossilisation (part of the field of study known as "taphonomy"), but she agreed to show us around Lothagam at the beginning of our first season of work there. Her excellent maps formed the basis of the work that Frank Brown's former student Craig Feibel compiled when he joined us for subsequent field seasons.

The problem for both Patterson's team and Kay was that they had enormous trouble finding suitable material for dating. Twenty-five years later, we encountered the exact same problem. The main marker bed, called the Marker Tuff, is a thick prominent bed of distinctive pinkish-red rock full of sparkling crystals with a secure date that showed the lower sediments were older than 6.54 million years. But other than this, dating material eluded everyone. When Craig later joined us, he searched high and low to find pumices suitable for dating but only found tiny bits of material that appeared to have been washed into small ponds. In desperation, he sent some of these off to Ian McDougall, an Australian geophysicist. Craig was banking on the slim chance that Ian might possibly be able to analyse these minuscule samples with newly available dating techniques and the new machines that had recently been installed in his lab. To our amazement and delight, Ian succeeded. He came back to us with a series of dates that indicated that the fossiliferous sediments at Lothagam range in age from more than 7.5 to less than 3.5 million years—exactly the time interval when the earliest hominins are thought to have split from our common ancestor with apes.

Armed with Craig's new dates and my newly trained eye, I would see a much clearer picture of Lothagam's geology were I to retrace my earlier steps with Kay and Patrick at the end of four field seasons. The site is an uplifted fault block about ten kilometres long and six kilometres wide. Two roughly parallel ridges oriented in a north-south direction protect lower-lying exposures in a central valley between them. These exposures slope gently down to the north and south from a sandy saddle of much younger deposits, which bridges the parallel hills to form a distinctive H shape. The site is named after the highest easternmost ridge, Lothagam Hill. In the Turkana language, *losthagam* describes something rough, varied, and heterogeneous, an apt description that eloquently captures the area's colourful assortment of rocks. There are extensive conglomerates, some with huge boulders and others with smaller stones, pebbles, and cobbles cemented together with sandstones. Basalt horizons cap the top and sandwich the conglomerate layers. One of these is a spectacular outcrop of columnar basalt, impressive in the perfect symmetry of the hexagonal pillars arranged in eerily neat rows.

Lothagam Hill is a horst, the geological term for a block of sediments left standing undisturbed while the sediments on either side are downfaulted. Consequently, although the horst is the highest topographical feature of the site, it is also the oldest, deposited roughly between 14.2 and 9.1 million years ago. Had there not been the disruptive tectonic activity, the oldest sediments would be underneath all the consecutively younger layers. But not at Lothagam, where oldest is at the top, and the subsequent layers of sediments are found not only at a lower level at the sides of the horst, but they are also tilted, split into separate units, and interrupted by erosional gullies. No wonder it was so confusing!

Craig renamed the ancient sediments older than nine million years the Nabwal Arangan beds after the *nabwal arangan* or "red waterhole," a muddy pool of startling rouge that sits in a gorge cutting through the horst. The only fossils recovered in the Nabwal Arangan beds were fossilised wood, so we did not spend much time prospecting there. The view from the top of the horst is spectacular, however, and sometimes we got up early on Sundays to climb it. The hill towers over the rest of Lothagam, and to the east, the lake glistens in the sun in ever-varying shades of jade, and on a clear day, one can also see the hills flanking the opposite shore. High, steep-sided sand dunes continually form along the base of the horst as the strong easterly winds dump sand from the vast surrounding desert. To the west is a panoramic view of the entire site between the horst and the parallel lava-capped ridge that forms the western boundary.

Cutting across the exposures to the west of the horst are successively younger exposures. The sediments are all tilted at a steep angle with long sloping pavements of hard rock terminating in almost vertical cliffs. Craig named the next oldest group of beds in the sequence the Nawata Formation. *Nawata* is a Turkana name for a distinctive type of grass that grows in the seasonal Nawata River, which drains the north-

ern half of Lothagam's central valley and cuts through these sedimentary deposits. The grass looks as though it is made from a cluster of
miniature bottlebrushes bound together at their base and mounted on
top of a long narrow stalk, the whole composition burnished a rich gold
by the desert heat. Apart from a few doum palms at the northern end
of the riverbed, the ornate *nawata* grass is virtually the only vegetation
in the hot, barren, sedimentary badlands.

Deposition of the Nawata sediments began soon after 9.1 million
years ago and continued for nearly four million years. Some eighty
metres above the base of the sequence, there is a distinctive layer of
bright brick-red sediment called the Red Marker. Craig divided the
Nawata Formation into two members, the Lower and Upper Nawata,
and picked the Red Marker to indicate the top of the Lower Nawata
sequence. The Upper Nawata begins with a layer of volcanic ash, which
is the Marker Tuff that we were following that memorable day with
Kay and Patrick, and lies directly above the Red Marker. The volcanic eruption that produced the ash of the Marker Tuff happened 6.54
million years ago, giving us lower (9.1 million) and upper (6.6 million)
chronological boundaries for the Lower Nawata beds. Another very
distinctive marker bed, the Purple Marker, defines the top of the Upper Nawata sequence. Ian McDougall's analyses gave us a good series
of dates for the Lower Nawata beds that allowed us to identify the age
of the fossils very precisely. But we had enormous trouble identifying
dating intervals for the Upper Nawata beds. Apart from the Marker
Tuff date of 6.54 million years at the base of the sequence, there are no
other secure dates for the Upper Nawata beds, including for the Purple
Marker tuff. This dearth of dates is terribly frustrating—all the more
so because the Upper Nawata was a time of great upheaval.

The sedimentary layers in the Nawata Formation are mostly conglomerates, mudstones, and sandstones. Tellingly, the sandstones are
"upward fining"—with coarser particles at the bottom and finer ones
at the top. This is the typical deposition of a slow-flowing river system,
where the largest, heavier particles settle out first. With the buildup of

sediment on the edges of this ancient meandering river that ran across Lothagam all those millions of years ago, the outer curves of the river-bed got shallower, and the river shifted course slightly as it cut a new and deeper channel. The water on the edge slowed down, and progressively finer materials settled on top of the coarse layers laid down by what was once the fast-flowing centre of the channel. As the river wove its way back and forth on the floodplain, it cut a new channel through sediments laid down earlier; the pattern that the sediments form is called braiding for its resemblance to a plait we might braid in our hair.

Differences in thickness and composition of the sedimentary layers show that there were fluctuations in the volume of water and the speed of the river throughout this time period. Massive fossil reefs of the Nile oyster, *Etheria elliptica,* laid down in sandstone channels are common throughout the beds of the Lower Nawata, and their presence demonstrates that the river complex was flowing all year round as these molluscs thrive only in well-oxygenated fresh water. We also found thin limestone layers with minute bivalve molluscs called ostracods in them. Interpreting the rocks, Craig told us that this used to be a lush riverine environment with broad shallow channels, back swamps, and oxbows formed as the meandering river cut new channels. The landscape was a mosaic of floodplain savannas dissected by gallery woodland tracing the river's path, which provided a rich and bountiful habitat that could support a large diversity of life. During most of this time, a large perennial river was flowing, but the fossil soils showed that during two intervals, around 6.7 million years ago and 5.2 million years ago, deposition of sediments slowed practically to nothing, which suggested intervals that were considerably dryer.

Above the Nawata Formation beds, a new set of sediments, the Apak Member, form the earliest sequence of the Nachukui Formation. The geology of the Apak is a bit more complex than that of the Nawata. Thick beds of multistory deposits, each characterized by coarse-grained sandstones at the bottom and built up with progressively finer materials and mudstones towards the top, tell their own story. This was also a river system, but one that was faster flowing and with less back swamps. The absence of the Nile oyster suggests that the river might have been seasonal, and Craig believed it represented a different drainage system from the Nawata, perhaps ancestral to the modern Kerio River that currently lies to the east of Lothagam.

At the end of the Apak interval, some 4.2 million years ago, there was a volcanic eruption that led to the creation of the Lothagam Basalt, which tops the easternmost of Lothagam's parallel ridges. Just before the volcano that spurted this basalt erupted, the river system that had persisted through much of the Apak was replaced by a lake. There are impressive beds of distorted, squashed snails where the red-hot ribbon of thick magma encroached inexorably on the lakeshore and preserved them for posterity.

As you clamber up the exposed lake beds at the top of the Apak onto the Lothagam Basalt, you are rewarded with a spectacular view across the younger beds to the west. The ancient lake, which begins in the uppermost beds of the Apak Member with the squashed snails, continues through the Mururongori Member. These lake beds are important because they correlate with the distinctive olive-green fine clays found all around the basin. They are all that is left of the huge Lonyumun Lake, which marks the beginning of the sediments in Omo-Turkana Basin as we know it today. I'll come back to the story of their geology later on for we come across the waxing and waning of a lake in the basin over and over again in the years to come. At any rate, with the geology of Lothagam unlocked at last, we were fast learning that most of the evolutionary action was going on before the vast Lonyumun Lake swamped the landscape some 4.1 million years ago.

The sediments in the Nawata Formation and Apak Member contain the majority of fossils, and the story that the fossils tell echoes Craig's interpretation of the geology. From the vantage point of modern-day Lothagam's bare and barren rocks, it is hard to imagine such a diverse and watery world. Hippos were everywhere—they turned out to be the most frequently preserved mammals when we tallied up the fossils at the end of four field seasons, making up 27 percent of the mammalian fauna. And no other East or Central African fossil locality rivals Lothagam for the diversity, abundance, or quality of preservation of turtles and crocodiles.

Turtles from the Nawata Formation include at least six different species. By far the most common was a brand-new genus and species of side-necked turtle, *Turkanemys pattersoni*, so named because these turtles fold their necks sideways within the protective shell rather than tucking them in as other turtles do. The type specimen for this turtle was found by Patterson's team. It is exquisitely preserved and almost complete, with the skull, mandible, and most of the skeleton tucked inside the shell. We found countless other specimens of this turtle and were initially puzzled by the fact that almost all of them were adults of roughly the same age. This pattern was very similar to that seen today at modern nesting sites, where large numbers of adult turtles congregate to breed, always returning unerringly to the same place year after year.

Five species of crocodilians also rubbed scaly shoulders, several of which grew to formidable lengths. The most common was a very large broad-nosed crocodile. The biggest skull and mandible we found turned out to be roughly a metre long when the pieces were reconstructed. Only the year before, Samira had been assigned a statistics project, and it just so happened that she had picked crocodiles as her subject. Studying different types of modern crocodiles, she worked out that the species all have similar ratios between head, body, and tail lengths probably because the weight must be evenly distributed for the animal to be able to move efficiently, especially in water. This entailed some rather perilous practical data collection perched on the walls separating the holding pens in the reptile park at the museum in Nairobi while juggling camera and measuring stick. My support for the project suffered rather grave doubts when she came to my office telling me how close she came to falling in when a crocodile lunged at her! We plugged our big specimen into her ratio to see how long it might have been. The result was a whopping twenty-one feet—about the same size as big male saltwater crocodiles in Australia, the largest reptiles alive today.

It is mind-boggling to think that this aquatic environment could support so many of these large formidable predators. They must have had a high degree of specialization in different ecological niches because there must have been a remarkably abundant and diverse range of prey for them to eat. Among this bountiful prey were large numbers of fish. In the Lower Nawata beds, the fossil fish were all small-sized species that eat other fish and vegetation and thrive in well-vegetated bays and swampy shallows. In the Upper Nawata, these fish become less common, and by the Apak, the fossil fish represent species that prefer more oxygenated water consistent with a faster-flowing river system. We also found a number of birds, which, apart from ostriches, were all species at home in an aquatic environment. The terrestrial fauna included many species typically inhabiting well-watered, well-vegetated habitats, and we were finding a considerable diversity of monkeys that would have thrived in the swamps and the gallery woodland along the river.

All the evidence pointed to an aquatic habitat heavily influenced by changes in its water system. All the Turkana Basin fossil sites are associated with an ancient water course. Without the water, there would be no sedimentation—and therefore no fossil record. But something bigger and more interesting seemed to be afoot at Lothagam. A pattern was unfolding and repeating itself across the main aquatic groups. We found archaic species, often making their last appearance in the Nawata, living alongside modern species that were making their debut and becoming more common in the later sediments. Kathy found that all the small-sized fish she recovered from the Nawata Formation were archaic species, whereas these archaic elements disappeared or were rare in the Apak Member; extant genera predominated there instead. The side-necked turtle that was using Lothagam as a nesting ground is another case in point. This curious-looking creature belongs to a lineage of side-necked turtles that is represented today by only one species on Madagascar, and until the Lothagam fossils turned up, nothing at all was known about its ancestry. *Turkanemys pattersoni* represents the last species we know of from this lineage on the African continent, and it lived alongside the earliest-known fossil of the modern soft-shelled turtle, *Cycloderma frenatum.* A second extinct soft-shelled turtle is also known only from Lothagam.

The pattern repeats itself with crocodiles. The only crocodile alive today in Lake Turkana is the Nile crocodile, *Crocodylus niloticus,* and we found the earliest-known record of this species at Lothagam, which was virtually indistinguishable from its modern descendants. The other modern crocodile that persists today only in Lake Tanganyika, *Crocodylus cataphractus,* also turns up for the first time at Lothagam. These two crocodiles lived alongside now extinct species, the long-snouted *Euthecodon* crocodile, the giant, broad-nosed crocodile whose length we estimated using Samira's ratios, and a new species of gavial, *Eogavialis andrewsi.* Gavials are found today only in India and Pakistan, and the Lothagam gavial represents a new geographic record for these crocodilians that were previously unknown in East Africa.

The bountiful hippo fossils also told an intriguing story. Modern hippos belong to two genera, both found only in Africa. *Hippopotamus amphibius* is the scientific name for the widely occurring sub-Saharan hippo. *Choeropsis liberiensis* is a forest-dwelling pygmy hippo now confined to the lowland forests of West Africa from Guinea to the Ivory Coast. In the common modern hippo, the eyes, ears, and nose are positioned high on the head, so when the hippo is submerged, they protrude just above the water line. It is this morphology that allows *Hippopotamus amphibius* to still be able to see, hear, and smell above the water. The most commonly occurring hippo in Lothagam is *Archaeopotamus harvardi.* This creature shows the beginning of the modern configuration: although positioned close to the top of the skull, its ear, eye, and nasal sockets are not as high as in modern hippos. A second much rarer and smaller hippo of the same genus, *Archaeopota-*

mus lothagamensis, also lived during the Nawata interval. The ancestral *Archaeopotamus* lineage eventually became extinct and was gradually replaced by a hippo that has long been thought to be the earliest known species of the modern genus: *Hippopotamus protamphibius,* which was first found at Lothagam in the Apak Member.

A massive change was clearly underway that caused a wave of extinction and a generation of new species. The turnover was not limited to a particular group of animals but repeats itself again and again. Why was this? What did such a change signify? We needed to systematically study all the groups of animals we were finding to see how widely the pattern was recurring and figure out what was driving it. The teeth from the *Archaeopotamus harvardi* specimens seemed to indicate that its diet might have been changing over time because more recent specimens had progressively smaller incisors. Might this be a clue? An even more intriguing question then arose — was this change caused by a local event, such as localised tectonic activity, or was some larger, more global change underway five to six million years ago? If this was global, then could it be the same change that led an early ape to habitually walk upright rather than on all fours?

Working at Lothagam with its logistical and geological complexities, and at a time of personal worries because of Richard's new political life, was daunting. Yet it was also exhilarating. Our discoveries proved to be critical to understanding the environment in which our early bipedal ancestors lived and when major upheavals in the environment of Turkana created opportunities for new species to evolve. With this new understanding, I could now look again at our maps of fossil exposures and plan our next step in search of the hominins themselves.

6

A BRAVE NEW WORLD

FROM THE OUTSET, IT WAS ABUNDANTLY CLEAR THAT I NEEDED reinforcements at Lothagam. New fossils were pouring in. We kept running out of plaster of paris to encase the fragile bones for their journey back to Nairobi, and were it not for Richard's visits to camp when he invariably departed in a considerably heavier aircraft than he arrived in, I honestly don't know how we would have safely transported them all back. The wealth of knowledge in these bones was astounding, and I needed help from specialists to sift through it all.

This was a distinction of scale rather than scope: from Richard's very first expedition in 1968 when he led the Kenyan team during the trinational exploration of the Omo region of Ethiopia, we had always benefited hugely from close collaboration with other scientists. But the possible roads of enquiry from Lothagam seemed almost limitless. Eventually, this translated into a decade of rigorous, collaborative research with experts in a wide array of subjects. I derived great satisfaction from the fruits of this labour — a big fat monograph of our findings inspired by (and, I like to think, fully worthy of) Mary Leakey's exacting standards in her own monographs of her work at Olduvai Gorge and Laetoli, the site where she found the world's most famous set of footprints.

Kathy Stewart first joined us in our second season in 1991 when she and I spent so many memorable hours battling the wind and the flies as we sifted through our bone fragments, and she returned each subsequent year in the field. Eleanor Weston, a PhD student at the time, first came to Lothagam in 1992 to study hippos. From the fossils laid out in the wooden trays under the shade of the mess tent, Kathy and Eleanor noticed the patterns emerging in the fish and hippo fossils early on. Other experts later scrutinised different aspects of the collection back at the lab in Nairobi. What we really wanted to know was: is the change so strikingly underway in the aquatic realm evident among the terrestrial mammals too? If so and if this proved to be a global rather than a local climatic event, it could point to the impetus behind our own evolution into bipedalism.

John Harris, a soft-spoken, hardworking, and careful scientist with a perpetual nervous energy that I suspect was at least partly fuelled by his fondness for endless cups of strong black coffee, joined us in these studies. He had worked with us from the very beginning—starting at Koobi Fora in 1968, where he later fell in love with my sister, whom he subsequently married. He and Judy are no longer together, but our scientific relationship endured. He worked with us all over the Turkana Basin and knows the fauna intimately, so he was ideally suited to the task. I had to work hard to persuade John to compile the Lothagam monograph with me, however; having helped Mary edit the Laetoli volume, as well as two of the volumes on our research at East Turkana, he knew precisely how much work was involved and had vowed never to do this kind of thing ever again. It was fortunate for me that he changed his mind. John not only studied the fauna and coauthored many of the chapters but his previous editorial experience was also indispensable as my coeditor.

John and I took on the pigs together, and he tackled his specialty, the bovids, along with some of the other herbivores. I shared the monkeys with Mark Teaford, an expert in the interpretation of primate dental wear and diets, and Carol Ward, who had worked with me previously in studying monkey skeletons. Sure enough, all these mammal groupings seemed to fit into a pattern. Species that were common at earlier

Miocene sites took their curtain call at Lothagam. And another telling denominator was also soon apparent: there were key differences in the teeth of the ancient and modern groupings. Through time, individual teeth of many of the herbivores became more complex with additional cusps, higher crowns, and more intricate enamel patterns. These adaptations all increase the resistance of the tooth to wear by abrasion, so they are typically found in mammals that eat grass. This is because grass contains minute silicon capsules called phytoliths. Phytoliths are extremely abrasive and wear teeth down with ruthless efficiency. Once an animal's teeth are worn out, it can no longer feed. So for grazers, adaptations that increase the longevity of the tooth are very important for the reproductive success and long-term survival of the species.

Now that we knew what we were looking for, it was obvious. Among the mammals, all the primitive representatives tended to be browsers feeding on leaves: a primitive elephant (*Stegotetrabelodon*), a hornless rhino (*Brachypotherium*), primitive browsing antelopes (*Tragoportax*), and a primitive giraffid (*Palaeotragus*). Living alongside these ancient creatures were much more modern-looking herbivores—most with teeth suited to eating grass—which appear in the fossil record for the very first time at Lothagam. The "moderns" are all close relatives of mammals familiar today on the East African plains. They included early elephants as well as the ancestors of modern rhinos, giraffes, and pigs.

The pigs (*Suidae*) gave a textbook illustration of an evolutionary trend in dentition towards increasingly complex, high-crowned molars. Several lineages of suids replace one another serially through the ages in the Turkana Basin, and pig teeth are so distinctive that they are excellent indicators of the age of newly discovered sites before they can be dated by more precise methods. In the various species within each lineage, the length and height of the third molars increase significantly with a corresponding reduction in the size of the premolars. At Lothagam, changes in pig teeth were particularly clear in the different species of *Nyanzachoerus*. The earlier species, *Nyanzachoerus syrticus,* is the most common in the Nawata Formation. Through time, its third molars become increasingly more complex, adding extra cusps to the back of the teeth and increasing the height of the crowns. In the later Apak Member, the common pig is *Nyanzachoerus australis.* This species has even longer, more complex, and higher-crowned third molars than *N. syrticus.*

Bryan Patterson found many mandibles of *N. syrticus,* both male and female. His team also found a complete and beautifully preserved male skull with enormous knobs on either side of its head and strong ridges along the top of the skull. The female skull remained elusive, however, and I was always very curious about whether the female skull would resemble the ornate male one. One day, the field crew told me that I should look at a large pig skull sitting on the exposures. Other than the complete hominin skull Richard and I found on our camel survey in 1969, this was the only time I ever saw a complete skull of a large mammal totally exposed on the surface with little damage to the bone. A female *N. syrticus* skull was lying on the rocky outcrop as if someone had recently put it there for us to find. And it lacked the striking decoration of the males.

These two key insights from Lothagam—that ancient animal species were being replaced by the ancestors of modern ones around five million years ago and that in many mammals this meant a shift from browsing to grazing, and thus the evolution of taller and more complex teeth that take longer to wear down—are beautifully reflected in the evolution of elephants. Pascal Tassy, a charming, soft-spoken Frenchman and the best expert to be found on elephant evolution, helped us to untangle the elephant story at Lothagam. Elephants have evolved their own elaborate system of tooth replacement, which uniquely solves the problem of how to prolong the longevity of their teeth against the daily grind of tough fibrous vegetation. Modern elephants have a single molar in each quadrant of the mouth that is composed of a number of parallel plates. After the eruption of the first tiny molar before the elephant is born, these teeth are replaced five times during its lifetime, each replacement tooth being larger than the previous one and with

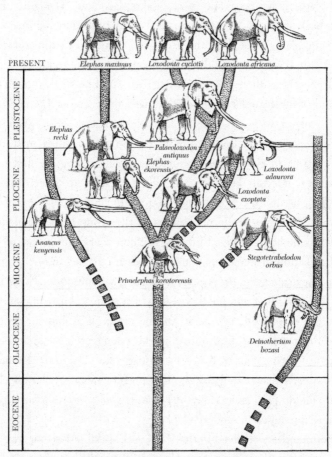

PRESENT — Elephas maximus — Loxodonta cyclotis — Loxodonta africana

PLEISTOCENE

Elephas recki

Palaeoloxodon antiquus
Elephas ekorensis

Loxodonta adaurora

PLIOCENE

Loxodonta exoptata

Anancus kenyensis

MIOCENE

Prinnelephas korotorensis

Stegotetrabelodon orbus

OLIGOCENE

Deinotherium bozasi

EOCENE

TURKANA BASIN DURING THE MIOCENE

At Lothagam, the evolution of proboscideans towards serial tooth replacement is clearly documented. This solution allowed for a change in diet towards tougher fibrous vegetation to meet the challenges of a changing habitat during the Miocene and ensured their survival.

more plates, in a process known as serial replacement. The teeth erupt at the back of the jaw and grow forward as the preceding tooth is worn down and plates break off at the front—so instead of wearing down all their teeth as they grow and age, elephants wear one tooth at a time, which lengthens their chewing lives. The evolution of this ingenious method of tooth replacement can be partially traced at Lothagam—and it evolved twice, an example of parallel evolution down two different families of elephantoids responding to the same adaptive pressures.

The story of elephant evolution begins long before the first sediments were deposited at Lothagam, and it is easiest to conceive in the form of a family tree like the one you might find in the front of old family Bibles. Ancestral proboscideans—the group to which modern elephants belong—evolved fifty-five million years ago as part of the great radiation of mammals that followed the extinction of the dinosaurs. These ancestral proboscidiens begat palaeomastodons, which in turn begat the first gomphotheres in Africa some twenty-four million years ago. Gomphotheres are an ancient group of proboscideans that had elongated jaws with primitive, thickly enameled teeth and a cumbersome arrangement of two tusks in each jaw. They are significant in our story because they are the first proboscideans to show a clear enlargement of the molars and the beginnings of an evolutionary adaptation that resulted in modern elephants.

Gomphotheres thrived in the Early and Middle Miocene (twenty-three to ten million years ago). Up until then, the most recent East African fossil (*Choerolophodon ngorora*) was thought to be from this time. Thus, to my very great astonishment, Pascal Tassy one day showed me a single very primitive and worn lower molar of a "trilophodont" (meaning "three lophs," the cross plates that make up elephantoid teeth) gomphothere from the Apak Member at Lothagam. This new Apak Member tooth is about five million years younger than the Early and Middle Miocene. It is now the very last known occurrence of a trilophodont gomphothere in East Africa and adds to the evidence of the late survivorship of ancient species in the Turkana Basin.

But this ancient gomphothere was not the only one of its kind at Lothagam. It was already competing with no less than four other proboscidean species. By far the most common elephantid at Lothagam was the primitive-looking *Stegotetrabelodon*. This curious creature had cumbersome tusks both in its upper and lower jaw like the gomphotheres, but its thickly enameled teeth seem to show the beginning of serial tooth replacement. The second group are the resilient deinotheres. These rather uncommon elephant-sized proboscideans had tusks in their lower jaws instead of their upper jaws and did not have serial replacement of their much simpler teeth. They are the exception in the Late Miocene elephant world and remained browsers throughout their evolution. But the real evolutionary challengers were the other two groups that had already evolved—in parallel—more efficient grazing teeth through serial replacement.

The first of these is a proboscidean called *Anancus*. This creature had teeth with very thick enamel and fewer tooth lophs than modern elephants but is otherwise very similar to them: the mandible is similarly shaped, and the teeth serially replace over the animal's life. The fact that *Anancus*, like modern elephants, also evolved a means of living longer by wearing down its teeth one at a time at the end of the Miocene shows how much grittier and tougher the diet of elephants at the time had become.

But the biggest discovery was yet to come. One day, I was lucky enough to find the earliest known specimen of an ancestor of modern elephants—known as *Palaeoloxodon*. I climbed a small peak towards the western side of Lothagam, where I came across the lower jaw of a young elephant partially exposed on the surface. I could see that it had been eroding from the sediments for some time as there were hundreds of broken fragments lying nearby. I was surprised that this jaw had teeth closely resembling those of a modern elephant. It had big molars with many plates in a short jaw showing obvious serial replacement. But it was also clearly from the Upper Nawata sediments. Over the following days, I carefully reconstructed as many of the fragments as I could before excavating the hidden parts. This mandible would turn out to be hugely significant because it showed that the modern group of elephants evolved in the Late Miocene more than 5.5 million years ago.

Lothagam has the youngest gomphothere, an ancient form evolving serial tooth replacement (*Stegotetrabelodon*), and two new lineages—*Anancus* and the oldest true elephant (*Palaeoloxodon*)—that had independently evolved teeth that allowed them to graze effectively. Amazingly, the precursor to the modern African elephant, *Loxodonta,* with its distinctive "lozenge" shape of the lophs on its teeth, makes its first appearance in the Apak Member—making Lothagam the site with the greatest diversity of known proboscideans. With the exception of the ecologically conservative and rare deinotheres, the diverse species of proboscideans at Lothagam had all found the same way to maintain their large bodies on a diet of abrasive grass. Quite independently, they evolved hard-wearing and serially replacing teeth. Once again, we were seeing the disappearance and replacement of primitive species by the earliest species of modern genera as well as the temporal evolution of increasingly high-crowned molars, thinner enamel, and complex occlusal enamel patterns that dramatically increase the cutting surface of each tooth.

But why were we seeing such a significant ecological shift with clear, widespread adaptations to grass eating instead of browsing? What happened to herald the end for so many Miocene mammals and usher in a brand-new era? And could the early evolution of our hominin ancestors be driven by the same adaptive pressures?

As so often happens in science, somebody else was puzzling over the same quandary but through quite a different line of enquiry. The man who had also picked up on this phenomenon was our old friend and colleague, Thure Cerling, an easygoing geophysicist with a vast and eclectic knowledge accumulated over years of fieldwork that has taken him to all corners of the globe. Our association with Thure stretches all the way back to Richard's first Omo expedition in 1968, and he is always a pleasure to work with in the field. He wrote to me in 1992 that he'd been "analysing carbon isotopes in fossil teeth fragments of herbivores. I have exciting evidence of a major change in the vegetation in the Siwaliks at about seven million years, and I want to see if I can detect similar changes in the isotopes at the same time in East Africa." The Siwalik deposits are in Pakistan, and although the majority are older than those at Lothagam, some cover Lothagam's sweeping age span. "Please do join us, as soon as you can!" I replied. "This is a perfect coincidence —we seem to be picking up a similar signal from the tooth morphology in many of the herbivores at Lothagam. But how on earth do you detect such a major change in habitat just from tiny little fragments of teeth?"

Taking me back to one of my first biochemistry lessons, Thure explained that all cells have a carbon base and that plants manufacture their carbon through a process of photosynthesis. Capturing carbon dioxide from the air and combining it with water, plants use energy derived from sunlight to split the oxygen and hydrogen in the water and reconfigure the atoms as oxygen and glucose (glucose being the simplest hydrocarbon, which is the building block of all cells). Dredging back to my days at the polytechnic, I did remember the basic formula for photosynthesis, $CO_2 + H_2O + \text{light energy} \rightarrow (CH_2O)n + O_2$. But it's not that simple! A number of intermediary steps are involved between throwing carbon dioxide, water, and light together and producing sugar and oxygen.

There are actually three types of photosynthesis adapted to different conditions, and as they have an important bearing on our story, they merit a bit of explanation. The most common type is the one I learned about in school—C_3 photosynthesis. With less machinery (fewer enzymes and no specialised anatomy), this process is the most economical: C_3 plants use the same enzyme, rubisco, to both latch hold of the carbon dioxide and process it during photosynthesis. But it turns out that rubisco is only efficient at getting hold of carbon dioxide in conditions of moderate heat and light. At higher temperatures, the enzyme also latches on to oxygen, and the reaction creates a different compound the plant does not need, causing the plant to waste precious energy.

C_4 photosynthesis is an adaptation to avoid this predicament in hotter climates. C_4 plants use a different enzyme to get hold of carbon dioxide, which it then delivers to the rubisco enzyme in an inner cell for the photosynthesis process. Because this enzyme is more efficient in carbon dioxide uptake, the plant can keep its pores open for a shorter time, thus cutting evaporation rates and saving on scarce water. A key difference in this process is that the intermediate step creates a four-carbon molecule as opposed to the conventional three-carbon sugars first created in the C_3 pathway. In cooler climates, it is most economical for plants to use the C_3 pathway, but above a certain temperature threshold, the additional expense of C_4 reaps greater efficiencies, which makes it worthwhile. Thure also told me about CAM, a third photosynthetic pathway that is ideally suited to arid conditions because the plant is able to shut down operations completely during prolonged dry spells. But because CAM is restricted to fewer plant species, mostly succulents, it needn't concern us here.

So how does all this biochemistry relate to the changes we were witnessing at Lothagam? Put very simply, we are what we eat. An herbivore that derives its basic building-block carbon from plant material will pick up the photosynthetic signature of the plant matter it ingests, and this is preserved in the tooth enamel, hair, and hooves. By measuring the relative composition of the carbon isotopes, we are able to discern the proportion of C_3 to C_4 plants in the animal's diet. Because most of the trees and shrubs in sub-Saharan Africa today are C_3 plants and the majority of grasses and hedges are C_4, this ratio neatly translates to the proportion of leaves versus grasses ingested by the animal. It turns out that this signature is preserved perfectly in fossil tooth enamel. Thure had been analysing the teeth from modern mammals coupled with dietary information established by direct observation as a control group for comparing the fossils.

This is just brilliant! I thought. A tiny fragment of fossil can now yield invaluable indicators of the animal's diet. Similar analyses of carbonates in fossil soils indicate the dominant vegetation at the time, and the ratios of two different oxygen isotopes can tell us about the sources of drinking water the animal used. All these analyses can greatly add to our understanding of the climate and habitat millions of years ago. Once I had grasped the basics, I couldn't wait to hear what Thure had found in Pakistan. I eagerly pored over his papers as I waited for him to join us in the field.

Thure's research in Pakistan documented a dramatic shift in the biomass away from plants photosynthezing with the C_3 pathway to a dominance of C_4 plants some seven million years ago. We don't know for certain what caused this change in vegetation, but we do know that the relative efficiencies of C_3 or C_4 pathways at different ambient temperatures is related to the concentration of carbon dioxide in the atmosphere: below a certain CO_2 level, the rubisco enzyme simply isn't very efficient at pulling carbon dioxide out of the air. If Thure's observations of a sudden expansion of C_4 biomass in Pakistan were repeated at Lothagam, this could indicate a warming event that was global rather than localised and could be the impetus that led our own ancestors to adapt to a changing environment. We know that CO_2 levels in the atmosphere had been steadily decreasing during the Late Miocene. If Thure was right, I couldn't help but wonder what this might mean for the havoc we are causing with our current runaway carbon emissions. Will this affect our global food supply? As a species, we are almost totally dependent on a terrifyingly small number of cash crops for the bulk of our food supply: just three cereals now provide about 60 percent of plant-based human energy intake. Over the last century, some 75 percent of plant genetic diversity has been lost. Out of some 250,000 to 300,000 known edible plant species, only 150 to 200 are used today. Metaphorically speaking, we don't have very many eggs in our basket!

Thure finally joined us for a couple of days in the field at the beginning of August in 1992, and I excitedly put the whole field crew onto finding tooth fragments for him. Focusing on the time interval when we expected to detect a change, we searched systematically, starting at the base of the Lower Nawata and walking up-section through each time interval through the Apak. Thure made detailed notes on the geology and provenance while I collected the specimens and plotted them on the aerial photo. He was amazed at the field crew's skill and how much they found in such a small time—after a day and a half, we had ninety-eight specimens! The sample provided identifiable fragments of the dominant herbivore taxa from each time sequence: elephants, rhinos, horses, giraffes, pigs, hippos, and antelopes. A fringe benefit of Richard's new position was that he now had easy access to bones of modern herbivores from a wide range of habitats across Kenya's network of parks—most notably from the mounting stacks of skulls and mandibles of elephants lost to poaching. A number of Richard's wardens kindly contributed to our research with tooth fragments that significantly bolstered our control sample of modern herbivores.

Thure worked with John Harris on the analyses back in Utah, and we couldn't wait to get our hands on the results. When they finally came in, they were fascinating. As Thure had suspected, they tallied neatly with what we were seeing in the morphology of the pig and elephant teeth. As in Pakistan, many of the species changed from dominantly browsing to dominantly grazing diets during the Nawata, and the analyses of the fossil soils showed a corresponding replacement of C_3 plants with C_4 grasses as the dominant vegetation. It was clear that there was a dramatic change in the vegetation at this time. Subsequent work in South and North America has shown the same signal in horses, so it seems that this change was indeed global in nature. Such a widespread change in habitat at low latitudes would have completely altered the world. And, as new feeding opportunities opened up and old ones diminished or disappeared, this would certainly have driven both the mass extinctions and the evolution of numerous species we were witnessing at Lothagam. Our analyses showed that the animal groupings did not all undergo this transition at once; it was a phased progression instead.

The charge was led by the horses and the proboscideans, with the horses the first to change to a largely C_4 diet. The pigs were the slowest to fully exploit the new C_4 dietary resource. In spite of the apparent changes in suid molars that would have favoured a grazing diet, pigs remained mixed feeders through the Nawata Formation. Not all animals became grazers, however. Extensive forests and woodlands remained, and some animals, like the deinotheres, continued to exploit these habitats. Among the species of rhinoceros represented at Lothagam, two strategies were successful—we find the earliest occurrence of the modern white rhino, a grazer, as well as the earliest occurrence of the black rhino, which remained a browser. This complex ecological partitioning of diets is best illustrated by modern bovids that show the whole range of feeding strategies. Hypergrazers such as the oryx, hartebeest, buffalo, and kob are at one end of the spectrum, and hyperbrowsers at the other extreme include kudu and duikers. Mixed feeders such as impala and gazelle successfully exploited both food types.

The implications of this change in habitat and the timing of it are mind-boggling. The massive faunal turnover occurred at the end of the Miocene between seven and five million years ago. This corresponds exactly with the time that the human lineage is estimated to have split from that of the apes. Could the change to a new "C_4 world" in Africa have provided the initial stimulus for bipedality?

I HAVE ALWAYS been convinced that the main driver for such a momentous change must be first and foremost related to feeding. Any slight variations that help give an individual animal a competitive advantage in finding food, defending against predators, and gaining a mate are strong selective pressures in evolution because individuals must reproduce for a species to continue. And those variations related

to feeding are paramount because individuals must obviously first survive to adulthood before they can breed. These selective pressures act
most strongly in times of stress and hardship because an animal that
cannot feed efficiently will not live long enough to pass on its genes nor
will it breed successfully during prolonged food shortages.

Charles Darwin did not believe that natural selection could be witnessed in real time as he was convinced that changes visible to the
human eye must accumulate over many generations beyond a human
life span. But in the famed "Enchanted Isles," where the beginnings of
Darwin's revolutionary theory of evolution were spawned, the extent
that climatic fluctuations exert selective pressures on a population is
crystal clear. In the Galápagos, there is a little island called Daphne
Major. Here, the drama of the struggle of life over death is played out
daily on a stage that is pared down to the minimum essentials.

On this lava-strewn crater of an island, there is a population of
finches small enough to count and tag but large enough to provide a
statistically robust sample. The finches are absurdly easy to catch because they have no fear of humans, and they have highly variable beaks
that are easy to measure within a fraction of a percent. The finches'
ability to forage successfully depends almost entirely on the shape and
size of the beaks that nature endows them with, and the beak a particular bird is endowed with is highly hereditary. Then there are the seeds
that the birds eat, which vary in size and toughness but belong to few-
enough species that researchers can easily recognise and count exactly
what type of seeds individual birds are eating. They can also relatively
easily calculate the density of seed mass on the island at any one time.
Through continuous meticulous research in this near-perfect natural
laboratory since 1973, Peter and Rosemary Grant and their team of
finch watchers have witnessed both natural and sexual selection unfold
before their very eyes in what is quite possibly the greatest show on
earth. They were able to prove through solid, unassailable data, what
Darwin could only impute by logic. Little idiosyncrasies, such as a tiny
variation in the shape and size of a bird's beak, can literally be the dif-

ference between life and death, survival and extinction. In Darwin's own words, "the smallest grain in the balance, in the long run, must tell on which death shall fall, and which shall survive."

The Grants' study of the Galápagos ground finches would later be eloquently described by Jonathan Weiner in his Pulitzer Prize–winning book, *The Beak of the Finch*, a riveting and crisply detailed account of evolution in action. As I read the Grants' scientific articles while I was pondering the complete transformation of Lothagam's long-ago climate, what struck me most was the irrefutable correlation they found between the size and shape of the finches' beaks with climatic fluctuations.

The Grants first arrived in the Galápagos during the wet season, and the finches were enjoying four good years of decent rainfall and plentiful food. The finches concentrated on two dozen species of seed and spent half of their foraging time eating seven favourite soft seeds and fruits. This confusing picture is what threw Darwin off the scent some 140 years earlier, for he also arrived during the wet season. If all the finches are eating the same food, why does the large ground finch, *Geospiza magnirostris,* have a large deep beak and the small ground finch, *Geospiza fuliginosa,* have one about half as deep and half as long? And why is there a medium ground finch with a medium beak, *Geospiza fortis,* occupying a middle zone? If Darwin is correct, all these variations should eventually zero out into an "average" middling bird with an average, medium beak?

The Grants got their first inkling of why this was not happening when they came back in the dry season. Now the birds were devoting only one-thirtieth of their time to the seven species that they had favoured in the wet season. On a "struggle index" of seed hardness and size, the average seed now ranked 6 compared to 0.5 three months earlier (the toughest seeds to crack approached 14). And the volume of food had decreased by 84 percent. With less to go around, the finches had all become specialists. Cactus finches now dined almost exclusively on cactus, and the large *G. magnirostris* focused on big heavy seeds.

Even within a single species, the medium *G. fortis* ground finch, the pattern was the same: each bird according to its beak. *Fortis* finches with relatively big beaks focused on big seeds, medium individuals ate medium seeds, and small-beaked birds ate the smallest seeds.

Then the study got really interesting. During the 1976 wet season, Daphne Major received 137 millimetres of rain, and the resulting density of ten grams of seed per square metre sustained a population of more than a thousand *fortis* finches and three hundred cactus finches. But in the following year, there was an epic drought, and only twenty-four millimetres of rain fell in the whole year. The cactus finches bred following the first rainfall, but not a single fledgling survived past three months. The *fortis* did not breed at all. The seed density dropped dramatically to three grams of seed per square metre. There was very little food, and between eleven a.m. and three p.m., the black sunbaked lava was too hot for the birds to forage. By January 1978, there were fewer than two hundred finches alive, a grim survival rate of one in seven. The Grants found that the oldest and the biggest birds had lived while the younger generations were practically wiped out. Surviving *fortis* were, on average, 5 to 6 percent larger than their dead brethren. Before the drought, the *fortis* population's average beak was 10.68 mm long and 9.42 mm deep. After the drought, the survivors had an average beak 11.07 mm long and 9.96 mm deep. Half a millimetre or less is the seemingly infinitesimal difference that decided their fate. One drought and one generation were enough to show that natural selection was much swifter and stronger than Darwin himself had ever imagined.

But Daphne Major had more rich surprises to offer the Grants. There was another key feature to the roster of survivors: many more males made it than females because males are on average bigger than females. When at last the clouds gathered and fifty millimetres of rain fell in a single torrential deluge, a breeding frenzy began. But each female had six males to choose from, and choose she did. The females deliberately picked the biggest birds with the biggest beaks as mates.

Their offspring had even larger and deeper beaks. Sexual selection accentuated the power of natural selection during the drought. For the next four years, this trend towards ever-larger birds with bigger and deeper beaks was perpetuated.

Then El Niño arrived in 1982. Every three to six years, the warming of the waters of the Pacific Ocean wreak havoc on worldwide weather patterns, and the Galápagos sit in the eye of this upheaval. On Daphne Major, sheets of water sluiced down the steep slopes of the volcano, and more rain fell than any time in living memory. The bare, depleted lava burst into leaf and flower, yielding such a bumper crop that there was twelve times as much seed as the year before. But it was too wet for the arid-loving cactus, so the big-seed crop crashed as the small-seed supply boomed. In such a time of plenty, this didn't matter overly much to the finches, who feasted indiscriminately. Their numbers climbed by over 400 percent, soaring to more than two thousand individuals by June.

The following year, only fifty-three millimetres of rain fell and a paltry four millimetres barely dampened the earth the year after that —and now the big-beaked *fortis* were the ones in trouble. The finches had overshot the island's carrying capacity, and small seeds were in greater supply than big ones. Imagine a room with grains of rice strewn across the floor. Give two teams a bowl and an implement to gather their dinner. You would sleep on a far fuller tummy if you were on the team armed with tweezers rather than pliers. And so it was for the finches. The biggest birds, needing the most calories to fuel their big bodies and having the least dexterity in their big beaks, were now at a disadvantage. Big-beaked big birds were dying, and small-beaked small birds were flourishing. And more males were dying than females. Nature had neatly and ruthlessly reversed the trend of the previous years. The "normal" distribution of the size of the finches and their beaks swung back again according to the whims of a fickle climate.

But what if the climate hadn't swung back again? If the dryer spell had persisted, perhaps the small-beaked finches would have become extinct. Might a new giant-beaked finch have taken its place on the famous family tree of thirteen Galápagos finches? This was the scenario at Lothagam between seven and five million years ago. There were extreme fluctuations in climate with dramatic swings between wetter and dryer times, the evidence for which we see today in the changing nature of the ancient river deposits of the Nawata and the Apak. But there was also an inexorable drying trend—the conditions never fully swung back. The implications of such a change were enormous and far-reaching—there would have been ramifications all the way down the food chain as new habitats and feeding opportunities presented themselves and were exploited by new species.

If, as the evidence suggests, a major radiation of C_4 grass species replaced many of the previously dominant herbs and shrubs, this would have created a large number of new feeding niches. Thousands of species of grass-eating animals, both large and small, would have followed. Although we find mostly large animals in the fossil record, this change would primarily have affected the many smaller creatures, including caterpillars, butterflies, beetles, birds, lizards, frogs, and many others whose remains would have decomposed before fossilisation could occur. But this radiation of new grass-eaters would have created new feeding opportunities for both omnivores and carnivores higher up the food chain—including our own ancestors. This could be the elusive catalyst that we had been searching for that might have stimulated a radiation of early human ancestors exploiting new open-country habitats and these new food resources.

To my enormous frustration, we found tantalizingly little evidence of hominins at Lothagam in spite of the abundance of other fossils. At 3.5 million years, we found two whole teeth and two fragments of teeth. After intensively searching for four seasons, we found only two more isolated teeth dated somewhere around five million years during the crucial Upper Nawata times. At this same age, there is the lower jaw fragment found by Patterson's team in the 1960s, but this specimen is too incomplete to assign taxonomically to either ape or human with any certainty. But there is no evidence whatsoever for hominins in the older sediments at the bottom of the Nawata—and I had chosen Lothagam specifically because this was the time frame where I desperately hoped to find them.

We can, nevertheless, speculate that, like the other fauna, these early hominins were exploiting the upheavals in the food chain to take advantage of new rich dietary opportunities. For instance, we know from the tooth morphology of more recent and better-preserved, hominins that early human ancestors probably ate fruits, insects, birds' eggs, and sometimes small mammals. But at five million years, when they first appear in the record at Lothagam, did they walk on all fours like an ape? Or were they the first intrepid explorers who stepped out of the trees on two legs?

BECOMING BIPEDAL is the pivotal event that enabled further evolutionary changes that set humans apart. No longer bound to the physical demands of walking quadrupedally, the forelimbs were free for other tasks, and this enabled the development of a dexterous hand over time. Manual dexterity vastly improved the efficiency with which an individual could gather food, enabling even a young infant to easily put berries and other hard-to-reach foods in its mouth—but, crucially, it also allowed for further developments later on, such as the manufacture of effective stone-tool kits and the ability to carry and store foraged food. All of these improvements in food uptake are what eventually enabled human ancestors to grow larger brains, which are among the most calorie-expensive organs in the body. Thus, bipedality was the key adaptation that led to the divergence of the ape and human lineages. Such an altered landscape does proffer the incentives for a momentous alteration in ape locomotion to an upright stance. However, without the fossil evidence, I could only conjecture that the sweeping changes in all the other fauna would have affected the apes too.

Modern apes are not very good at walking on the ground on all fours. For example, modern chimpanzees, who walk on their knuckles, expend some 35 percent more calories getting from A to B on the ground than a similarly sized quadruped such as a large dog. So it is probably not very surprising that they traverse the ground for less than a mile or so each day. But the changing habitats we were witnessing in the vista of Lothagam's geology must have significantly reduced the available forest cover in East Africa, gradually transforming the landscape into the mosaic of floodplain savannas dissected by gallery woodland along the rivers.

A consequence of such a widespread reduction in forest cover would have been a significant decrease in foraging opportunities in the forests. If our early ancestors were arboreal and moved on the ground inefficiently like modern apes, then the ability to better cover open ground would have had vast potential benefits. Moving into the newly evolved grasslands would have dramatically increased their foraging opportunities, but these earliest hominins would still have needed to be able to retreat rapidly back into the trees to escape dangerous carnivores. Thus, as forest area gradually diminished, there would have been a strong selective advantage to moving efficiently in the increasingly wide-open spaces—in other words, to walk bipedally.

So why were we not seeing any hominins in Lothagam before the five-million-year mark? I could think of only two possibilities. It simply couldn't be that my team had missed the evidence, after searching so hard. They were too good for that. One possibility was that the hominins first evolved to suit the new landscape somewhere else and only colonised the area around Lothagam at about five million years. After this time, we begin to find increasing evidence of their presence elsewhere in the Turkana Basin. The other possibility was that they were at Lothagam all along but only in small numbers because they were prey to voracious carnivores that kept their population small and that scavenged most of their remains. Maybe the evidence is now all but gone even though the hominins were there.

There were indeed several large and ferocious predators during the Nawata. The carnivore I found in that scorching bend of the river on our very first 1989 survey is a good example. When we excavated this specimen, we unearthed a near-complete skeleton of an animal similar to a wolverine. This would have been a fearsome, strong, and muscular predator. There were also several species of sabre-toothed cats at Lothagam that would have made easy pickings of ape-sized prey. The most common of these cats is well-known from another skeleton that we first spotted in 1990 while walking up the dry sandy bed of the River Nawata. All that was exposed on the surface were several foot bones encased in extremely hard rock that were barely protruding out of the cliff face high above us. A large part of the base of this cliff had broken off and provided a wall that protected the fossils from being washed away by fast-flowing floodwaters whenever it rained. In 1992, we excavated and sieved the sediment that was trapped by this wall. We were rewarded with many bones of the skeleton as well as some broken pieces of the skull.

This made us even more certain that we would find more remains if we could follow the hand bones into the cliff with an excavation. But how could we do this? It would mean erecting scaffolding, removing the top of the cliff, and digging into the hard rock. In 1993, Alan Walker supervised this challenging excavation. What was most interesting about this leopard-sized felid was the large claw on the forepaw that presented a formidable cutting weapon. When Lars Werdelin eventually studied the skeleton, he gave it the name *Lokotunjailurus emageritus,* a name derived from the Turkana words for "cat" and "claw."

This sabre-toothed cat would certainly have given the early hominins at Lothagam good reason to fear exposure on the ground. A possibly even more serious threat was *Dinofelis,* a similar-sized felid with a smaller canine and less specialised chewing teeth that crops up throughout the Nawata and Apak members. All in all, there must have been strong evolutionary incentives to risk becoming an easy meal by leaving tree cover.

Numerous theories have been put forward to explain the strategic change in the locomotor pattern of our ancestors to bipedality. The new stance permitted a greater variety of sexual and aggressive displays, and allowed our ancestors to see over high grass and other vegetation. Hard-to-reach foods that were unavailable to competitors could be accessed from the higher vantage point of two legs: by standing upright, our earliest ancestors could have reached farther to pick fruits, berries, and seeds and search for insects or pluck eggs from birds' nests. Strong seasonality was a feature of the weather patterns at this time, and seasonality leads to times of food scarcity. Individuals that had access to more food in times of drought would have had a strong selective advantage (witness the change in a few generations of finches on Daphne Major). Adaptations to avoid falling prey to the formidable assembly of carnivores must have played their part as well. Moving upright allowed early humans to regulate their body temperature by reducing the surface area exposed to the hot African sun—but this thermoregulatory advantage only kicked in once they were above a certain height. In all likelihood, it was most probably a combination of many factors that led to the evolution of bipedality in human ancestors. These theories are hard to prove, but it is likely that feeding was a key factor and that the opening of new feeding niches through habitat change was a significant driving force.

Even without much evidence of the hominins I badly hoped to find, Lothagam was a palaeontologist's paradise. It gave us a remarkably comprehensive snapshot of a period in our prehistory that was hugely significant in our evolution. Over the course of five years, we found fossils that provided an exquisitely detailed faunal record, and we deciphered the geology and the corresponding evidence for habitat change, which must surely have played a key role in opening new feeding niches for the earliest hominins to exploit.

But in spite of all this, I was on the clock, and I was running out of time. My five years of funding were nearly up, and I badly needed to find some of our ancestors. Our donors like "sexy" finds, and none of the fabulous wealth of knowledge we gained at Lothagam was head-line-grabbing material. "Mass extinctions seven million years ago; forests turn to grass" really didn't cut the mustard! (Although now, given that a new wave of extinction is upon us, this might be an easier story to sell.) There are few alternative sites of a similar age in Kenya, and none of them are as richly fossiliferous as Lothagam. I therefore began to consider sites close to Lothagam of a slightly younger age, hoping that one of these might shed further light on the emergence of bipedality. But as I pondered where to go next, I was interrupted by another huge setback in our personal lives that would change things forever.

7

||||||||||||||||||||||||||||||||||||||||||||||||||||

NINE LIVES

WEDNESDAY, JUNE 2, 1993, DAWNED MUCH LIKE ANY OTHER DAY in the field. We wiped away the remnants of our breakfast and set off into the exposures of the Apak member just as the flies awakened and the last tinges of pink and orange faded from the sky. I shook off the unpleasant vestiges of a bad dream that, unusual for me, I could still remember in the morning. Just before the field season began, I bought Richard some new fancy swimming fins to use when we went snorkeling during our regular holidays to our home in Lamu on Kenya's northern coast. They were bright yellow and black, and he really liked them — but in the dream, he couldn't use them because he hadn't any feet. I am not superstitious, and I don't believe in prophetic dreams. I only remembered this peculiar and unpleasant coincidence long after the horrors of the days to come when I was back in Nairobi gazing miserably at these gaudily cheerful pair of flippers.

The morning passed with nothing more remarkable than the discovery of a nice horse tooth. I busied myself at one of the carnivore excavations after lunch. But later in the afternoon, the droning hum of a distant plane materialized into something completely unexpected — a KWS aircraft buzzing the camp to alert us of its imminent arrival at the airstrip. My heart sank into a leaden ball in my stomach; only bad news could have brought the plane here. Communication through the radiotelephone was extremely inefficient, and more often than not, we failed to make contact with Nairobi. If anyone had to get a message to us in a hurry, the only sure way was to fly into camp. The first shivers of premonition had me rushing to the car and over to the airfield. Premonition turned into outright alarm as I saw the KWS chief pilot, Phil Mathews, and another pilot and good friend, Sacha Cooke, climbing out without Richard and lifting worried faces towards me.

It turned out that Richard's engine had cut out on a short flight to Naivasha, and he had crashed his plane over a thick forest. He had managed to find one small clearing to land but had snagged a large mango tree as the plane glided headlong toward the tiny gap in the trees. He was alive, as were his passengers, Phil and Sacha reassured me, but they told me that I must immediately return to Nairobi. All they would say is that he had a broken leg and a broken nose. I rushed back to camp to gather a few belongings and scribble a note to Kamoya, and I was soon on my way back to Nairobi.

Phil and Sacha remained tight-lipped, but as I pried a few more details out of them and news filtered through over the KWS radio, the accident began to sound more and more serious. Yet I still could not bring myself to believe that Richard might be badly injured. Time and again he has been likened to a cat with the proverbial nine lives, getting himself out of seemingly impossible situations and reaffirming my deep-rooted faith in his ability to survive. This wouldn't be any different, I told myself firmly, as the fault lines of the Rift Valley unfolded far too slowly below us. Arid dusty plains gradually changed to steep escarpments and then, at long last, to the green patchwork of fields and dotted huts on the fertile highlands that give way to the development on the outskirts of Nairobi. The familiar landmarks had never, ever passed this slowly before.

We finally landed just before dark, and I rushed into the hospital as Richard was about to be whisked away into surgery. The surgeon spoke to me as he scrubbed up, briskly informing me that both Richard's legs were severely damaged although he did not think there were other serious injuries. No, I couldn't see Richard yet. "Go home," he advised. He was planning to clean the wounds that were still full of earth and grass. I could phone the hospital or come back in four hours as they certainly wouldn't be through with the surgery before that. Hugely frustrated at not being able to see Richard or learn the true extent of his injuries, let alone make any informed decisions for him, I reluctantly left and went to see Mary at her home in Langata before rushing home to relieve Sacha, who had been valiantly manning the phone that was ringing off the hook.

When I finally got to see Richard later that night as he was being wheeled back from the operating theatre, I was shocked. He was in intensive care, his face a swollen mess, and his head and legs heavily bandaged. At least I could be thankful that his lungs, ribs, kidney, and spinal cord were undamaged. His four passengers had varying and relatively minor injuries but none as serious as Richard's, but his bodyguard did have a broken arm. Richard had already been given a blood transfusion, which worried me as the hospital had not asked us for blood. Although the blood was screened, I would have preferred them to use blood from a close friend or relative as is common practice to this day in the country given the high incidence of serious diseases like HIV. Infuriatingly, my own blood, correctly typed and safe, was sitting unused in the hospital's blood bank! I had donated some as a contingency for Joyce Poole, an elephant researcher, neighbour, and close friend of Richard's, who had given birth shortly before this.

At last, the doctor came to talk to me about Richard's injuries. He didn't mince any words. He blandly described Richard's right leg as "pulp" and added that both his ankles were very badly broken. His left foot had been dislocated, and many of the nerves had been severed. During the surgery, the doctor had put a screw into the left talus (anklebone) to stabilize it and removed dead bone from the right leg. I was thankful at least for my anatomy training as I could follow the full extent of these horrible injuries. The next couple of days confirmed my initial perception—that the doctor's bedside manner was simply dreadful. Adding to my worries, a number of people privately expressed doubts about his being the right surgeon to tackle Richard's injuries and strongly urged me to get a second opinion. As I questioned people further, it became evident that there were tensions between this doctor and other members of the medical team. While none of the medics questioned the doctor's ability, they clearly did not get along well. I had deep qualms about whether the team could keep a patient's best interests paramount if they were not sharing information with one another but constantly bickering. Then the hospital—and the whole country—ran out of the pain-killing drug Pethedine (Demerol). Richard was incredibly sick and he was in unspeakable pain. For the second time in our married life, our fate was in the hands of doctors and nurses, and in Richard's sheer determination to persevere against the odds.

The days passed in a slow-motion blur. We turned away a constant stream of visitors and fielded endless phone calls that Richard was far too sick to take. Flowers from well-wishers piled up outside the door as they couldn't be brought into the sterile IC unit, and we sent them on to other patients. As they had more than a decade earlier, friends and colleagues rallied around, even offering to raise money to help offset our mounting hospital bills. Richard's brother Philip and I privately consulted with Dr. David Silverstein, a well-respected and prominent physician at the hospital who was also the president's private doctor. He was emphatic that he would seek the best trauma centre in the world if he were in our shoes—and that the level of care and expertise was simply not high enough in Nairobi to treat Richard's extensive injuries.

Despite pleas from the whole family, Richard was adamant about wanting to stay. Not only was he worried about the cost of travelling, but he also wanted to continue working. He insisted that all the things he had been building at KWS would crumble if he left. He started to refuse pain medicine so he could keep a clear head, and the hospital room became jumbled with office equipment as he valiantly tried to run KWS from his bed. For a man like Richard who was used to being in complete control, his total dependency on the hospital staff not only rankled, it was also unbearable. Continuing to work was his way of fighting back and a key to his very survival just as it had been more than two decades previously.

Nevertheless, he was truly sick, and I worried about his insistence on staying put. I despaired at my own lack of control in the situation. How would I be able to make important decisions when I lacked the medical knowledge about the true extent of his injuries? I worried that, in his state of great pain and dulled by drugs, Richard might stubbornly keep trying to put the country's interests before his own, which would not be in anybody's interest if it ended up killing him.

At the hangar containing the wreckage of the plane, Sacha was shocked and puzzled. The engineers were still trying to identify what had gone wrong. Sacha had thought that he would be the next person to fly the plane when he was scheduled to fetch me from Turkana later that week. The plane had just passed its service with flying colours, and Sacha had checked out the plane two days earlier with a veteran pilot who worked for the flying doctors at the African Medical and Research Foundation. The two of them had put the engine through all its paces, doing spins, turns, stalls, and all the other things pilots do to check an engine thoroughly. Everyone remarked that this fantastic little engine was working perfectly. So how could it abruptly die only seven minutes out of Nairobi?

Richard's antipoaching drive and strong stance against corruption had earned him any number of enemies, and we were both aware of the death threats that had been made against him. The spectre of sabotage loomed large in our minds, and we were desperate to find a simple mechanical fault to set our fears to rest. Although the result might be equivalent, somehow a nasty accident was far easier to stomach than a failed sabotage attempt. Many man hours were spent trying to figure out what went wrong, and parts of the engine were sent to specialist engineers in the United States. But because we were unable to rule out sabotage, we moved Richard to a more secure wing at the hospital, and bodyguards were deployed outside his room.

We never did find out what caused the engine to fail over the forest. Phil later told me that the AMREF pilot who had first located the crash site had told him that the plane was so badly damaged that there was no hope of anyone coming out alive. The first onlookers at the scene had apparently concurred—everyone's possessions, including Richard's gold Rolex watch, were stolen and their unconscious, broken bodies left unattended until a good Samaritan took most of the passengers to the nearest hospital in his pickup. Richard was evacuated in a helicopter, and I shudder to think about the pain he endured on that trip to hospital. But before his evacuation, Richard also rose to the occasion through his shock with his usual grit and determination, calmly directing people to get his staff to the hospital and arranging everything as though he were fine. Looking at the photos of the wreck later, I had to marvel that everyone had survived. The photos showed a twisted, gruesome tangle of metal, with one mangled wing still suspended from the gnarled old mango tree that had slowed their fall. This was all that remained of the faithful little workhorse that had clocked most of its miles running supplies and scientists to Turkana.

Things came to a head on June 8. Richard was now out of intensive care, and I was sleeping in a cot beside his bed. He had a terrible night after reacting to some sleeping pills he'd been given. He'd been hallucinating and trying to clamber out of the bed by throwing everything out of his way and disconnecting the cannula in his wrist. "Where am I, why am I here?" he kept mumbling. My efforts to soothe him seemed to make him even more agitated. "Why are my legs so heavy, and why can't we cut them off?" he presciently demanded, and he wanted to know why he was tied to the bed and whether or not he was a prisoner even though all that restrained him were bandages and IV tubes. It was simply dreadful to listen to him and be unable to help. The nurses eventually got hold of a doctor who prescribed a tranquilizer. This stopped him from trying to break free of the bed although his deep distress and endless questions continued. It was one of the longest nights of my life, equaled in length only by one some thirteen years earlier when Richard clung precariously to life as his body fought to reject the newly transplanted kidney given to him by Philip.

The president of Kenya had planned to come by to see Richard and discuss some pressing matters about KWS early the very next morning. But I knew that in his current state, Richard would not be able to have a coherent conversation, so I asked Charles Njonjo, one of our closest and oldest family friends and Richard's greatest confidante, to arrange for the appointment to be postponed. But perversely, when Richard woke up later, it was with a clear head. He remembered nothing at all of the night or his earlier state. He was absolutely furious with me! Never in the course of our whole marriage, he said, had I done anything so inimical—he needed that meeting for reasons I could not possibly begin to understand. Then, to make a thoroughly horrid day far worse, he started to have temperature spikes. After his kidney transplant, we had been warned by his doctors in England that we should watch for such signs as they were possible indications of serious problems with his kidney. Both of us worried that a severe infection might cause complications and irreparable damage.

When the president later came to visit, he personally assured Richard that he would not allow people to take advantage of his absence to push through untoward dealings or undo the progress he had been making at KWS. I was forgiven, and Moi's public support, together with our fears about his kidney, finally persuaded Richard to accept the generous help we were being offered from friends. Without further ado, he telephoned Prince Claus of the Netherlands, a dear friend. The prince arranged for an English surgeon, Chris Colton, known for his successful treatment of a polo injury sustained by Britain's Prince Charles some years earlier, to come to Nairobi to assess the situation. Colton flew to Kenya on Friday, June 11, and after a short surgery that he oversaw with the local surgeon, he insisted that Richard be flown to Nottingham without delay. Monday found us bound for England after Richard's stretcher had been forklifted into the plane and bolted down over six economy seats at the very back. All through the long journey, he kept sliding toward the rear because of the flight angle, bumping and jarring his broken legs. I felt dreadfully sorry for him, but I was thankful beyond measure that he had at last agreed to seek the best care. My fieldwork was far from my mind, but Louise, on summer break from her undergraduate studies at the University of Bristol, had gone to Lothagam in my place. With Alan and Kamoya's help, she would ensure that the season was not wasted.

The days at Nottingham melded together into endless consultations with specialists and operations every three or four days. The Coltons kindly lent me a room in their house where I could change and bathe, and their garden offered some respite during the long string of operations that Colton performed on Richard's legs. On more than one occasion, I fell asleep in their bathtub, awaking with a start in stone-cold water and stared at the soapy rim that had congealed along the waterline until my dazed and scattered thoughts returned me to the present. Although my communications with Louise and the field crew were few and far between, Richard remained obsessed with the necessity to communicate frequently with his office, and once again, fax machines and other office paraphernalia competed with nursing equipment in a cluttered hospital room. He also insisted on a decent supply of good wine. This problem was also solved by the ever-generous Prince Claus. By royal command, the Netherlands ambassador in London was dispatched to personally procure a number of fine wines —a most unusual mission for him, I'm sure. The nurses insisted that Richard not drink alcohol, and I brazenly assured them that the empty bottles they encountered daily at our doorway were entirely due to my own overconsumption. But they didn't buy it. Against this backdrop, the fax machine spat out reams of paper and the phone rang off the hook in a tiny hospital room packed with office equipment and papers. The hospital began to complain that there were too many phone calls and that we were blocking up the phones since they had only two lines!

Richard came back from the first surgery with metal struts and bolts all over his legs and pins through his bones. Colton pegged the odds of keeping each foot at fifty-fifty. Because of the septicemia that had set in and nearly killed Richard in Kenya, bone grafts and transplants were not an option. The infections also led the doctors to quarantine Richard.

There was a big sign on the door that announced BARRIER NURSING. Anyone who entered had to don gloves and aprons, and whatever came in couldn't be taken out again. The whole affair was surreal and made me feel as though we had landed in a nightmare from outer space.

It eventually became clear that the left leg could not be saved. Mary took the news badly. I also called Samira. She had been travelling in Asia at the time of the accident, and after tracking her down, I had encouraged her not to come home, minimizing how bad things actually were. She appeared at the door soon after. Our eyes met over the metal bolts and struts of Richard's bandaged and scaffolded legs, hers as wide as saucers and silently accusing me of keeping the seriousness of the situation from her. I was too weary and had not the words to explain that admitting the severity—and asking for support—would have made it all too hard to bear. Samira threw herself into the task of nursing me while I nursed Richard. She took over all the shopping and laundry, and brought enticing home-cooked meals to the hospital every day, always extracting me for a few hours of relief. Joyce Poole flew to Nottingham with her baby girl for a few days, and George Bronfman, another of Richard's closest friends, came frequently, always bringing a delicious picnic to tempt Richard to eat. I left Richard in their care with some peace of mind when I snatched a few hours of respite. Anna, Richard's eldest daughter from his marriage to Margaret, was also a constant and welcome visitor at weekends. Messages and gifts from friends and well-wishers poured in, and many close friends took time out of their busy schedules to travel to Nottingham to drop in. Once again, we were overwhelmed and grateful for the outpouring of support and generosity.

The kindness of one friend in particular was very bittersweet. Renata Williams, who attended the University of Bangor with me and was known in our university days as a leading light in the debating society, rang me out of the blue one day. Renata was living in a village close to Nottingham. While she and her family went on holiday, she opened her home to Samira and trusted her with her car. A kitchen, laundry, and wheels made looking after Richard a great deal easier. But Renata was living her own tragedy. She had battled cancer for years, and she would lose this fight the following year.

On my birthday, July 28, Richard's left leg was amputated. Soon afterwards, it became apparent that his other leg would probably have to go too. Richard was very down and was finding it harder and harder to recuperate from the repeated surgeries and anaesthetics he had already undergone. We decided that he should go back home for a fortnight to recuperate before they did anything to the right leg. The idea of being in Kenya again did much to boost our damp spirits. Richard was fitted with a wheelchair and a temporary prosthesis, and by August 12, we were on our way to Kenya. We took the trip in grand style, thanks to more generosity from an old school friend of Richard's, Geoffrey Kent, and his wife, Jorie. The Kents arranged for us to fly home first class after a night in Claridge's, one of London's finest establishments.

THE FIRST DAY BACK was difficult and emotional for us all. We were painfully aware of Richard's disabilities and his despondency. But we soon adjusted to a new rhythm. We put a comfortable old armchair into the back of a KWS van, and Richard wheeled his chair up the ramp and travelled in relative comfort to his office where he received a hero's rapturous welcome.

Louise had been supervising the fieldwork at Turkana, and we all felt it was important that I go to Turkana and see what they had found. I taught Joyce how to dress Richard's wounds, and with his encouragement, I left him in her care and took off to Turkana. I was accompanied by Louise, Kamoya, and Bob Campbell. Bob was a longtime friend who had worked with us at Turkana on and off as a photographer since the very early days. Bob's interests were wide-ranging and included a vast knowledge of mechanics from his days running the Jaguar maintenance shop in Nairobi. He was also well seasoned in the bush, having spent most of his career as a wildlife photographer far off the beaten track.

Since we had no direct means of communication except on Louise's occasional trips to Nairobi for news and supplies, I thirstily absorbed a backlog of all the camp news. They had much to show me, and for the first time in two months, I relaxed a little.

Bob and Kamoya were anxious to show me a hominin mandible Nzube had found on a foray they had made to the Nakoret exposures across the Kerio River in a bleak range of lava called the Loriu hills, which was roughly thirty miles away from our camp at Lothagam. I had planned this trip before Richard's accident, and they had gone ahead in my absence. There was, however, one considerable obstacle: the Kerio River, which can contain a substantial flow of water whenever rain falls some 150 miles away in the Elgeyo-Marakwet escarpment, the main source of the river. If it were to rain in the hills while they were across the river, the team could be stuck for many days. Richard's daughter Anna and her fiancé, Ambrose Langley-Poole, had found a clever solution to this problem in the form of some large heavy-duty plastic drums known as bog cogs. These bolt onto the outside of the Land Rover's wheels and provide increased traction, thereby greatly improving the chances of achieving a safe transit.

Even though I couldn't be there myself, I had asked Bob to join the expedition and lend his expertise to Louise for the crossing. Fording any river of size by vehicle requires a measure of knowledge and assessment of risk, especially when the base consists mainly of sand and soft mud. Bob, Louise, and Kamoya had tried once and failed. They made their second and successful attempt on July 27. While we were preparing for the amputation in Nottingham, they had set off with basic camping gear, water, food, and the bog cogs and a full complement of the fossil hunting team. A hot wind was blasting from the southeast that lifted skeins of fine sand above the scrub of the arid countryside and completely obliterated their tracks as they made for the distant line of green that marked the river. With some judicious cutting, they managed to clear a way through the dense thickets along a cattle and goat trail and break out at a low bank only to find the river still flowing strongly. But a hundred yards away, the far bank looked reasonably suitable for an unobstructed exit.

Bob waded out with some of the others to test the bed, finding mainly sand but also many patches of soft mud. The water at the bank was two or more feet deep; farther out lay some partially exposed banks with channels of unknown depth on either side. There was little chance that their laden four-wheel-drive Land Rovers could cross unaided. The bog cogs were put to test. They worked superbly, their width giving plenty of traction over the soft spots and making short work of the exit at the far bank. The team excitedly headed southeast towards the dark Loriu hills.

Searches of the slopes confirmed that there were very few fossils in the pale sediments on the hillsides, but out in the low-lying ground to the west, there were large quantities of fragmented fossil bone. Assembling for lunch the following afternoon and seeking refuge from the blistering sun under the thin shade of an acacia tree, the team waited a long time for Nzube to appear. Just as worries that he had either lost track of time or run into difficulties were surfacing, he appeared. Exhausted and stone-faced, he headed straight for the water bag and drank long and deeply. He listened as Kamoya, Louise, and Bob berated him for causing much worry. Mildly chastened but clearly unperturbed, he suddenly grinned broadly. He had found a mandible — a hominin mandible!

It was to the site of this mandible that we eagerly retraced the team's steps eighteen days later, with Louise remaining in camp at Lothagam with a radio as a backup should anything go wrong. But the river was a whisper of its earlier flow, and we forded it with ease without a thought about bog cogs. A blazing sun accompanied us across the many sections of soft sand; the wind blew strongly out of the east and thermals created frequent dust devils. We picked up the helpful survey line and made good time, reaching the site of Nzube's rare find before dark. We camped that night in a sand river, listening to the high whine of the sand flies and mosquitoes. After so many nights of hearing hospital cries, groans of distress, and beeping machines, this annoying din had never sounded so welcome.

• • •

MY VISIT PASSED far too fast, but I left feeling refreshed and reassured that the expedition was running smoothly. Under Louise and Kamoya's guidance, our work at Lothagam would be completed by season's end. We were back in Nottingham by the end of August, and Richard soon decided that a second amputation was preferable to a painful and gammy leg. I found this second one much harder than I had expected. It was completely galling to think of the many reconstructive surgeries he had already undergone to save the foot—but it was futile to dwell on this wasted time, pain, money, and emotion. My more immediate concern was how Richard would cope. He had always been so active —how would he adjust to the loss of both his legs and the resulting disability? We did not speak of this.

Richard threw himself into learning how to walk and amazed the nurses with his tenacity and the speed with which he was up on his temporary new legs. "I used to be a master on stilts as a boy. It is like riding a bicycle—once you have learnt, you do not forget how," he told them, brushing off their praise and modestly playing down the huge challenges he had overcome. By October 26, he was back in Nairobi walking unassisted on his permanent prosthetics. He passed his medical and went flying on November 12. Life slowly returned to "normal" or perhaps to a new "normal." Richard was determined to see past his disability, and because he did, others soon saw him as he sees himself, undiminished and fighting for what he believes with determination and grit. Richard resumed his hectic schedule at KWS, and I began to plan my next field season. The wheelchair was relegated to collect dust in the attic, where it would sit for decades. But if his new career had cast doubt on our ever working together again in the field, the accident made this possibility even more remote. We never spoke of it, but beneath cheerful façades, we both deeply mourned all that was lost that fateful day the plane came down.

8

A NEW EARLY BIPED

"I'VE DECIDED THAT WE ARE GOING TO KANAPOI NEXT," I AN-
nounced brightly to Richard one morning. Much as I expected, his
response was a mere grunt. Some rather inauspicious incidents had
firmly prejudiced him against the site, but my gut instinct was to give
it a go.

Richard had first attempted to survey Kanapoi in 1985. He and
Alan Walker had flown into a makeshift airstrip near the site, where
they met up with Kamoya, Nzube, and Wambua Mangao to survey the
area on foot. Wambua, like Kamoya and Nzube, first worked for Mary
at Olduvai before joining Richard's team in the early 1970s. Wambua is
a quiet, solid, and very strong man with an uncanny ability to find the
most beautiful and completely concealed specimens. The trio had cut
across country from the excavation of the Turkana Boy many miles to
the north and had worked hectically since early morning to ready the
airstrip and build a temporary camp. The construction of the airstrip
proved more difficult than expected, and they were already weary by
the time Richard arrived. He immediately marched them off on an
extensive and exhausting search of the exposures. Their weariness only
intensified as mere scraps of fossil bone were all that could be found.
Worse, Alan was feeling very off-colour, and by midafternoon, Richard
realised that he needed medical help and decided to fly back to Nairobi
before dark.

Kamoya recalled that he had never felt so worn out in his life when, caked in a chalky layer of fine dust, they returned to the camp they had hastily erected that morning. The plan was to continue surveying for a couple of days. However, their efforts came to an abrupt halt that very same day when a group of bandits sauntered into camp. This in itself was not enough to warrant undue alarm since armed men frequently passed through. Although they often positively bristled with an assortment of battered weapons that probably dated back to World War II, they were usually content to move on after filling their bellies with a good hot meal, fresh water, and some strong sweet tea. They were interested in stealing livestock or carrying out revenge attacks, not interfering with a curious collection of outsiders scrabbling around and intently scrutinizing rocks. But on this day, Kamoya noticed that the local youths employed as camp help were unaccountably nervous and flighty. The boys spoke the same language as the interlopers, and when Kamoya pressed them, they nervously admitted that they had overheard plans to do rather more than share a meal with the team. Under the keen eye of a group of armed men with malicious intent, what could be done?

Outnumbered and unarmed, Kamoya opted to wait for the cover of darkness. They carried on their usual activities as casually and unselfconsciously as possible. The traditional mancala board game *bao* was played, cigarettes were smoked, tea was served, firewood was cut, and dinner made. The minutes and hours dragged painfully by. When night finally fell, they left the food conspicuously on the fire until the last possible minute as they slashed the guy ropes holding up their tents and surreptitiously gathered what belongings they could in the darkness. Then they piled everything and everyone into the car, and tore out of camp as fast as possible, a scalding hot pot of goat stew bouncing at their feet.

As a result, Kanapoi remained a question mark over the years. Whenever Richard and I had occasion to fly over the site on our way farther north, my questions were invariably dismissed. Apart from the fact that Richard had spent very little time surveying there, I had other reasons to hope for better things from Kanapoi: Bryan Patterson, of Lothagam fame, had also been to Kanapoi some thirty years earlier, and he had found a hominin humerus in 1965. On our own brief survey of Kanapoi in 1989, we had not even begun prospecting when Wambua called out to stop the car and pointed to what looked like one of the large boulders marking the edge of the road—it was a complete and beautifully preserved elephant mandible. If Richard and the crew had not found anything, it did not necessarily mean that there was nothing to find—I now suspect that on top of having scarcely any time to look for bones, Kamoya and the team were also looking in the wrong place. The fossil layers at Kanapoi are extremely localised. Indeed, years later, I would ask Kamoya, Nzube, and Wambua which of all the sites they had surveyed did they think would be the most worth returning to. "Kanapoi," these three veterans of the hunt instantly replied in unison.

The deposits at Kanapoi contrast greatly with those at Lothagam even though Kanapoi is only about fifty-five kilometres away. Where Lothagam is all treacherous steep slopes and eroded canyons, its ancient rocks sculpted into all manner of curious shapes in a kaleidoscope of rich reds, greens, and browns, Kanapoi's deposits are much gentler in incline and lighter in tone, with buff shades of grey, brown, and white brightly reflecting the sharp desert sun. It is correspondingly less hot. Our camp was set attractively under shady acacia and palm trees on the banks of a small tributary sand river called the Atalomeyan. Conveniently situated in the middle of the exposures, it allowed us to walk to work most of the time, saving precious fuel. And there was good drinking water just a few miles from our camp as Christian missionaries had installed a hand-operated water pump in the nearby village of Kaesamalit. A mere flick of the wrist to turn a large wheel brought forth the miracle of pure, delicious water! To top it off, the site was also several hours closer to Nairobi and accessed by a relatively good road.

The only drawback was that the last part of this road, a dirt track leading east from the tarmac trunk road from Nairobi, crossed some fairly large seasonal rivers, including three crossings of the broad, meandering Kalabata River close to camp. The dreadful drought that characterized our time at Lothagam ended while we were at Kanapoi, and on more than one occasion, we made the thirteen-hour journey from Nairobi to find the Kalabata in full spate.

This happened the very first time I made the trip, with Samira, at the beginning of the first field season in 1994. We crossed the river once with ease. It was flowing but not deep. Halfway across the second crossing, the river deepened, and we suddenly found ourselves in dangerous fast-flowing water. It was too late to reverse, so we kept going and had one nasty moment when we were not sure we would make it safely across. At the third crossing, the river was unquestionably deep and flowing fast. In my mind's eye, I could picture the car tilting and sinking into the sandy, muddy bottom—water running through the doors and getting into the fuel tanks—before rolling over and being swept downstream. No. Heavily laden with supplies, it would be folly to try to cross. There was nothing for it but to settle down and wait patiently for the river to recede. In the gathering darkness, we set a stick in the riverbank to mark the water level, and then snacked on the remnants of our lunch. We laid down our *kangas,* the brightly printed rectangular cloths that Kenyan women put to any number of uses, and wrapped up against the mosquitoes while the level of the inky water dropped inch by inch through the night.

All in all, however, conditions at Kanapoi were far easier than at Lothagam—and when the first hominin came in at the end of only the second week of work, morale shot up. Wambua found this first treasure—a lovely piece of upper jaw (maxilla). It came from an individual who lived to a ripe old age and had worn its teeth down through years of chewing to the point that you couldn't tell too much about it. This precious fossil was also covered in matrix that would have to be meticulously removed once we got back to the lab in Nairobi, a process that takes endless days of patient work with an air drill under a microscope. Nevertheless, after five long years of painstaking searching at Lothagam that yielded only a paltry collection of isolated teeth, some of the men were starting to doubt their ability to find respectable hominins at all. Wambua's success injected new determination and a healthy spirit of competition back in the crew. And I allowed myself to hope that Kanapoi would indeed have more of the hominins that Lothagam had so sorely lacked. Having persuaded the National Geographic Society to give us a grant for that first year at Kanapoi, I was desperate to find some hominins to ensure that the organization would continue to fund the project.

We didn't have long to wait. We put some of the crew to sieving the surface material around Wambua's maxilla while others continued to prospect. The sieve turned up several other identifiable fragments of hominin, including the other side of the recently discovered upper jaw. This was the first of many sieves we conducted at Kanapoi—and they were some of the biggest and most tedious sieving jobs we did anywhere because the gently sloping topography had scattered the fragments over a huge surface area.

What would turn out to be the most significant find of the season did not come from a sieve, however, and it came at a most inopportune time: on the first day of a very brief and rushed visit from Craig Feibel when Kamoya and I had planned to devote ourselves exclusively to studying the geology with him. We had invited Craig to do the Kanapoi geology after the fantastic job he did at Lothagam, but he was working at Koobi Fora that season and could get away for only three days, which was barely long enough for him to show us the lay of the land. So we had very little time to gain critical initial geological insights. As we stood on the side of a small buff while Craig studied the topography, Kamoya casually pointed to the ground between us. "Look at that nice piece of tibia," he said, understatedly. Both of us immediately thought that it might be hominin—but bones that look like hominin tibiae have a nasty, disappointing habit of turning out to be something else altogether. The tibia Kamoya had discovered was the proximal (top) end, with the articular surface (the top of the bone that fits with the bottom end of the thigh bone) quite severely eroded, which made it even harder to identify. It also seemed to be implausibly big. So I replied equally casually, "Mmm. Very nice. We must be sure to return to collect that some other time." We built a little cairn and took careful mental note of where we were.

For the rest of Craig's visit, Kamoya and I privately pondered the piece, saying nothing to anyone, while we learnt as much as we could about the geology of Kanapoi. In camp, we found ourselves furtively poring over the cast of another tibia found more than two decades earlier. It was a great deal smaller than the piece we had marked with the cairn on the hillside. But there was no mistaking the strong resemblance. The plane arrived to pick up Craig from an airstrip at Lokori, a small village to the south, just as he finished hastily scribbling some notes and diagrams for us that would faithfully guide us through the rest of the field season. We thanked him, waved good-bye, and watched the little plane recede hazily into the distance. Then we rushed back to the exposures to collect the piece of tibia and bring it back to camp to study properly. The whole crew unequivocally pronounced it hominin!

The cast we compared our tibia to belonged to Lucy, who was discovered by Don Johanson at a site called Hadar in Ethiopia. Don, like many others, had cut his teeth in the Omo Valley as a student and later had the great fortune to be invited to Hadar by Maurice Taieb, a French geologist who had noticed the richness of fossils littering the site. Taieb's invitation established Don in one of Africa's especially rewarding sites—Lucy is one of the most spectacular hominin finds of all time and all the more remarkable for her 3.2 million years of age. With what Johanson estimated to be 40 percent of her skeleton intact, including valuable diagnostic bones like the pelvis and femur, she was the most complete example of any species of early hominin found at the time. This may not sound like much, but to a palaeontologist, such completeness is nothing short of miraculous, the sort of thing dreams —and careers—are made of. A veritable mine of information was contained in her ancient remains. Furthermore, Don's other discoveries at Hadar, including the remains of multiple individuals he dubbed the "first family," provided an astounding wealth of additional evidence about this early species, *Australopithecus afarensis*.

Not much of Lucy's skull was found, but there was enough to show that her brain was tiny—estimated to be comparable in size to that of a modern ape's—and her jaw was more V-shaped than our own U-shaped one. Far more telling were Lucy's well-preserved hip and leg bones, essential in the interpretation of locomotion. They left no doubt that she was bipedal. In a quadruped, such as a modern-day chimpanzee, each hind leg descends vertically from the hip socket to feet placed quite far apart on the ground. But in bipedal locomotion, this arrangement would cause the centre of gravity to shift wildly on each supporting hind leg, making for clumsy and inefficient steps and the side-to-side "wobble" that chimpanzees have. In humans, the femoral shaft is angled relative to the condyles (knee-joint surfaces), and below the knee, the tibiae descend straight to the ground. This "carrying angle" between the long axes of the femur and the tibia allows the centre of gravity to move forward in a straight line as we walk. Simply put, bringing the knees closer together than the hips is what allows the biped to balance on one leg at a time without waddling or toppling over while striding forward.

Lucy's distal femur and tibia exhibited this carrying angle just like modern humans do. To prevent the knee from dislocating due to the angle of the leg shaft, there is a prominent lip on the patella (kneecap). The condyles are large and the bony shaft is robust, so the limb bone is built to withstand the added strain of bearing weight on just two limbs. The pelvis has also evolved to accommodate an upright stance and the need to balance the trunk on only one limb with each stride forward. The talus (anklebone) shows evidence for a convergent big toe, which increases efficiency in bipedal locomotion (at the expense of agility and gripping power in the trees). And Lucy's vertebrae share our spinal curvature, which is necessary for a permanently upright stance. This intriguing picture laid to rest a long-running debate about the sequence

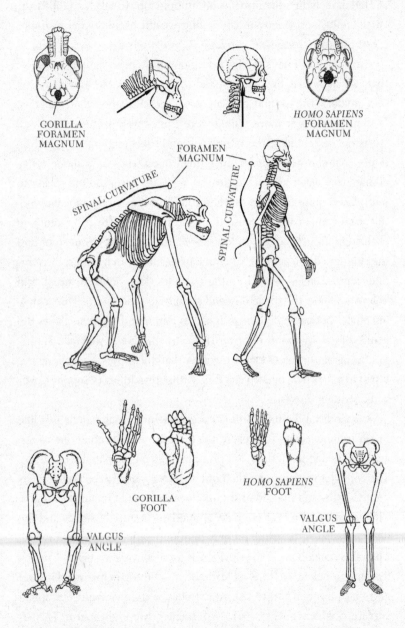

GORILLA
FORAMEN
MAGNUM

HOMO SAPIENS
FORAMEN
MAGNUM

FORAMEN
MAGNUM

SPINAL CURVATURE

SPINAL CURVATURE

SPINAL CURVATURE

GORILLA
FOOT

HOMO SAPIENS
FOOT

VALGUS
ANGLE

VALGUS
ANGLE

Becoming bipedal is the pivotal adaptation that set humans apart from other apes (not drawn to scale).

in which the key adaptations that define human evolution appeared. Contrary to the views long held by some palaeontologists, bipedalism evidently fully developed long before any evidence of enlarged brain capacity.

For all these human traits, Lucy was also very primitive. Although she was certainly moving bipedally, her body proportions were quite different from ours. Standing only about one metre tall, she had a long trunk and long arms relative to the length of her legs. Her long torso and arms, her rather short legs and broad pelvis, and her somewhat curved fingers and toes strongly suggested to me that she retained considerable agility in the trees. This school of thought is not one that everyone agrees with, however. Some scientists assert that the morphology suited to tree climbing is a relic from her past and that she was fully bipedal and never retreated to the trees.

In 2000, long after we had left Kanapoi, a new fossil find near Hadar would shed light on this very question in the form of a juvenile *A. afarensis* from Dikika. This gorgeous specimen includes an almost perfect skull with its brain cast still enclosed and much of its skeleton. It took many years for its finder, Zeresenay Alemseged, to clean the brittle, priceless bones. They included the complete scapula, a real rarity because this bone is so delicate that it is almost never preserved and found. The scapula looks remarkably like that of a gorilla's—and not at all like ours. Another fragile bone was also preserved—the bony labyrinth of the inner ear. We use our bony labyrinth for balance, which means that it looks very different in bipeds than it does in quadrupeds. This telltale bone also strongly suggested that *A. afarensis* had the ability to move through the trees.

These features support our hypothesis that the earliest bipeds would still have sought refuge in the trees from time to time in order to escape the many dangerous predators that shared their environment. *A. afarensis* was both a bipedal and an arboreal ape—with a very different locomotor repertoire from that of modern humans.

Don Johanson found Lucy in Hadar, Ethiopia, in 1974 (and gallingly for us, the very day *after* Richard, Mary, John Harris, and I visited Johanson and Maurice Taieb's camp to see some of their initial discoveries). Mary was working at Laetoli on the edge of the Serengeti plains at the time, where she had recovered fragmentary remains of a human ancestor similar to those from Hadar. Her team would also make a momentous discovery in 1976. Unlike the extensive barren badlands of Hadar and Turkana, the sediments at Laetoli are exposed only in limited eroded areas as much of the site is covered in tall grasses and thick stands of acacia. These were frequented by bad-tempered Cape buffaloes and were rich with other game, which together supported such an inordinate population of ticks that they needed to be brushed off several times a day rather than picked off individually. Mary also remarked on the extraordinary number of deadly and highly camouflaged puff adders around camp as well as the "considerable nuisance" made by elephants chasing members of her team away from their work. In spite of these menaces, it was to the elephants that Mary would owe the most remarkable discovery of her career.

According to the now-fabled legend, some visiting scientists were engaged in some rather unscientific slinging of elephant dung in an impromptu battle. Ducking down in a flat gully to replenish his arsenal of the large fibrous balls, palaeontologist Andrew Hill noticed numerous prints of birds and other animals preserved in an exposed layer of fossilised volcanic ash. The game forgotten, everyone began to look for evidence of prints. Having found hominin fossils in the area, Mary immediately put the whole team to work on the task, hoping to find evidence of human prints. The ash layer was exposed in eighteen different sites, and on the largest of these, Mary and geologist Richard Hay calculated that approximately 18,400 tracks crisscrossed the surface. They eventually identified twenty different species that ranged from an insect to an elephant. Even without hominin prints, the other tracks, combined with the fossil remains of many of the animals at Laetoli, represented a mass of new information.

It was a long shot, but hominin prints were eventually found in

1978, surpassing Mary's greatest hopes. After many weeks of excavation over two field expeditions, a trail of more than twenty-seven metres (eighty feet) was exposed. Two individuals walked across the ashen landscape side by side, with a third smaller one trailing closely behind and carefully placing each foot in one of the prints left by the individual in front. A serendipitous sequence of events preserved this mesmerizing snapshot. The simmering Sadiman, a volcano that flanks the Serengeti plains, first gently showered a layer of ash on the ground. Before the trio set off across the ash-covered surface, a light rain fell. Due to the unusual chemical composition of the ash, the added moisture gave the layer a consistency akin to that of wet concrete. Like concrete, this layer hardened as it dried, preserving the indentations (there are even slight impressions of raindrops recorded). Then a subsequent ashfall from the volcano, followed by more light rain, blanketed and sealed the prints, effectively safekeeping them for Mary's team to uncover some 3.6 million years later. This trail vividly conjures an evocative, uncannily humanlike image — my own children would often amuse themselves the same way when Richard and I took them for a walk along the sandy lakeshore at Koobi Fora after a hard day's work. The Laetoli footprints, set in stone in exquisite and graphic detail, put firmly to rest any lingering doubts that early hominins walked on two legs. Fossils, however complete, can only be interpreted. Here, the behaviour was actually recorded — and no one could argue with that. Bipedalism had been pushed back in time from Lucy's age of 3.2 million years to 3.6 million years ago.

In the first few years at Hadar, Don Johanson would come through Nairobi with his fossils on his way back to America and would stay with us in our home. For security reasons, the Nairobi National Museum technicians made casts of his specimens before he flew to America with the originals. While he waited, we would all eagerly compare his finds with our own from Koobi Fora and those from Tanzania, and spend many enjoyable hours discussing what they might mean. One hot topic was the question of just how many species all these fossils represented.

Not only were the Laetoli bones thought to be some half a million years older than those from Hadar, but within the entire sample, there was considerable variation in the size and shape of some of the bones. Was this variation within a normal range for a sexually dimorphic species that had much larger males with some prominent defining features that were different from their female counterparts? Or were there two entirely different types of creatures?

Mary invited Tim White to study the Laetoli material. Tim, now acclaimed as a meticulous scientist, was a student at the University of Michigan. Fast gaining a reputation that has endured for being intolerant of bad science, Tim was already outspoken and frank in conveying what he believed, although he was charming and a gentleman in less formal situations. In 1977, he and Don Johanson teamed up to study the Hadar and Laetoli fossils. They set about assessing just how different the individuals were, comparing them to a chimpanzee and a human, and taking key measurements on all the specimens to compare how much they differed from one another. Johanson and Taieb's initial paper in *Nature* describing the Hadar finds from 1973 and 1974 had concluded that two or three species were represented at the site. But the detailed analysis by White and Johanson suggested that this was not so. Based on an exceptionally large sample of this human ancestor (which now, after more than forty years of fieldwork, numbers more than one hundred individuals) and after much thought and careful study, the two scientists concluded that the entire sample represented a single species. In 1978, they named this species *A. afarensis* (southern ape-man from the Afar). They attributed the large variation to a high degree of sexual dimorphism, with males being considerably larger than females.

Johanson and White did not only conclude that this was a single species—they went further. Contending that this was the only species living at the time, and the earliest known biped, they also asserted that *A. afarensis* was the common ancestor of *all* later hominins. They boldly drew a family tree that depicted *A. afarensis* on the main trunk, with all other known hominins branching out from this stem species. Although the type specimen for *A. afarensis* hailed from Laetoli, Lucy came to represent this single ancestor, having attained a sort of celebrity status among fossils for her completeness and her endearing nickname. She would reign unchallenged for twenty-five years in this position until our discoveries from Kanapoi because there was little or no evidence to prove that she was not alone.

However, there were divergent opinions as to the correct interpretation of these Hadar and Laetoli fossils. Mary, who had been the one to invite Tim White to study the Laetoli specimens, was dismayed by the two scientists' choice to merge the hominins from both sites, and name the species by the Afar site but select the holotype (the specimen that best describes the attributes of the species) from Laetoli. This choice —to derive a name from one site and choose a holotype from another site—was unprecedented and taxonomically unacceptable to many. A great many miles and, it was believed at the time, nearly three-quarters of a million years separated the two sites (more accurate dates would later prove that the Hadar sediments were older than previously thought, narrowing the age gap between the two sites). Arguably, since Lucy was the most complete specimen known, she was a logical choice for the holotype as she allowed for an exceptionally detailed description. Johanson's reasoning for countering this argument was that Lucy would be inappropriate for a holotype because she was not an average representative of the sample because she one of the smallest of the bunch. In contrast, LH-4 (Laetoli Hominin 4), the mandible they chose, was closer in features to a mathematically derived "average individual."

But Mary vehemently disagreed with Johanson and White's decision to assign all the fossils to the australopithecine genus and a single species because she believed that her fossils should have been assigned to the genus *Homo*. She disassociated herself from the paper they submitted, and her previously close relationship with Tim White became strained. Meanwhile, Richard's own relationship with the pair was also breaking down because Richard refused to accept the single-ancestor theory, saying only that Lucy may or may not turn out to be an ancestor. This was falsely interpreted by Don as a failure to acknowledge the enormous importance of his spectacular find. But Richard refused to be drawn into the fray, insisting that only the discovery of more fossils would provide the answers as to what Lucy really represented. The media made much of the disagreements, painting in highly colourful, personal terms what was, at least to us, a professional difference of opinion.

The controversy about whether these specimens should be attributed to the genus *Australopithecus* has since largely been settled, although how to best interpret the huge amount of variation in the sample remains controversial. A partial 3.6-million-year-old skeleton discovered in 2005 at Woranso-Mille and nicknamed Kadanuumuu ("big man" in the Afar language) has stretched the range of variation in *afarensis* even further to encompass individuals who have gorilla and humanlike shoulder blades and may be more than two feet taller than one another. But, independent of these arguments, I have always disagreed with Johanson and White's other central assertion, namely that A. *afarensis* is a common ancestor for all later hominins. When seen in context of other mammalian species and the generally accepted evolutionary theory, Lucy should not have been alone. It just didn't make sense.

After the appearance of a major new adaptive feature such as bipedality, a radiation of species would be expected to have followed with a plethora of different hominins evolving to fill the new feeding niches that bipedality opened up. The evidence from Lothagam had pointed overwhelmingly to huge habitat change—a strong impetus for just such a dramatic evolutionary stride. In this context, 3.2 million years is relatively recent—if our ancestors split from the ape lineage approximately six million years ago, then why was there no radiation of hominins for three million years after this initial divergence that corresponded to the change in climate? Across the animal kingdom, no other newly emerged lineage with a comparable record of three million years of unilinear evolution after the appearance of a significant novel trait is known.

All this was very much at the forefront of my mind as I looked at the Kanapoi material and revisited the casts of bones from Hadar and Laetoli. Although larger, our tibia was similar in shape to Lucy's and almost identical to the larger tibiae from Hadar. The Kanapoi tibia presented incontrovertible evidence of full bipedality more than four million years ago, pushing the date back in time more than half a million years from when the Laetoli trio stepped out across the ash-covered plain. Bipedality is the hallmark human trait that set hominins on a separate trajectory from apes. No doubt about it, Kamoya's tibia was the most significant find we'd had in a number of years.

But little else could be told from this isolated piece of tibia and Wambua's old, worn maxilla. The new ancestor remained an enigma. What did our mystery hominin eat? The teeth were simply too worn to tell us much about its diet. And what did it look like? Although we knew it was definitely bipedal, how dexterous were its hands? Was it *A. afarensis* or one of Lucy's ancestors? Perhaps it was an ancestor of an altogether different lineage, which would demonstrate the diversity at that time that I had long suspected but not been able to prove? We desperately needed more fossils to tell us about this creature, and the team set about finding them with even more determination than ever.

1994 WAS A GOOD YEAR for us. The great drought that had parched the landscape and cloaked it in dust when we worked at Lothagam was replaced that year at Kanapoi by a beautiful feather-light sprinkling of lush green that slowly turned to yellow gold as the rains faded. On some days, the heavens opened, flooding the river where we were camped and grounding us. The normally hard exposures softened like putty, and walking across them would risk snapping fragile fossils as our feet sank in the mud. With the overcast sky hanging low and pewter, we used these days to transfer our field catalogue of fossils to the computer. The crew caught up on laundry and their correspondence home, and enjoyed a well-earned rest while I waited impatiently for the ground to dry out enough to safely walk on the exposures again.

Even better for us, our hominin drought also ended. Soon after Wambua found his maxilla, Nzube found three beautiful unworn and isolated hominin teeth: an incisor, a premolar, and a huge canine. They were in pristine condition and must have come from a really large individual. This latest find further infused the team with competitiveness, and the mood in camp was very upbeat in spite of the fact that each of these finds added another bout of seemingly endless sieving. This particular sieve was especially rewarding and produced all the missing teeth. The result was a hardly worn, beautifully preserved, and astonishingly complete lower dentition.

Not everything turned out to be what we hoped, however. On July 28, which happens to be my birthday, Wambua found what initially looked like a marvelous birthday present, a potential hominin skull. But something about it wasn't quite right. As I excavated this fragile specimen, I thought at one moment that it was a hominin, and the next that it was not. Finally, we lifted it out and found that it was even more strange-looking underneath. I had no idea what it could be, and Kamoya promptly nicknamed it the *bahasha* ("envelope" in Kiswahili) for its flattened appearance. The specimen had fallen out of the bank in some relatively recent rain, and as we sieved, we retrieved a number of fragments of tooth roots scattered among the stones and sand in the gulley, all with very fresh breaks. When we finally managed to extract it, we immediately saw that we had been misled. Our odd birthday find was an enormous knob from the skull of an ancient pig, which was used for display to attract a mate!

Kamoya returned one day to the sieve site of his tibia and came back to camp in the late afternoon pleased as punch with the distal (lower) end of the bone. This second discovery made the find doubly important and raised high hopes of finding the middle section. To our enormous disappointment, we never did find this piece despite extensive sieving.

Samuel Ngui, who had helped Kathy Stewart search for fish at Lothagam, also found his very first hominin at Kanapoi that year. Like Nzube, Ngui specialised in finding monkeys, which are often impossibly small and even harder to find than hominins. He spotted two matching pieces of a mandible from the left and right sides with the crowns of the two teeth in each fragment broken off. The following year, after six endless weeks of sieving this site, we had an almost complete beautiful lower jaw with all its teeth.

LATE IN THE 1994 field season, on September 10, Richard and some friends, including the photographer Bob Campbell who'd been with Louise that last field season in Lothagam, joined me in camp for a long weekend. Richard now came to the field very infrequently as he found the extreme heat and uneven terrain very difficult. Not only was getting around a chore, but without legs, his body could not stay cool in the high desert temperatures because of its decreased surface area. But he couldn't have picked a better weekend to visit.

Kamoya and I took our guests on a tour of the exposures, stopping first at Nzube's sieve site so I could get Richard's advice on how much farther to go. He agreed with my assessment: we could finish up for the season, but it would be worth our while to extend to the flanking slope the following year. We then proceeded to an elephant skull that we had just finished excavating. Richard had promised to oversee the delicate task of plastering and removing this in one piece. If not done correctly, the weight of such a large specimen can result in its falling apart, so I was much relieved to have his advice. Amid much banter and talk of the old days, we finished plastering the top half of the elephant and had turned it successfully on its side when I urged Nzube to join the sieving team. But he did not get very far. "Meave, come and see what I've found!" he cried with a smug, delighted grin. "You are not going to believe your eyes," he added with great certainty. Although his expression usually gave him away, Nzube was usually much more taciturn and noncommittal in the way he spoke about his discoveries. Needing no further encouragement, I rushed after him, leaving a rather frustrated Richard sitting on the ground beside the elephant.

It was indeed incredible. A beautiful mandible was concealed among the pebbles. All that could be seen on the surface was an almost complete set of teeth glittering like a string of jewels in the soft morning light. Nearby lay a temporal bone from the skull that included the external opening for the ear and the bone surface that articulates with the lower jaw. Three incisors were missing from the mandible, and the third molars were not exposed. But the skull fragment indicated that more might be there if we were lucky. We knelt and stared at it in disbelief. There is nothing quite like the thrill of being one of the very first to set eyes on an ancient link to our past, and this one was just exquisite. Unlike the maxilla Wambua had found, these teeth were immaculate and relatively unworn so we would be able to learn much from them. Bob snapped away busily with his cameras while I rushed back to tell Richard.

We retreated to camp during the heat and glaring light of midday, elated by the tremendous discovery. In the afternoon, we returned to the site, and Richard excavated the mandible while Kamoya and I began to move the larger stones in preparation for sieving. For one blissful afternoon, with Richard by my side and the rest of the old gang working in companionable silence and the thrill of an important discovery humming in the air, it felt just like old times.

THIS STOLEN AFTERNOON was all the more precious because the new bent of Richard's interests did not sit well with the government and I continued to worry for his safety. While he was in hospital in Nottingham, jealous colleagues took advantage of his absence by accusing him of mismanagement and corruption at KWS, just as they had done during his kidney transplant. Once again, an enquiry was held, and no evidence whatsoever of corruption was found. Amid mounting political pressure and interference upon his return, it became increasingly impossible for him to do the job as he wanted, and he resigned in early 1994. He was briefly reinstated, but a few months later, he was forced to resign again. His experience at KWS had led him to believe that while strong institutions could be built in Kenya they would only endure with a solid and more mature political foundation. He then threw himself wholeheartedly into politics, founding a new political party in May 1995 called Safina ("ark" in Kiswahili). But if Richard and his ardent supporters believed they could rescue Kenyan politics and sail off into a democratic future, the current government believed otherwise. The party members were thwarted at every turn, their licenses denied and their members detained. A year after we sat peacefully excavating Nzube's hominin would find me back at Kanapoi sitting anxiously by a crackling radio listening to Richard warning me that I might hear of a "little trouble" he'd had in Nakuru.

The little trouble transpired to be a run-in with an extremely hostile rent-a-mob, where he was beaten over the head with batons and severely lashed with a hippo-hide whip at least thirty times across his back. Among the other Safina members present, two were hurt so badly that they were in intensive care. Because the foreign correspondent Louise Tunbridge was also badly beaten at the rally, we heard a terrifying graphic description of the incident on the BBC World Service that evening. Richard later admitted the severity of the situation to me and revealed that his greatest fear had been losing his footing on his prosthetic legs, because if he had been jostled to the ground, he would assuredly have been trampled to death. With the main roads blocked off by police, they escaped through the safety of the slums, where Safina enjoyed widespread support, in the shell of Richard's vehicle, its windscreen and windows completely smashed.

NZUBE'S FOSSIL came out beautifully that magical afternoon in 1994 at Kanapoi—a complete lower jaw that we immediately suspected

was not the same as A. *afarensis*. We hastened back to camp to compare it with the Hadar casts and Nzube's beautiful set of lower teeth.
Talk about closing the season the following week quickly turned to talk
about how long it would take to sieve this new specimen. Starting at
the bottom at the edge of the sandy streambed, we steadily sieved the
slope. We were very excited, especially after we recovered an isolated
molar tooth at the bottom of the slope. But over the succeeding days,
our hopes faded as we proceeded closer and closer to the top of the
slope, and it became clear that nothing further was going to emerge.
We would have to wait until the following year to excavate this bed and
see if there could be something more emerging from the layer where
it was originally buried. We swept the site clean, placed two rows of
protective stones to prevent erosion, and finished up on the last day in
the late morning.

Despite the frustrating days sieving, I could barely believe my luck.
In just one season, I had enough fossils to answer some of our most
pressing questions. Plus, the cache of exciting hominin finds was exactly what I needed to assure the National Geographic Society that
they should continue to support the research. I simply could not wait
to get back to Nairobi to compare them all in detail and enlist the expertise of my colleagues. How different will our hominin be from A.
afarensis, and where will it belong on the family tree?

The crew began dismantling the camp, eager to see their families
again after a long field season. In no time at all, most of the tents were
down, and we continued loading by moonlight until our home for the
past three months was securely stowed and tied down in the lorry, and
the stash of fossils carefully wrapped and arranged in cartons and pad-
ded trays in the back of the other vehicles. We slept on mattresses laid
out on tarpaulins, rising early to begin the long trip back to Nairobi.
Little did we know that one further adventure awaited us before we
would get home.

In the morning, try as we might, we could not persuade my normally reliable Land Rover to start. We tried all the usual tricks—we

cleaned out the fuel filter and the fuel lines; we looked at the fuel pump and the spare-tank switch. The Land Rover was a Defender model with a diesel engine. These workhorses can almost always be started by towing, but by midmorning, we had towed it all over the place to no avail.

"We're just going to have to tow it to Lokichar to find a mechanic who can help," Kamoya decided. Lokichar is the town at the junction with the tarmac trunk road to Nairobi and is a great distance away. We set off gingerly, using an old and worn tow rope that looked rather feeble. We did not get past the first small gulley before it snapped. The same thing happened again and again, and the tow rope was getting shorter and shorter. The first Kalabata crossing was difficult, but we managed to navigate the soft sand without having to do more than push in several places. The second crossing was more challenging and required plenty of pushing as well as lining the ruts in the road with palm leaves to provide traction against the soft sand. We got to the third crossing at noon and did not make it to the other side until two o'clock. We were exhausted and hot after unloading all of our carefully packed equipment from the car. We had hoped to have been well on our way to Kitale by now, a town halfway home.

"We won't even make it to Loperot at this rate!" I exclaimed in despair as the tattered tow rope broke yet again. "We have to, so we will" was the unflappable Kamoya's reply. Loperot, halfway between our camp and Lokichar, was unlikely to have a mechanic, assuming we got that far. We sent Benson—an excellent fossil hunter and jack-of-all-trades—ahead to Lokichar in the lorry so he could call our friend Bob Campbell. Bob's immense knowledge of engines and bush mechanics seemed like our best hope of starting the obstinate engine.

Bob racked his brain to think of something that would help Benson trace the fault, the faint line crackling and the coins dropping away as the minutes ticked by. "We already tried that!" Benson replied in frustration to everything Bob suggested. The best Bob could think of was to make sure there was no air in the fuel lines and that fuel was definitely reaching the injector pump. Benson's disappointment was

palpable through the crackling interference as he broke the connection. Moments later, a thought flashed through Bob's head. Could the
Land Rover be the victim of an electrical fault? He suddenly remembered that when the spark plugs are not igniting the petrol mixture,
diesel engines are shut down by cutting off the fuel supply either by
mechanical means or an electrical cut-off switch. Bob tried calling the
operator to locate the call box, but when he finally got through, nobody
was there to answer.

Benson was already on his way back to our convoy. He had found
somebody with two steel rods we could use to jury-rig a tow bar, and
he had found a mechanic. The mechanic proved to be singularly un-
helpful. He had made only a few cursory checks before a Toyota heading back towards town approached. The mechanic looked at his watch,
hailed the vehicle, and jumped in, announcing that he had to be back in
Lokichar right way. The tow bar improved matters considerably, however, and by five p.m., we were back at the pay phone in Lokichar.

"Hello! Bob?" Tired and discouraged, I was calling in desperation
to ask if he could suggest anything that we could try next. We'd spent
the entire day towing our lifeless Land Rover a mere thirty miles—
over sandy roads where even fully functional four-wheel drives tended
to bog down. With night fast approaching, the thought of towing the
vehicle to the nearest garage at Lodwar was not a happy one. Lodwar
lay fifty miles north in the opposite direction of home. The nearest
town on the way south was Kapenguria, 110 miles away—but that
route takes you from the floor of the Rift Valley up the western wall of
the rift through the Marich Pass, a long winding climb of four thousand
feet that would surely tax the energy of the towing vehicle.

Bob immediately broke in before I even had a chance to speak.
"I've thought of something else that might work. Have a look at the
injector pump and see if there is a wire connected to it. If there is, then
disconnect it and search for a length of wire to reach from the positive
battery terminal down to the connection on the pump." At least we
had something new to try. We followed Bob's instructions cautiously

—but to our immeasurable relief, the engine suddenly sprang to life. It transpired that the starter switch controlling the lifeline to the injector pump was the culprit, a very unusual fault. Tired but happy again, we could head for home first thing in the morning.

I spent our last night much as I had the first—lying impatiently on the ground beside the vehicle beneath a full moon. I was too worn out to notice that this time it was the rumbling of heavily laden trucks belting by on the main road rather than the flowing Kalabata River that lulled my dreams.

ANOTHER PIECE OF THE PUZZLE

WE RETURNED TO NAIROBI FLUSHED WITH SUCCESS AND WITH-
out further mishap. We were harbouring a strong suspicion that we
might have hominins that were not only older than Lucy and her kin
but that were also something completely new, and I couldn't wait to
study them further. The very first evening upon my return, friend and
science writer Virginia Morell telephoned me. "Have you heard the ru-
mour that Tim White has found seventeen hominin skeletons in Ethiopia?" she asked excitedly. "Apparently, it's something even older than
Lucy!" she continued. I had mixed feelings as I fumbled for a reply. If
Tim really had found seventeen skeletons, what a boon for our science.
But it would make my precious haul seem rather paltry, and I wished
we hadn't made our discoveries at exactly the same time. It was very
discomfiting to think that my potentially new species might have already been discovered — and named — elsewhere, and I was most anxious to learn if we had found the same thing.

It eventually transpired that Tim had not found seventeen skeletons. Shortly after Virginia's call, he published descriptions of seventeen
specimens, mostly teeth and pieces of skull and mandible. He named
a new species, *Ardipithecus ramidus. Ramid* means "root" in the language of the local Afar people, and Tim believed that his 4.4-million-
year-old fossils were ancestral to Lucy and the root of the human evolutionary tree. I immediately contacted Tim to congratulate him, and

sharing my own news, I suggested that we meet to compare our latest finds. Tim agreed that this would be a good idea, and I arranged to visit Addis Ababa after his return from the field the following January. In the meantime, we set about cleaning our fossils and making plaster of paris replicas in preparation for my visit to Addis.

Tim was a most gracious host, and throughout my visit, he showed me around the Ethiopian collections, and whenever time allowed, he enthusiastically joined me in making comparisons between our bones. "Look at how human that upper canine looks," I exclaimed. "Yes, and see how it occludes against the lower premolar," agreed Tim, as we pored over one of the first specimens he had discovered, which included a very primitive apelike milk molar from a baby as well as some adult teeth including canines and part of the back of a skull. In most higher primates other than humans, the long sharp canine is rather like a dagger, continuously self-sharpening against the lower premolar in a configuration known as the C/P3 complex. Tim was pointing out that *Ardipithecus* did not share this trait and was rather more human-looking.

"And see how the foramen magnum is in a really forward position," he continued, pointing to the opening at the base of the skull through which the spinal cord leaves the braincase. The position of this hole is an indicator of whether the individual walked on two legs or four because the angle that the head is held varies in different postures. These two features are what had led Tim to cautiously conclude that *Ardipithecus ramidus* belonged on the human rather than the ape ancestral line. "Well, you need to find some more specimens to tell us if it truly was bipedal or not!" I cheekily suggested, with little inkling of what he was going to show me next.

Tim then revealed his next and most spectacular surprise of all. He had just returned from an exceptional field season, bearing many parts of a very fragile skeleton that could answer this very question about bipedality. However, the protective plaster casing obscured all but the top of the matrix-encased bones, which were too delicate to handle without preparation. "The bone is so brittle that a mere drop of Beda-cryl, carelessly applied, can cause the bone to crumble," he explained. "I've had to use the fine point of a small syringe needle to squeeze tiny amounts of the Bedacryl onto it time and again until it is hard." I could only peer at the top of the fragile bones peeping through the plaster. It was immensely frustrating, but I was curious and eager.

Some fifteen years later, Tim published his findings on this skeleton —which included much of the skull and the incredibly rare hands, feet, limbs, and portions of the pelvis. Much could be learned from *Ardipithecus* and it held a few surprises. The most telling of these was how distinct it was from chimpanzees, thus upturning a long-held assumption that chimpanzees represent the primitive condition from which we evolved. In fact, both modern African apes and humans have embarked on separate evolutionary trajectories from our last common ancestor.

Ardipithecus was able to climb along tree branches on all fours but also walked upright on two legs. It had long relatively dexterous fingers with flexible joints that allowed the body to be supported on the palms while moving through the trees. It lacked the morphology that allows the "knuckle walking" characteristic of chimps and gorillas. Its feet, however, retained the opposable, grasping big toe that would be useful for climbing, which has disappeared in fully bipedal hominins.

From my time in Addis with Tim, it quickly became obvious that our Kanapoi hominin was not the same as *Ardipithecus ramidus*. The latter was a small, primitive, apelike species with very thinly enameled teeth and relatively large canines. These contrasted with the thickly enameled and relatively smaller canines of the Kanapoi specimens. But were our bones from Kanapoi the same as *Australopithecus afarensis*? This was the question that I now most wanted to answer, and I discussed it at length with Tim.

But further study to decide if our fossils represented *afarensis* or a new species would have to wait. I was leaving almost immediately for a short field season on the east side of the lake with Alan Walker and the crew. While I chafed at the delay, I was somewhat consoled that at least we would be looking for fossils of a similar age to Kanapoi. If I was lucky, we might turn up still more answers.

IN 1987, several hominin teeth had been discovered by John Kimengich at Allia Bay, which lies to the south of Koobi Fora and marks the southern boundary of Sibiloi National Park. They were washing out of an ancient bone bed that was chockablock with fossils, so there was a good chance that more hominins might be found. In 1988, we had put in a small excavation and had indeed recovered more hominin teeth and a piece of mandible. Moreover, the Allia Bay exposures had been dated at 3.9 million years, right between the older Kanapoi and *Ardipithecus* fossils and the *afarensis* material. Thus, the site was of huge significance; these were the oldest hominins that we had from the east side of the lake, and they were only slightly younger than the ones from Kanapoi. By 1995, the site was in danger from erosion and it urgently needed excavating before it washed away completely. I had been worrying about how to find the time and the funds for this until Alan Walker provided a solution when he raised funds from the American National Science Foundation for this task. For the next three years, we spent January and February at Allia Bay and June to August at Kanapoi. This hectic field program left little time for catching up with the inevitable backlog of office work and admin, let alone time for studying the fossils.

The fossils in the Allia Bay bone bed were entombed in a cement-hard matrix and had to be extracted at a painstakingly slow pace. Although much of the bone was rolled and fragmented, we recovered some well-preserved hominin specimens, including several isolated teeth and a maxilla. In the three years we spent working this site, we excavated 115 square metres and recovered eighteen hominins—that's a surprisingly high density of one hominin for every 6.4 square metres. These were jumbled up with a plethora of other bones belonging to crocodiles, bovids, pigs, and a host of other animals. This rich assemblage of animal bones again confirmed that by four million years ago the transition that we'd seen starting at Lothagam had resulted in a modern-looking, grazing-dominated animal community.

When we returned to Kanapoi in June of 1995 for our second season, I embarked on a large sieving enterprise in the face of near universal opposition from my crew. The year before, Ngui had spotted the two fragments of a lower jaw that I had instantly recognised as hominin. Nobody else — including Richard — believed me, however, thinking instead that they were bovid bones. This unprepossessing piece of jawbone showed only the outline of teeth that had broken off from the roots. Nevertheless, trusting my instinct, I hoped we would find more by screening the long shallow streambed that led down a gentle incline from where Ngui had found the fragments. All the sand in the channel had to be removed and sieved, which involved a great deal of arduous and tedious labour because the channel was more than sixty metres long. It felt as though the sieve would go on forever. Lasting more than six weeks with the whole crew working on it for most of the time, this was one of the largest and most taxing sieves we ever did. Nevertheless, we were rewarded as more and more tiny fragments of jaw and teeth came out of the sieves. All my detractors and naysayers were silenced as it became apparent that I was right. Tiny fragment by tiny fragment, we pieced together a large unmistakably hominin mandible with enormous canine sockets that clearly must have belonged to a male. It dwarfed some of the other specimens we had found and led us to question whether it belonged to the same species as the others.

IN 1995, Alan Walker, Carol Ward, and I began to study both the Allia Bay and the Kanapoi material. Carol had been a student of Alan's, and her chief interest was hominin postcranial fossils. We initially compared the specimens from Kanapoi with those from Hadar. When Don Johanson and Tim White had examined the Hadar collection in detail years earlier, they were surprised at how primitive and apelike the specimens were. In turn, we were surprised to find that the Kanapoi specimens were even more apelike than the Hadar ones in some characters. For example, the small piece of skull that preserved the external opening of the ear showed the ear to be extremely small, like that of a chimpanzee, and much smaller than any known from Hadar. The mandible that Nzube had discovered at the end of the 1994 field season was strikingly different from that of *A. afarensis* because it had almost parallel tooth rows like a chimpanzee or gorilla has. In addition, the outer surface of the jaw below the incisors (where modern humans have a chin) also differed. Similarly, when we compared the relative sizes of the canines with the sizes of the cheek teeth, the Kanapoi canines were larger and different in shape. The lower premolars immediately behind the canines (P_3 and P_4) were more primitive too.

On the other hand, the Kanapoi material was more similar in other characters to *afarensis* than to modern chimpanzees and gorillas. The similarities in the two collections included thicker tooth enamel than is typical of modern African apes. This was not unexpected, since it most likely reflected dietary trends associated with the long-term changes from closed to more open habitats that we had revealed at Lothagam.

Like *A. afarensis* (and modern African apes), the Kanapoi hominins show a large difference in size between males and females. It was the marked size difference that had led Johanson to initially think that he had more than one species at Hadar. But when he looked at the size difference between male and female modern gorillas (the most sexually dimorphic modern African primates), the range of variation in the Hadar collections was not exceptional. When we compared the enormous sockets for the canines in the lower jaw of Ngui's mandible with the small canines of Nzube's beautiful mandible, we also initially wondered if these specimens could be from more than one species. The same comparison with modern gorillas, however, convinced us that we had only one highly sexually dimorphic species.

A trend of evolving characters can be traced from the Kanapoi material to Lucy's kin. In *afarensis,* the canine sexual dimorphism is reduced, but the high body-size dimorphism is retained. This suggests a social system characterized by high levels of male-male competition that did not depend upon the use of fearsome-looking canine teeth. This could be explained by the move to bipedality, which allowed the development of alternative means of display, such as chest beating or branch flourishing. But it made our fossils that still showed a high degree of canine dimorphism all the more interesting.

Although there were many characters that clearly linked them, the two collections were readily distinguished due to the more primitive nature of the Kanapoi fossils. Our next step was to check these features on both the specimens from Laetoli, which at 3.6 million years are slightly older than those from Hadar, and the new Allia Bay fossils, which at 3.9 million years are slightly younger than those from Kanapoi. Grouped this way, the evolutionary changes sprang even more clearly into focus. We were looking at fossils from four populations of decreasing geological age that displayed a distinct trend towards less apelike and more hominin-like characters. Should all these specimens be included in one species or should we name a new species for the earlier specimens since these were so clearly distinct from A. *afarensis*?

Our findings confronted us with a pressing dilemma that is a recurring source of much of the controversy in palaeontology. When do you decide to name a new species, and when do you lump things together? In reality, there is no clear line drawn in the sand to divide one group from another. Evolution works more gradually than that. Modern gulls in Northern Europe give a particularly elegant example of this. Two distinct species—the herring gull and lesser black-backed gull—both trace their ancestry to a Siberian species. This original bird spread slowly to the east and the west, all the while diverging into a chain of subspecies. Each one of these subspecies can breed with those adjacent to them. The two extremes of the populations, the herring and lesser black-backed gulls, overlap in Northern Europe but they cannot interbreed. These birds form what evolutionists call a classic "ring species," for the birds from all the subspecies represent a continuum of traits. They demonstrate just how arbitrary a dividing line between species can be. We faced a different sort of arbitrary division with our hominin fossils, a continuum of traits across time rather than across a geographical space.

We agonised over how we should interpret these fossils. Lucy and the other younger fossils from Hadar clearly bore the most differences to our fossils at the older sites of Kanapoi. These differences were testimony to the million years that separated the majority of these specimens. The Allia Bay and Laetoli fossils, intermediate in age, showed intermediate forms. If our fossils had been around when Johanson and White did their analysis, their decision to pick an intermediate "average individual" for their holotype for A. afarensis might not have made much sense to them. But significant finds are often made years or even decades apart, and previous names hold until they don't make sense anymore. We would have to build on what had gone on before and decide how the Kanapoi hominins differed from the Laetoli holotype of *Australopithecus afarensis,* LH-4. After much study and careful thought, we took the plunge and decided that all our fossils, including those from Allia Bay, were sufficiently different from all the A. *afarensis* material to deserve being called a new species that was ancestral to *afarensis* in the genus *Australopithecus.* We published a preliminary paper in *Nature* in 1996 that described and named the fossils. We picked the name *Australopithecus anamensis (anam* means "lake" in the Turkana language). At that time, *anamensis* represented the oldest known biped, and we received much publicity and acclaim around the world.

IN 2019, the publication of a new and fascinating discovery from the Godaya Valley in Ethiopia challenged our interpretation that *anamensis-afarensis* is indeed a single evolving lineage. The largely complete 3.8-million-year-old cranium was found by Yohannes Haile-Selassie in 2016 and is informally dubbed MRD as a shortened version of its full accession number in the National Museum of Ethiopia. It shows for the first time what an adult male A. *anamensis* could have looked like.

Haile-Selassie and his colleagues suggest that, despite being younger than the material from Kanapoi and Allia Bay, some of MRD's features are more primitive than those in these earlier A. *anamensis* fossils, which would be inconsistent with their belonging to a single evolving lineage that would become *afarensis*. Most striking is the forward position of the cheekbones. As further evidence to debunk the single-lineage hypothesis, the authors also present a second 3.9-million-year-old frontal bone (part of the forehead) that they assign to A. *afarensis* rather than A. *anamensis*. If they are correct, there was a period of more than 100,000 years when the two species coexisted, which would render it impossible for them to be a single lineage. However, several aspects of these finds still need to be clarified. MRD's cheekbones are very distinctive, but the authors used a high degree of digital reconstruction to correct distortions and estimate missing parts, so the forward-projecting cheekbones are notable for how smoothed they are, with little sign of the original bone surface. While such digital reconstructions are excellent ways of visualizing what is missing or distorted in a fossil, interpretations based on them always deserve stronger scrutiny. Moreover, these findings are reliant on a single specimen so any variation within the species cannot be assessed, and as we have seen in the Kanapoi, Allia Bay, and Hadar fossil collections, not all individuals from these ancient species looked alike. Lastly, the frontal bone is not a good diagnostic bone, with one looking much like another, so establishing the contemporaneity of *anamensis* and *afarensis* on the basis of this fossil fragment will need corroboration. To my mind, this discovery is enormously significant and of huge interest in and of its own. It is consistent with the theory I have long held that there is likely to be more rather than less diversity in a radiation of early bipeds. Yet for the time being, I remain open-minded as to whether the weight of evidence that they present is sufficient to definitively debunk the theory that *afarensis* evolved from *anamensis*.

. . .

MEANWHILE, BACK IN 1995, I was extremely curious about the environment our new *anamensis* lived in. In great contrast to Lothagam, the sediments we were working in at Kanapoi covered a mere snippet of time in geological terms—just 100,000 years. Craig Feibel, who had made that flying visit during our first field season, joined us to study the geology, and he named these sediments the Kanapoi Formation.

A river system believed to be ancestral to the modern Kerio River snaked its way northwards through the ancient volcanic landscape for fifty thousand years some 4.1 million years ago. Conveniently, two volcanic eruptions sandwich this time interval and allow us to have remarkably accurate dates for it. The enormous Lonyumun Lake then flooded the area as it expanded southward, and we can see this in the claystones and siltstones that make up the middle of the sequence. These oldest riverine layers correlate with the uppermost sediments in the Apak Member at Lothagam—remember where the squashed snails at the Lonyumun lakeshore were sizzled and distorted into a permanent record by the erupting Lothagam basalt?

The top third of the Kanapoi sequence records the establishment of a second river system as Lonyumun Lake receded, and the Kerio River delta advanced northward apace with the retreating lakeshore. A third ashfall, the Kanapoi Tuff, dated at 4.07 million years ago, was deposited in this deltaic complex and filled the bays and flood basins between the many distributary channels of the northward-advancing delta. This ash layer is clearly visible today where it blankets a large area of deposits to the north and east of our campsite.

Not far from Kanapoi, the modern Kerio River is a verdant ribbon slicing through the barren desert plains. While working at Kanapoi, we occasionally visited the Kerio on Sundays. Along with plenty of birdlife and smaller mammals, we always found prodigious quantities of fresh elephant dung underfoot even though the pachyderms have not been seen outside the green riverine belt in years and the river is strongly seasonal. The extreme contrast between the dry barren landscape at Kanapoi and the lush vegetation along the riverbanks vividly demonstrated to us just how localised the conditions we reveal at particular fossil sites could be—only a few miles away, conditions support a completely different biota. This could have been equally true in ancient times, for the fossils that we discover are always tied to a water system that is responsible for the sedimentation process preserving them.

Fossil soils, or palaeosols, add another dimension to our understanding about the past climate and habitat. In the aftermath of a storm, angry rivers, red with silt and tossing about logs and other vegetable detritus, are a force to watch. When the rivers swell beyond the narrow confines of their usual course, they burst their banks, and the furiously flowing water slows as it spreads out over the floodplain. Their burden of sand and silt dumps out of the water and forms a new layer of soil. This is quickly colonised by a new season of grasses, shrubs, and trees, and when these die and decompose, rich organic matter is added to the topsoil before a new rainy season begins the cycle all over again. In this way, soils develop, layer upon layer, at an incredibly slow pace, and the creation of a well-formed palaeosol takes hundreds of thousands of years. This means that the soils serve as an excellent proxy through time of the conditions under which they were created, although, as with all other geological records on land, there are large gaps from all those periods when the sediments were blown away by winds or carried away by water before hardening or being capped by rocks and sandstones.

Jonathan Wynn, whom I never once saw in the field without his trademark red bandana bound about his long and shaggy brown hair, is passionate about palaeosols. Jonathan first joined us in 1995 to study the palaeosols at Kanapoi when he was still a graduate student at the University of Oregon. Jonathan found that in general the Kanapoi palaeosols ranged from poorly drained vertisols (clay soils that crack open into a patchwork surface of yawning gaps in the dry season) that supported grasslands to well-drained alluvial soils that supported gallery woodland along the courses of open streams. In other words, the habitat was a mosaic setting of open and closed habitats, not just a dry savanna plain. Kanapoi at that time probably had a semiarid climate that averaged somewhere between 350 to 600 millimetres of annual rainfall, and the landscape was probably similar to that found today in the vicinity of the Omo Delta at the north end of Lake Turkana. The palaeoenvironmental information from Kanapoi provided the earliest evidence of hominins venturing into relatively open habitats.

The climate was also not unlike that at Kanapoi today. We know this from a comparison of modern and fossil rodents. One of our best hominin specimens was recovered in a horizon that had numerous fragments of the tiny skulls, teeth, and skeletal elements of bats and mice. This was Nzube's mandible that Richard helped excavate and what we eventually chose as the type specimen for *Australopithecus anamensis*. These small mammals are particularly good palaeoenvironmental indicators, so to recover a sample in the same horizon as a hominin specimen is exceptionally fortuitous. Kiptalam Cheboi was a crew member famous for his skill at finding the smallest of specimens, and he devoted countless hours to sieving for these beautiful minute bones. Many years later, our colleague Fredrick Kyalo Manthi sampled the modern rodents and shrews from Kanapoi for his PhD dissertation in 2006. When Kyalo (pronounced *Cha-lo*) compared his modern collection with the fossil sample, he found many resemblances that suggested they lived in similar habitats.

But before then, John Harris once again joined me in the study of the Kanapoi fauna. He shared my burning curiosity to know how the Kanapoi faunal assemblage would differ from Lothagam's. We found the Kanapoi mammals to be remarkably diverse—during four field seasons, we collected more than fifty species. Although few of these were new to science, the assemblage was instructive. While Lothagam witnessed the demise of many Miocene genera and the first appearance of several new creatures, a suite of these more modern species carried the day at Kanapoi. Indeed, none of the primitive large mammalian genera that characterize the Late Miocene at Lothagam were recorded at Kanapoi. Thus Kanapoi provides a link between the first appearances of more modern species in the upper levels at Lothagam and the more progressive species found in later sediments elsewhere in the Turkana Basin.

All told, the picture we built at Kanapoi revealed a mosaic habitat of woodland flanking the river and in the fan of the fertile river delta that gave way to semiarid grassland. This mosaic setting was frequented by species that are precursors of many of the African game animals today. The herbivores were much as we would expect following the huge shift in climate witnessed at Lothagam: when tallied up, the number of grazing species outnumbered browsing species by a ratio of two to one. The grazers were also more common: we found more specimens of grazing species in a ratio of three grazers to each browser, which suggests a far greater abundance of the grass-eating mammals. Assuming that the dietary adaptations of these herbivores are similar to those of their modern counterparts, their fossils indicate a relatively dry climate and a mixture of woodland and open grassland, again confirming the general trends of opening and drying environments observed at Lothagam.

The Kanapoi pigs, like those in the upper horizons at Lothagam, were grazers too. And our discoveries upturned conventional wisdom on the classification of one of these pigs, *Notochoerus jaegeri*, which had hereto been thought a completely different genus, the earlier *Nyanzachoerus* common at Lothagam. Pigs are excellent markers of evolutionary changes through time, and having accurate interpretations of their diversity backed up with good diagnostic fossils is extremely important.

The lower jaws of *Notochoerus* resemble the shovel of a mechanical digger. They have small widely separated peglike incisors protruding from the front edge of their rather broad and flat mandibles. In contrast, the lower jaw of *Nyanzachoerus* is more like a garden trowel: large closely aligned incisors form a sharp scooped edge ideal for rootling. A series of well-preserved mandibles and a single skull from Kanapoi clearly demonstrated that the previous generic affiliation was incorrect and that *Nyanzachoerus jaegeri* should be referred to the genus *Notochoerus*.

This skull very nearly never saw the light of day. It was already an incredibly fragile specimen when Nzube found it close to the hominin tibia during sieving for the missing piece of leg shaft. His discovery coincided with a visit from Richard, who was very frank about the extremely poor shape it was in. "You are wasting your time!" was his blunt assessment, and while he gave me some helpful advice on how to proceed, he wanted no part of the excavation. Undeterred, I devoted several days of meticulous excavation to the skull. Then I gave it a liberal application of Bedacryl to ready it for plastering and carefully covered it with rocks and branches from wait-a-bit thornbushes to protect it from the marauding efforts of hungry hyaenas as it dried overnight. The next morning, I was dismayed to find that a hyaena had indeed tried to get at it. The persistent beast had managed to remove much of the protection before giving up. It caused considerable damage to this crucial specimen, which I needed if I was to persuade the pig fraternity that the earliest notochoere was actually *Notochoerus jaegeri*. Still, the shape of its cheek teeth showed conclusively that this reclassification was needed.

Higher up the food chain, and therefore much rarer, are the carnivores, and we collected an impressive assemblage at Kanapoi that was larger and more diverse than any found at other East African Pliocene sites—eight species in eight genera that represented five families. They beautifully document the first post-Miocene radiation of endemic African species. In contrast to other carnivore assemblages of the same age, the Kanapoi carnivores include only species whose immediate forebearers are found in Africa. These formidable predators thrived off rich pickings, which included *anamensis*.

BUT WHAT EXACTLY was this early hominin, and how different was it to Lucy's kin, *Australopithecus afarensis*? Most intriguing was the question of how the *anamensis* hand would differ from a chimpanzee's. This is critical to establish when our ancestors acquired the dexterous hands that are so essential to all the fine and fiddly tasks that were pre-requisites to our ancestors' ability to make stone tools, which in turn fuelled the greater brainpower that led to the progression of technological advancements that brought us to where we are today.

In the early days of human palaeontology, scientists assumed that our large brains were the key feature that set us apart from apes and, fooled by a forgery—Piltdown Man—thought that they had the fossil evidence to prove it. Discoveries of bipedal hominins with small brains in South Africa showed how wrong they were while *Australopithecus afarensis* eventually proved that the big brain came after bipedalism and increased dexterity. But the idea that manual dexterity is a result of freeing the hands from the constraints of walking on all fours remained an assumption with no fossils to show for it.

All primates have a manipulative hand—but ours is by far the most mobile. We owe this dexterity in large part to the shared evolution of the primate order. For example, the mobility in our shoulders and elbows originally evolved for swinging through the trees. And our five-digit configuration of strong, flexible fingers evolved for grasping branches. But because these traits were also very useful for manipulative tasks, we improved upon them. To these shared primate advantages in our upper body configuration, we added another, by freeing the hand from its locomotive constraints.

Such a high degree of dexterity boils down to three principal features: the proportions of our fingers, an opposable thumb, and the mobility afforded by the morphology of the joints at the wrist, elbow, and shoulder. The African apes tend to have long curved fingers and reduced thumbs because this configuration gives the best grip on an assortment of different-sized branches. A colobine, the most skilled suspensory arboreal monkey, barely has a thumb at all. Conversely, monkeys that spend most of their time on the ground tend to have reduced fingers like baboons and patas monkeys do. The relative proportion of the thumb to the index finger is part of what allows for a precision grip between the tip of the index finger and the tip of the thumb. Be-

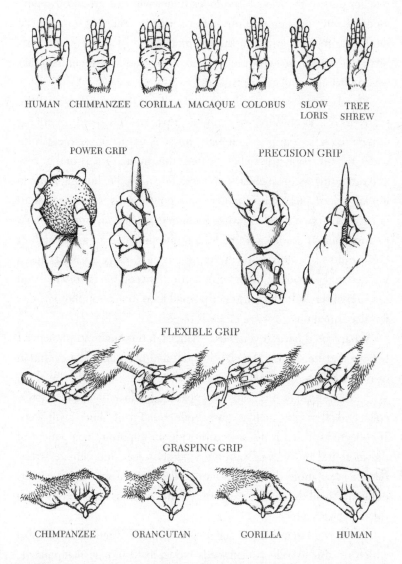

Manual dexterity vastly improved the efficiency with which an individual could gather food and also allowed for further developments such as the manufacture of effective stone-tool kits and the ability to carry and store foraged food (not drawn to scale).

cause an ape's thumb is short relative to its other digits, the thumb only reaches part of the way up the index finger whereas ground-dwelling monkeys have digit proportions better suited to achieving some sort of modified precision grip. But the full precision grip also requires flexibility in the joints to allow the rotation of the thumb—something that we excel at above all other primates.

The ability to rotate the thumb and some of the other fingers is the secret to our vastly superior array of grips. We can perform, among others, the scissor grip (which many humans use to smoke a cigarette with), the precision grip (where we use the thumb against one or several other fingers for such tasks as threading a needle, holding a key to unlock a door, picking up a pencil), and power grips (which we use to grasp a ball, open a jar, and wield a hammer). The only grips available to apes such as gorillas involve folding the fingers over an object—a hook grip to brandish a branch (or, in our case, carry a basket) and a less accurate version of the precision grip between the thumb pad and the side of the index finger to grasp small food items. Another key difference is that our grips are much stronger.

What would have driven these changes in our hand morphology? I believe that here again the clue is food; like bipedality, the evolution of manual dexterity was driven by selective forces related to feeding. With a dexterous, flexible hand, it is possible for an animal to effectively collect small objects such as grass seeds, small fruits, and tiny insects. If collected efficiently, these can provide an important component of the diet, and in times of extreme food shortages and climate stress, minute adaptations that permit more efficient food acquisition can literally make all the difference between life and death, as we saw with the Galápagos finches.

There is a fascinating analogue among modern monkeys, many of which are able to collect small seeds and fruits with their manipulative, dexterous fingers. Few can do this as fast and efficiently as modern gelada baboons. These baboons live only in the Ethiopia Highlands and feed exclusively on grass. After the rains, they eat the new green shoots; later, they eat the grass flowers; and after the grasses have flow-ered, they eat the seeds. When the grass becomes dry and withered, they dig up the grass tubers and feed on these. So they have a contin-uous source of food regardless of seasonal changes. Among modern primates, the geladas have hand proportions closest to our own and can almost achieve a precision grip. Ancestors of the modern geladas were common in the Turkana Basin and elsewhere in the Late Pliocene and Pleistocene. Their dental morphology and analyses of the stable car-bon isotopes in their fossilised tooth enamel show that these ancestral geladas, like their modern counterparts, were committed grazers. They had already evolved hand proportions that are similar to those of ge-lada baboons today, and manual dexterity would have been an essential component in their successful feeding.

The flexibility that provides the ability to rotate the fingers depends on the shape of the collection of bones that make up the wrist joint, so it is these bones, together with the finger bones, that we are most interested in finding to learn about the evolution of manual dexterity in hominins. Over time, we would expect fingers that originally resembled an ape's long curved phalanges to become shorter and straighter, which would reduce the difference in proportions between index-finger and thumb length. And the tightly interlocking wristbones that character-ize an ape should give way to wristbones that allow some rotation and flexibility. The problem is that hand bones are often destroyed before they are fossilised, and their pebblelike shapes are notoriously hard to discern, making manual dexterity an extremely elusive trait to chart through time.

Incredibly, almost all the hand bones of A. afarensis are known from Hadar. These bones show that the proportions of the finger bones (phalanges) were quite close to those of our own hands but that the fingers, although much shorter, were strong and quite curved like a chimpanzee's. Australopithecus afarensis had strong flexible fingers with which to firmly grasp branches. On the other hand, the bones of the wrist show that some movement between the joints was possible, although it is less than for humans, so the power grip would have been only weakly developed in *afarensis*. In other words, the fingers became shorter before major changes in the morphology of the wrist.

Although the hand of A. *afarensis* was considerably less dexterous than that of later hominins, its shorter fingers and partial wrist rotation revealed that the changes in hominin hands leading to greater dexterity had already begun a million years before the earliest stone tools known at the time, 2.6 million years ago. I was deeply curious to know whether the hand of A. *anamensis* was as manipulative as that of A. *afarensis* or if it would be more primitive. If we could possibly find a capitate (the most diagnostic of the wristbones), what would it look like? Against all odds, we had found one of these very bones at South Turkwell in my first year in charge of the fieldwork. It was Wambua who spotted a number of hand bones that included a beautifully preserved capitate. This looked exactly like the A. *afarensis* capitates, and at 3.5 million years, it is the same age as A. *afarensis* at Laetoli.

I had a second unanswered question regarding a tooth that could lay to rest any lingering doubts I might have about whether the Kanapoi fossils justifiably deserved their own separate species. When Tim White initially found *Ar. ramidus,* one particular specimen had convinced him that *Ar. ramidus* was much more primitive than A. *afarensis*. This was the small and rather insignificant-looking fragment of a baby's jaw that contained a deciduous (milk) lower first molar and caught my attention on my visit to Addis for its remarkable similarity to a chimpanzee's. The Ethiopian collections also included deciduous lower first molars belonging to A. *afarensis,* and these looked quite distinct. I was curious to know whether the lower first milk molar of A. *anamensis* would be more like that of A. *afarensis* or *Ar. ramidus.*

To answer both these questions—what the *anamensis* deciduous first molar would look like and how dexterous its hands would be— I needed fossils that are both rare and difficult to find because they are very small and fragile. I thought it most unlikely that I would be this lucky. The chances of coming across one are remote, and our best hopes lay in discovering one (as Tim had) in the small and delicate jaw of a baby. Hand bones are also small, and the even smaller wristbones are very rarely recovered as they are hard to recognise. A single phalanx that we found in 1995 showed us that the *anamensis* fingers were similar to those of *afarensis*. This phalanx was slightly curved, not very long, and had well-developed ridges for the attachment of strong flexing muscles; *anamensis*, like *afarensis*, had a hand that was still capable of firmly grasping branches when climbing. But we still had no idea of the morphology of the wrist, and finding any of the hand bones, let alone the exact bone we wanted, was like looking for a needle in a very large haystack.

When we returned to Kanapoi for the 1996 field season, it was these two elusive bones that we sought above all others. Taking a leaf out of Tim White's book, we decided to try his "hill crawl" method on a particularly prolific area where we had found so many isolated hominin teeth that we dubbed it the "graveyard." This area was too large to excavate and sieve given the time and funds we had available. Clutching numbered collecting bags, we spread out in a line on our hands and knees on the stony ground about half a metre apart. Then we inched forward through the morning, eyes glued to the ground as we scoured every centimetre and collected every single bone fragment we could find. Back at camp, I sorted through the fossils. A hill crawl enabled us to localise our sieves and excavations around points where we found hominins, which saved us from having to sieve the whole area. One patch on the top of a small hill looked particularly promising. Several unworn and apparently unerupted *anamensis* teeth were scattered on the slope of the hill, seemingly coming from a horizon not far from the top. We hoped that if we made an excavation here, we might be able to determine whether the bones were coming from a discrete soil layer. If this were so, we could concentrate our searches on a much smaller area. As we dug deeper into the hill, the compacted soil became extremely hard, and we doused it in precious water in the evening to soften it for removal the following day. It turned out that the fossils were not so localised, but to our amazement and disbelief, our excavation was rewarded with the discovery of one of the very bones we were looking for.

After hours of tedious digging through very hard rock, we began to uncover more isolated adult teeth as well as the milk teeth of a young juvenile. As more of these tiny, delicate, beautifully preserved teeth were recovered, I hoped against hope for the elusive lower milk molar. We found the tiny, erupted, and worn deciduous teeth including incisors, canines, and molars from the upper and lower jaws as well as unworn, exquisitely preserved adult teeth that had not yet erupted. We couldn't believe our luck when we also found a lower deciduous first molar!

Even more astonishing and improbable was Wambua's finding the jaw of another baby at a different site some distance away—and the jaw had a deciduous first molar firmly rooted in it! We could now compare our baby first molars with Tim's *Ar. ramidus* specimen. As we peered through the microscope back in Nairobi, we noticed that ours were intermediate in shape between *Ardipithecus ramidus* and *Australopithecus afarensis*, although they resembled *afarensis* more than they did the more primitive *ramidus*. This was reassuring, because we had postulated that *anamensis* belonged to the australopithecine genus, not *Ardipithecus*.

When we subsequently and unexpectedly found a capitate while sieving an area close to the main graveyard site, I was ecstatic. The capitate also confirmed what our initial analysis had told us. While the capitate of *afarensis* was in many ways more similar to that of modern humans than that of chimpanzees, the new *anamensis* capitate from Kanapoi still retained some chimpanzee features not observed in the younger *afarensis*. In particular, its morphology suggested that very little movement would have been possible between the wristbones—and much less movement than *afarensis* had. In many ways, the capitate had a morphology that was halfway between a chimpanzee's and that of *afarensis*. Now we could say with some certainty that the changes in hand morphology that ultimately led to a uniquely dexterous hand came gradually after bipedality was firmly established.

Our last field season at Kanapoi concluded in 1997 and coincided with the news that NASA had successfully landed on Mars, and once again, Nzube was transfixed and disbelieving. The Pathfinder Mission was even more incredible to him than Armstrong's moon landing three decades earlier. Parachutes to slow the descent as the projectile hurtled into the Martian atmosphere? Airbags to help *Sojourner* bounce to a safe landing? He merely chuckled at our descriptions. And then images were transmitted directly into our primitive camp on a laptop screen. Such technology seemed all the more impossible! It was an endlessly fascinating subject, and we soon found that we were all caught up in following *Sojourner's* progress through the field season while we gazed up at the stars above our camp. Everyone was enthralled by the ancient Martian floodplain, Ares Vallis, where the rover landed, which was not too different from our own remote surroundings where we were engaged in a similar task of imagining a past landscape by deciphering the fingerprints of rocks. We greedily waited for more images to be downloaded in Nairobi and brought to camp by a visiting scientist. Inspired by the spirit of exploration, we bent more determinedly to our own mission of discovery of the beginning of manual dexterity—the very adaptation that enabled modern humankind to develop the technical capability for space exploration.

OPEN-COUNTRY SURVIVORS

AFTER THE WONDERFUL DISCOVERIES OF *Australopithecus ana-mensis* at Kanapoi and Allia Bay, I was faced with a decision—should I look for the very rare Late Miocene sites of Turkana that preserve sediments older than five million years, such as Lothagam, to try to find even earlier hominins from whom *A. anamensis* may have evolved? Or should I look for new sites between four and three million years and resolve the question of what *A. anamensis* evolved into? Either plan would bring me closer to the issue of whether or not there was as much early hominin diversity as I believed there must have been. But this was not an easy decision, not least because a lot was being discovered elsewhere in the meantime. Fascinating new fossils were emerging from different parts of Africa that were shaking all our views.

At the heart of most of the ensuing controversies lies the question of early hominin diversity, made all the more intractable by the rarity of any hominin fossils older than four million years and the even more striking rarity of any ape fossils for the last six million years. There is far more known about hominin ancestry than about the equally long parallel lineage that gave rise to the chimpanzee—we only have one such example of fossil ape from the whole of Africa. This singular specimen is of a chimpanzee that lived somewhere between 545,000 and 284,000 years ago. I would guess that the reason for this incredible paucity is because of the different habitats of the two groups: the fossil sites we sample in the East African Rift are either riverine forest or more open woodland and savannah. None of them come close to the thickly forested habitats preferred by the modern apes of Central Africa. Nevertheless, I would dearly love to find an ape of equivalent age to any of the early hominins. But even more, I would love to find the ancestor of *A. anamensis*.

While we have good reason to think that *A. afarensis* probably evolved from *A. anamensis*, it is far less clear where *A. anamensis* came from. Tim White's *Ardipithecus ramidus*, the surprising 4.4-million-year-old fossils he showed me on my visit to Addis Ababa with my new Kanapoi discoveries, is one of the currently known possible contenders. *Ar. ramidus* is, as we would expect, more apelike and primitive than *A. anamensis*. It has smaller molars with much thinner enamel and longer canines than *A. anamensis* has.

Since his early work with Don Johanson that joined the Laetoli and Hadar fossils into the single species *Australopithecus afarensis*, Tim has strongly argued that there was one single main lineage of early hominins. In 2006, Tim and his colleagues described new fossils of *A. anamensis* from two sites in Ethiopia, Assa Issie and Aramis (the same site as *Ar. ramidus* but in younger sediments). At a similar age to Kanapoi, these fossils added nine new individuals of this species and extended its range by one thousand kilometres. Because they were found at Aramis in the same area as *Ar. ramidus* and *A. afarensis*, these three species represent a time-successive series from one place. This poses the question of whether this is one evolutionary lineage — or whether *Ar. ramidus* represents the diversity I keep expecting to see at this early age. It is certainly Tim's view that *Ar. ramidus* is the direct ancestor of *A. anamensis*. However, for this to be true, a lot of evolutionary change would have had to happen in a very compressed amount of time — only 200,000 years. *Ar. ramidus* is found at 4.4 million years, and the oldest specimens of *A. anamensis* lived at approximately 4.2

million. Yet the changes in the canine-premolar complex between the two are quite dramatic, and I remain unconvinced about their phylogenetic relationship.

But what about fossils much older than *Ar. ramidus*? Since the late 1990s, three new candidates for an earlier potential australopithecine ancestor have been discovered. These show a hodgepodge of ape and human traits. At this early age, so close to the split from our common ancestor with chimpanzees, it is incredibly difficult to distinguish from the fragmentary bones available whether they represent ape or hominin. And in any case, what are the defining characters that set hominin apart from ape? Bipedalism, an obvious divider, would not have arrived in an instant. Similarly, the canine would have reduced in size only gradually.

In fact, we do see this gradual process among the early australopithecines. The male canine has already evolved to the point that it occludes (bites down) directly on the lower premolar rather than overlapping outside this lower tooth (honing) as it does in apes and monkeys. Nevertheless, reflecting its ancestral condition in apes, an early male australopithecine like *A. anamensis* still has a rather large set of canines. Moreover, although the canine is no longer honing, the lower premolar still has some enamel extending down the outer surface of the root. When the canines hone, as they do in apes, they are continually sharpened as they wear against the first lower premolar, and in order to preserve this tooth as a sharpening device, the enamel surface extends partway down the outside of the root. In the more recent australopithecines, the extension of the enamel on the lower premolar recedes as the canine decreases in size, and its crown gets bigger as it increasingly takes on a chewing function.

These are the two defining characteristics—bipedalism and small canines—that can be argued about as defining a hominin versus an ape. Yet, like the ring species of herring gulls I mentioned in the last chapter, the line between the early differentiating ape and hominin lineages must surely be an arbitrary one. Nevertheless, the scientists who discovered these key specimens emphasize the hominin attributes, perhaps thinking that this would somehow be more interesting. Yet an ape at the threshold of this momentous split would tell us what the ancestral condition actually was. How egocentric of us humans to assume that while we have undergone sweeping evolutionary changes the apes have muddled along without changing much at all!

AMONG THE CONTENDERS of would-be earliest hominins is *Ardipithecus kadabba,* an older species presumed to be an ancestor of *Ar. ramidus.* This was found by the same Yohannes Haile-Selassie who recently published the cranium of *A. anamensis* from Ethiopia. What is most interesting about *Ar. kadabba,* dated to between 5.8 and 5.6 million years, is that it has a partially honing canine that self-sharpens against a thickly enameled lower premolar. Nevertheless, the cross section and shape of the upper canine is closer to that of a hominin than an ape. It would seem to be a transitional species, but there are frustratingly few bones and individuals represented in the sample.

A similar-aged fossil, called *Orrorin tugenensis,* was found by Brigitte Senut and Martin Pickford in the Tugen Hills near Lake Baringo, Kenya. The tragedy of this potentially fascinating find, nicknamed Millennium Man, is that few scientists have had the opportunity to study or even set eyes on the fossils. They are said to be locked away in a bank vault in contravention of Kenya's antiquity laws. Nevertheless, from the published descriptions, *Orrorin* would appear to have had fairly primitive teeth. It has small molars, and its front teeth are similar to those of a modern female chimpanzee's. However, its enamel is as thick as it is in hominins—perhaps another example of transitionary characteristics. A further interesting aspect of Millennium Man is its partial femur. From the shape of this femur, Senut and Pickford initially concluded that it walked bipedally. More recently, two other scientists have had the rare opportunity to study the femur. They found that it resembles that of australopithecines, pointing to the likelihood of Millennium Man being bipedal. *Ar. kadabba* also has a partial femur and a single toe bone, the shape of which led Haile-Selassie to reach the same conclusion.

The most riveting of these early discoveries that have been published is from Chad, where the French palaeontologist Michel Brunet found an even earlier, beautifully preserved, and remarkably complete skull and mandible from two different individuals of another half-ape, half-manlike creature. This find is all the more spectacular for the extreme conditions that Michel and his team had to operate in. The blasting hot winds and searing temperatures of Chad's Djurab Desert make Turkana seem like a positively mild working environment in comparison. Brunet came to Lamu with his then-wife Agnès to spend Christmas with us shortly after his sensational find in 2001. They brought with them a cast replica of the skull and a laptop loaded with digital images of their fieldwork. For several evenings, we were regaled with astonishing stories about their camp life. Brunet had to fly all his water (along with his wine!) to his remote camp, relying heavily on the French army to resupply him. Sediments laid down along the lush, well-vegetated shores of ancient Lake Chad now lie below a moving sea of sand dunes, which are propelled over the underlying bedrock by the force of the wind. Brunet's sites are ephemeral, here one year and gone the next when they are reburied beneath a mountain of sand—decreasing the chances of finding such a stunning fossil a thousandfold.

Because there are no datable layers in the sandy desert, the only way Brunet could date his fossil skull and mandible was to infer a date based on the fossils of the fauna that lived both in Chad and somewhere else where their age is known. This puts his fossils reliably between six and seven million years old, pushing the evidence for the date of the split between humans and apes further back. Lothagam was one of the sites where Brunet found similar fauna—so we were looking at the right age ourselves. Since then, Brunet and his team have applied a new method for dating to the material from Chad—cosmogenic nuclide dating—and the age they obtained (between 6.8 and 7.2 million years) is the same as they had predicted based on the fauna. Brunet's find is still more unique for its location far to the west of the Great Rift Valley. It serves as an important reminder that our ancestors were spread over large parts of Africa.

Since the initial discovery, Brunet's team has also found additional pieces of lower jaws and several isolated teeth. The well-preserved skull has several large cracks that have caused some distortion, but CT (computer tomography) scanning technology has enabled the research team to reconstruct its original form. The results are astonishing. It had a tiny brain—the smallest capacity of any adult hominin, yet it didn't look anything like an ape. Most noticeably, it did not have the long snout that early australopithecines share with apes. The canines were rather small, and they were nonhoning, a surprisingly "modern" character for a hominin that age, particularly when considering that it is much older than *Ar. kadabba,* which has partially honing canines. And the forward position of the foramen magnum and the orientation of the eye sockets also follow the requirements for bipedality and differ from that of apes. So Brunet concluded his fossils belonged to a biped and a hominin. That said, the incompleteness of the other specimens he found meant that comparisons could only be made of the teeth. These are sufficiently different that Brunet named both a new genus and a new species for his fossils, which he called *Sahelanthropus tchadensis.*

It is a real stretch, even with the fragmentary nature of *Orrorin tugenensis* and *Ar. kadabba,* to force these three—from widely different geographic areas and different features—into a single evolving lineage. It involves bunching thick and thin enamel and honing and nonhoning canines, for starters. It is far more likely that this is the evidence of early hominin diversity we have all been looking for. Any one of these, or an as yet undiscovered creature, could have been the stem species that evolved into the australopithecine line. While a tantalizing picture of a chronologically, geographically, and morphologically diverse set of potential ancestors for *Australopithecus anamensis* begins to form, the question of who its descendants were also remains unanswered.

• • •

THE TREND that we begin to see in the *anamensis-afarensis* line could be expected to continue in later species since the climate continues to dry over the next few million years. The most obvious of these is one that has to be related to diet — in response to the increasingly abrasive and tougher food found in open-country vegetation, the molars and premolars steadily increase in size, the enamel becomes thicker, and the incisors and canines reduce in size. Correspondingly, the skull morphology changes to accommodate the attachment of bigger muscles for more energetic and prolonged chewing as well as for biting with greater force. Since the point of having a long snout is to be able to bare the impressive canines by opening the mouth as wide as possible, the constraints on the size of the snout disappear if you no longer have large canines to display. If your diet involves a prodigious amount of forceful chewing of tough foods, a shorter snout provides a better platform for large molar teeth and strong muscles. As we would expect, the snout gets shorter in more recent australopithecines with less need of a display function.

This means the beginning of thicker enameled and larger chewing teeth (which scientists poetically call "megadonts") and bigger and better surfaces to attach muscles to on the skull ought to become more pronounced through time. That other hallmark of the hominin line, manual dexterity, which had already begun in *anamensis,* could also be expected to become more manifest in progressively younger hominins. These trends are indeed apparent in some later hominins — but other than manual dexterity, they are notably absent in our own *Homo* line. Did we really arise from the australopithecines? The jury is still very much hung on where the australopithecine lineage leads. But before delving into the twinned problem of their evolutionary destiny and the origins of our own genus, we need to consider the important insights from the hominins that existed after *A. anamensis.*

THERE IS A GRATIFYINGLY large sample of another well-known australopithecine. This species includes the very first hominin fossils ever found in Africa. Serendipity led to this first discovery from limestone cave deposits in the Transvaal. A student of Raymond Dart's had previously brought him a skull of a fossil baboon blasted from the Buxton Limeworks near Taung, which had piqued his curiosity, and Dart had requested any further fossils the owner of the quarry might find. Nestled in the jumble of limestone rocks and fossils that were subsequently delivered, Dart recognised the cast of a perfectly preserved brain and also found the skull that had enclosed the brain, and he would spend many months chipping away at the limestone matrix encasing it until he was able to reveal an almost perfectly preserved skull and mandible of a young child. Dart, an anatomist by training, was uniquely qualified to recognise the "convolutions and furrows of the brain" complete with the blood vessels of the skull. He had looked at enough human brains to see the unmistakably human attributes of what lay before him, and he was broadminded enough to see the implications. Too many differences separated the Taung Child from the baboon he had expected to find. Most telling of all was its diminutive humanlike milk canines and the position of the foramen magnum, which was far enough forward to suggest bipedal locomotion.

One problem was that the Taung Child was discovered in 1924, a time when many in the palaeontological establishment accepted as a matter of course the alleged supremacy of Europeans over other races. The fossils met with an extremely frosty reception when Dart hastily penned off a paper to *Nature* in which he enthusiastically announced his ape-man from Africa, *Australopithecus africanus*. For the palaeontological community, the very notion that human origins arose in Africa was anathema. Because of the prejudiced insistence that humans must have originated in Europe and the prevailing view that our large brain evolved before other adaptations such as bipedalism, Dart's tremendous small-brained African discovery was ridiculed and dismissed as merely some kind of juvenile ape.

Many more finds followed, including an adult female skull, nicknamed Mrs. Ples, found twelve years after the Taung Child by Robert

Broom, a maverick Scot who contributed much to the discovery and interpretation of the early South African hominin collections. But it wasn't until several decades later that Dart's discovery and interpretation were at last accepted.

A. *africanus* is a slightly built creature that in many ways resembles A. *afarensis*. Along with its small canines, it has megadont cheek teeth that, like A. *afarensis,* have thick enamel and are enlarged from the primitive condition. It was most certainly a biped, but until now, nothing could be said about how dexterous its hands were. This was about to change—for one of the greatest discoveries ever made will include the very bones we needed. This sample of *africanus* fossils was retrieved from the Silberberg Grotto, a cavern deep in the Sterkfontein cave system in Krugersdorp. They included a collection of foot bones belonging to an adult individual. But as so often happens in palaeontology, it turned out not to be quite that simple!

In 1978, Phillip Tobias, an eminent and highly respected palaeoanthropologist from the University of Witwatersrand, asked his field director, Alun Hughes, to look in the older and lower levels that are exposed in the Silberberg Grotto. Many of the fossils recovered from these caves were retrieved from lumps of limestone called breccias, which were blasted out by miners using explosives. They raised the blocks of breccia and rubble from the cave floor and extracted predominantly monkey fossils. These bones were stored in labelled boxes in the university's anatomy department and a work shed at Sterkfontein. Nearly two decades later, Ronald Clarke was puzzling over the bones at the University of Witwatersrand, where he works in the anatomy department. Ron is an old friend from when he worked at the Nairobi National Museum for Mary and Louis as an extremely skilled preparator, cleaning many of the Olduvai fossils and trying to stick the thousands of pieces of hominin fragments back together. Among the Silberberg fossils, Ron had noticed a surprising absence of antelope bones. He was looking for the missing bovids in the boxes in the Sterkfontein work shed one day when, to his astonishment, he noticed several bones belonging to a foot that was unmistakably hominin. These beautifully preserved and hitherto discarded fossils represented the first known *A. africanus* foot. Because of its small size, it soon became known as Little Foot.

In a completely separate coincidence, Ron was later looking for some fragments of a different hominin skull in the safe in the anatomy department when he noticed another box labelled as containing monkeys from the Silberberg Grotto. Seeing this, he immediately took the box with him to have a closer look and was rewarded with more foot bones that surely belonged to Little Foot. A fragment from the lower end of the tibia was also in the box, and this discovery spurred him to scrutinise the box's contents further. Sure enough, in a nondescript bag labelled BOVID TIBIA, Ron found a much more complete piece of hominin tibia that articulated perfectly with the left talus (the ankle-bone), one of the original four foot bones.

Ron next tried a literal shot in the dark. He dispatched two of his men with a torch to the grotto, tibia in hand. The cave walls and lumps of breccias are all studded with numerous protruding pieces of broken-off bone from the miners' blasting some sixty-five years earlier. After only two days of searching, they were able to match the broken-off surface of the left tibia to a piece of bone embedded in the breccias in the cave. And another break next to it matched up with the piece of the right tibia. The odds were that there could be far more of Little Foot's skeleton concealed in the breccias, and Ron has recently disclosed that this remarkable specimen now includes almost the entire skeleton, including a full set of hand bones. This has to be the most complete early hominin skeleton ever to be discovered, and its hand will surely shed light on the crucial question of how manual dexterity evolved.

But major controversies surround Little Foot. One relates to its age, the other to its identity among australopithecines. A big problem with the fantastic fossils emerging from South Africa is that they have all been found in limestone caves. Caves present notorious dating problems because the sedimentary layers get jumbled up as water erosion causes parts of the cave to fall in and disrupt the order of the sequence by mixing older bones with newer ones. Both stratigraphic dating methods and faunal comparisons are subject to a large degree of uncertainty—so large that two different initial interpretations put Little Foot at 1.07 million years old and 3.3 million years old. Since then, Ron and his team have applied the same novel dating technique that Brunet used to date the *Sahelanthropus* site—and proposed that the fossil dates to 3.67 million years ago. This makes it the same age as Lucy and only slightly younger than *anamensis* and, most important, significantly older than the *Australopithecus africanus* fossils found in other parts of the same cave at Sterkfontein that date between three and two million years ago. Most intriguing, Little Foot shows many primitive features that have led Ron Clarke to give it a new species name, *Australopithecus prometheus,* with possible closer affinities to the *anamensis-afarensis* fossils from East Africa than later South African remains. These proposals remain contentious, and more research is needed before we can establish with certainty how these South African hominins relate to their East African cousins.

These dating difficulties have also been the predominant challenge in sorting out how the different hominins from South Africa relate to one another. Other types of australopithecine have been found in other caves in the Transvaal—at Kromdraai, Swartkrans, Drimolen, and Malapa. One of these is relatively unique. While clearly an australopithecine, it shows a number of derived features that buck many of the trends observed in other australopithecines. Its name is *Australopithecus sediba,* and it was discovered by Lee Berger in the form of six partial skeletons at Malapa. The fossils, which include adults and infants, are two million years old and offer us insight into the range of adaptive opportunities hominins were facing at that time. Particularly interesting among these is the hand of *A. sediba,* which with its short fingers and long thumb suggests a precision grip. However, its wristbones, stunningly complete in one of the two adult skeletons, show primitive features suggesting little rotation and flexibility—a true evolutionary mosaic of features.

The other type of australopithecine has been known since the very early days of Broom and Dart, and is so different that most scientists now believe it should have its own genus. This creature is a true megadont: its molars are massive. Yet the incisors and canines aren't nearly as impressive—and the male canine is so small that it would barely scare away a mouse! In keeping with the nondescript front teeth, the snoutlike face of earlier australopithecines has also been replaced with a dish-shaped, snub-nosed face—as though somebody forcefully punched the nose in. The reason for this effect is that the architecture of the skull is all about chewing. To power the huge molar teeth, the face has big flaring cheekbones that provide room for seriously big chewing muscles and give the punched-in look. Because the brain is still too small to allow much surface area to attach muscles to, the skull has a very distinctive and prominent crest running along the midline at the top—a sagittal crest. This extends the needed surface to attach the muscles to.

This robustness lends itself to the hominin's species name, *robustus*. Originally assigned to the genus *Australopithecus,* these southerners are now considered to belong to a separate genus, *Paranthropus* ("next to man"). Through comparisons of the fauna and because the robust creature is never found in the same cave deposits as the more gracile *A. africanus,* we know that the robust hominin is the younger of the two. What we don't know is what it evolved from. For years, scientists argued whether *robustus* evolved in South Africa from *A. africanus* or from the lineage of *anamensis-afarensis* in East Africa where *P. robustus* had a relative that strongly resembled it: an even more impressive megadont.

The East African "robusts" had even bigger cheek teeth, an even bigger crest along the top of the skull, and even more sweeping cheekbones that are adapted to a similar type of diet requiring prolonged heavy chewing. When Mary found the most famous and first of these East African megadonts, she and Louis gave Dear Boy the species name *boisei* in recognition of Charles Boise, who supported them financially for many years. And my very first encounter with a hominin discovery was the *Paranthropus boisei* skull KNM-ER 406 that Richard and I came face-to-face with in the riverbed on our first memorable camel safari in 1969. The robust australopithecines were clearly common in the Turkana basin, where *P. boisei* crops up in most of the sediments younger than two million years until it died out sometime less than one million years ago. Although the hyper-robust males crop up again and again, the slighter females occur rarely. Only one partial skull has been found in East Africa. This female, KNM-ER 732, lacks the sagittal crest and massive muscle attachments of the male counterpart. We found it shortly after Richard and I found the *P. boisei* male skull in the riverbed.

The most impressive megadont of all is nicknamed the Black Skull for its gorgeous dark colouration. This impressive specimen was discovered in West Turkana in 1985 in 2.5-million-year-old sediments at a site called Lomekwi. This was soon after the Turkana Boy was discovered at Nariokotome and Richard had shifted focus from the east side of the lake. The field crew was exploring sediments about halfway up the lakeshore roughly opposite Koobi Fora. It was Alan Walker who found a pile of scrappy-looking bones that several of the crew had seen and dismissed as an unremarkable bovid, which naturally caused them considerable mortification when they realised what they had missed. Although they assumed at first that they were fragments from an *Australopithecus* skull, by the time Richard returned to Nairobi with the news the following day (happily for me, on my birthday), he had already realised that he had "some odd hominins" to show me and that he "might have to change his theories a little."

Pieced back together into an almost complete gorilla-like skull, the individual (probably male) epitomizes many of the robust characters of its line. It had a reduced dish-shaped snout (although it still protruded in contrast to the later *Paranthropus*), and an enormous sagittal crest that allowed masses of room for muscle attachments and compensated for the lack of space in a very small brain. Although the Black Skull is missing all but one tooth, the broken-off roots show that the cheek teeth were huge. These big molar teeth and the prominent bony ridges for attaching big muscles are the defining features of the Black Skull and, like those of *P. boisei* and *P. robustus,* are indicative of a diet of tough fibrous foods that required plenty of chewing. The Black Skull is named *Paranthropus aethiopicus* not because Richard chose it but because the species had already been found. In 1968, Camille Arambourg and Yves Coppens discovered a large mandible that they named *aethiopicus*—which remained largely ignored in a museum. This mandible shares the same robust features as the Black Skull and also lacks teeth.

HOW CAN WE interpret this megadont version of hominin? It is universally accepted that the differences between *Australopithecus* and *Paranthropus* are driven by dietary changes—Mary and Louis's Dear Boy skull was also known as Nutcracker Man. But it is only comparatively recently that anyone took a long and close look at the megadont's teeth using new technology and refined techniques that allow us to reconstruct in far greater detail the actual composition of its diet. The results from two of those techniques—examining the microscopic scratches and the chemistry of the teeth—have given us much to think about.

Scientists have long scrutinised teeth under a microscope to count the scratches and pits that scar the enamel from chewing particular types of food as a way to identify diet. But nowadays, powerful machines can do this tedious job of counting and measuring with ease, precision, and speed. Sure enough, compared to *A. africanus,* the greater number of pits and the larger and more numerous scratch marks suggest that *Paranthropus* must have been chewing on harder foods. But the chemistry of their teeth reveals a more complicated picture.

The isotopic analysis pioneered by Thure Cerling and his colleagues that proved so instructive to us at Lothagam has also been used to help explain why a group of hominins evolved such megadont teeth. These new methods are so precise that by taking a series of microscopic samples from different levels of the surface enamel Thure is now able to distinguish the isotopic composition of the enamel at different points in the individual's life. These analyses showed that the isotopic values were surprisingly variable, changing markedly both seasonally and in different years. This is consistent with a highly variable climate, but *Paranthropus,* with its whopping teeth, had such extreme and specialised chewing adaptations that everyone had assumed their energies were focused on one particular type of food. But Thure concluded that *Paranthropus* was not a dietary specialist and was instead eating a large variety of savannah-based foods, including a variety of grasses and sedges. In fact, *Paranthropus* had an extremely flexible diet. In Thure's words, these counterintuitive results "may indicate that its derived masticatory morphology signals an increase, rather than a decrease, in its potential foods." In plain English, becoming a successful generalist and broadening one's diet in a highly fluctuating climate is a sure way to maximise the chances of survival!

My favourite fossil monkey shows an interesting parallel evolution —the ancestors of the Ethiopian gelada baboons that, using their remarkably dexterous fingers, dine exclusively on all that grass has to offer in the Ethiopian Highlands. The geladas' ancestors are called *Theropithecus,* a ground-dwelling baboon that found itself in much the same predicament as the australopithecines as the climate dried and habitats opened between three and two million years ago. *Theropithecus brumpti* lived between 3.5 and 2.5 million years ago and preferred closed woodland. The males sported fearsome daggerlike honing canines more than an inch long in a long snout—the typical sexually dimorphic primate pattern. But its later relative, *Theropithecus oswaldi,* had to make some serious compromises to survive. And survive it did: *T. oswaldi* is the most common and most widespread fossil monkey known in East Africa. Its range extended from South Africa to Europe and across to India, and it survived for more than 1.5 million years until

extreme climate fluctuations and competition from diverse grazing bovids eventually drove it to extinction.

The evolutionary secret to this phenomenal success shows many
parallels with *Paranthropus*. Through this 1.5-million-year period, the
snout is reduced, the cheek teeth are enlarged, and the canines become quite small for a monkey. The lower incisors of a late *T. oswaldi*
are so reduced that they are usually missing altogether in older individuals. Since this is a baboon, not a hominin, the canines still hone and
are not reduced to the same degree—but considering the continuing
need for canines in sexual display, the changes to the snout and canines
are remarkable. The molars adapt to the selective pressure to withstand more abrasive material and more chewing in a different way from
Paranthropus because the starting shape of the tooth is very different.
So instead of developing thicker enamel on a flat grinding surface, *T.
oswaldi*'s molars resemble those of grazers such as hippos: with an increasing number of convolutions and indentations and a higher crown
to resist wear through the animal's lifetime. To power the grinding in
this evolving jaw, *T. oswaldi* has a sagittal crest running along the top
of the skull that is bigger and better than that of *P. robustus,* and it ac-
commodates the need for muscles to attach to a skull housing a smaller
baboon-sized brain. The youngest *T. oswaldi,* less than a million years
old, are also very big animals—comparable to female gorillas. This increase in body size is probably a response to predatory pressure in open
country, and there is much evidence that *T. oswaldi* was indeed heavily
predated upon, including by hominins.

We also find evidence that *Paranthropus* society was male dominated, which suggests that alternative methods of sexual display and
defence must have evolved to compensate for the puny canines brought
about by this dietary adaptation. Males in highly sexually dimorphic
primate societies (gorillas and baboons, for example) have a problem:
only one alpha male enjoys the undivided sexual attentions of a group
of females. For the rest of the males, life is about flying under the

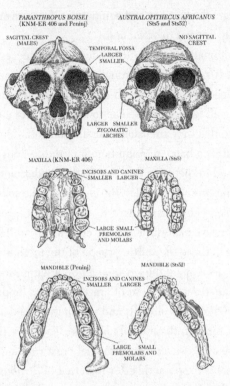

Among the australopithecines, illustrated here by KNM-ER 406 as robust and Sts5 as gracile, the robust forms show extreme megadont adaptations involving the size of the teeth and the flaring of the cheek bones and crests for the attachment of extra large muscles. Not drawn to scale.

radar, sneaking sexual favours when the dominant male is distracted, and constantly jockeying with other potential rivals to become the next kingpin. Upon reaching their prime, male gorillas become significantly larger and more powerful, weighing up to a staggering 275 kilos (more than 600 pounds). Their huge broad shoulders and their beautiful coats crowned by the distinctive silver hair on their powerful backs make these imposing ubermale individuals impossible to miss in the crowd. Although an average adult man is bigger than many female gorillas, the silverback is far heavier and ten times stronger than the biggest foot-ball players and wrestlers. These great beasts tower over females and nonalpha male gorillas, who view them with a mixture of respect and terror, deferring to them in everything and proffering preferential access to every single resource. A world with too many silverbacks would be brutish indeed. Nature's solution to this highly stressful situation is ingenious: males have their development arrested. While female gorillas reach sexual maturity by about age seven, males remain sterile until they mature into the silverback state between the ages of eleven and thirteen. But in the interim, blackbacks are highly vulnerable, living either on the periphery of the group or in smaller groups without the protection of a silverback, so predation is much higher.

The late Charlie Lockwood, who had an abiding and intense curiosity about the South African hominins, published a compelling study that shows that *P. robustus* probably also followed a pattern of extended male development like gorillas do. Lockwood went through the South African collection to pick out adult specimens with pieces of face and jaw. From the degree of wear on the cheek teeth, he estimated the age at death of each individual and found that old adult males were significantly larger and more robust than younger males and adult females. His conclusion was that while females attained their full size upon sexual maturity, males followed a pattern of extended development, growing larger after attaining dental maturity like modern male gorillas do. Lockwood also found many more males in his sample, which suggests more males fell victim to predation. When remains are found in caves, as they are in South Africa, the individuals were usually dragged there by a carnivore, as evidenced by the damage found on many of the bones.

One big and unsettled question about the robust hominins is how the South African megadonts tie into the well-dated australopithecine story in East Africa. Two different scenarios are perfectly plausible from an evolutionary standpoint. The first posits that the robust form in the east and south of the continent could have evolved in parallel from different stem species, each taking advantage of similar feeding niches of tough, fibrous vegetation. Other examples of parallel evolution abound. The venomous sting has evolved separately at least ten times: in jellyfish, spiders, scorpions, centipedes, insects, cone shell molluscs, snakes, cartilaginous fish (stingrays), bony fish (stone fish), mammals (male platypuses), and plants (stinging nettles). On an even grander scale, the radiation of pouched mammals (marsupials) in Australia parallels the evolution of placental mammals elsewhere in the world to an extraordinary degree and fills many of the same niches. Although many of the placental mammal and marsupial pairings look quite different, their adaptations are often very similar.

Today, the genus name *Paranthropus* is frequently applied to all the robust australopithecines, which implies that the two robust forms share a single ancestor—most probably the East African *anamensis-afarensis* lineage. This is the nomenclature we have followed in this book, but it is important to point out that the true evolutionary path from the early gracile australopithecines to the robust forms remains yet another conundrum in need of more fossil data.

If we look at all the robust australopithecines from East Africa in chronological sequence, we see that they exhibit increasingly megadont cheek teeth with corresponding adaptations of the skull. The later specimens have a slightly larger brain, and thus the sagittal crest is never as extreme as in the Black Skull since there is a larger surface area for attaching the large chewing muscles.

The reason that the dietary changes of the robust australopithecines is so significant is because you simply can't be gutsy and brainy at the same time as brains and guts are both metabolically expensive to maintain. Ruminants have incredibly specialised digestive tracts to break down the tough fibrous walls of plant cells. A cow has four chambers to its stomach—a veritable factory for fermentation—where billions of bacteria, protozoa, moulds, and yeasts do most of the heavy lifting. In the first and biggest compartment, the rumen, every millilitre of rumen fluid contains an extraordinary number of microbes: some 25 to 50 billion bacteria and 200,000 to 500,000 protozoa. These organisms digest the plant fibre and produce volatile fatty acids that are absorbed directly through the rumen wall into the bloodstream and converted into glucose in the liver. They supply 60 to 80 percent of the energy needed by the cow. The reticulum, with its honeycomb-like lining, is the chamber involved with rumination. When cattle ruminate, or chew their cuds, they are regurgitating a bolus of incompletely chewed feed that they then rechew and process a second time in the rumen. The fourth chamber, the abomasum, is the "true stomach," which functions much like the human stomach, producing acid and some enzymes to start protein digestion. The cow's small intestines function much like ours, absorbing the digested nutrients (except for the volatile fatty acids already absorbed upstream).

When I first started working for Louis at the Tigoni Primate Research Centre, I couldn't help noticing the colobines version of this process. Compared to the other monkeys, the *Colobus* were my hands-down favourites. I liked them the most because they were more relaxed and laid-back, placidly chewing leaves and sitting around with their large bellies protruding as they belched contentedly. Their behaviour was in sharp contrast to the lively, more aggressive *Cercopithecus* monkeys. Colobines don't have four stomachs—but they do have two. The first compartment, or fore stomach, functions the same way as the cow's rumen, where cellulolytic bacteria degrade the fibre. For both ruminants and leaf-eating monkeys, breaking tough fibres into digestible components takes time and requires the food to pass slowly down a long digestive tract. The *Cercopithecus* monkeys, in contrast, eat a lot of fruit and don't have the fermentation fore stomach or a long digestive tract because fruits are made of simple carbohydrates that are readily digestible with enzymes available in the acid stomach.

But switching from a diet comprised of mostly plant fibres to one with higher energy yielding fruit can only reduce the gut so far. The australopithecine strategy of becoming bipedal in order to move between the trees and the opening grasslands while foraging on fruits, plants, tubers, eggs, and insects worked well when it first evolved four million years ago. But obtaining sufficient energy from these foodstuffs would have been getting more and more difficult two million years later. Not only were a great diversity of competing species moving into the same feeding niche, but as we shall see, this kind of lifestyle was becoming a risky business as greater seasonal variability and longer-term extreme fluctuations in climate came into play.

THE AUSTRALOPITHECINES responded to the increased competition and changing climate by specializing in at least two different ways —one was to consume tough fibrous vegetation, for which they had to grow robust features (big chewing cheek teeth, large muscles for mastication, and probably a long gut). As we would expect, however, the brain size of these megadonts didn't get any bigger than that of *afarensis*. We already had a glimpse of the second alternative trajectory among late australopithecines in the nonrobust adaptations of *Australopithecus sediba* in South Africa, whose hands had evolved a precision grip. In East Africa, something similar may have been happening.

In 1997, Yohannes Haile-Selassie found yet another important fossil in Ethiopia's Afar region—a 2.5-million-year-old skull. This fossil is puzzling. It is conceivably a possible candidate for the female counterpart to the Black Skull, but it is more generally believed to be evidence of further diversity among the late australopithecines. Because the skull lacks the pronounced sagittal crest of the Black Skull, Haile Selassie's colleagues, including Tim White, attributed this specimen to a new species of *Australopithecus*, *A. ghari* (*ghari* means "surprise" in the Afar language). The combination of large cheek teeth with moderately large canines and the small sagittal crest characterizing *A. ghari* might arguably be expected at 2.5 million years. Although it is not yet at all clear where this hominin fits, it does emphasise the general trend of increasing megadontia seen in the australopithecines.

Equally interesting is evidence of the use of stone tools close to the age and site where *A. ghari* was discovered. Although no artefacts were actually found, several broken antelope long bones show percussion marks that indicated they had been smashed for the extraction of the bone marrow. Cut marks on these bones and on an antelope lower jaw clearly indicate that meat had been cut from these bones using stone flakes. This is some of the earliest evidence of stone tools, which strongly supports the suggestion that the hands of some East African late australopithecines had become more dexterous just as the hands of the young *sediba* had.

Even more tantalizing is the discovery of the very earliest stone tools currently known. These 3.3-million-year-old tools were found by Sonia Harmand in 2011 at Lomekwi on the west side of Lake Turkana. They are surprising for their degree of complexity. Not only would the maker of these tools have to have had significant motor skills and manual dexterity, but the tools demonstrate that their makers must have had a strong understanding of how the stone fractures because they were using a combination technique of core reduction and battering. This Lomekwian tool industry is astonishing and has pushed the earliest known tools back in time by 700,000 years.

SOME OF WHAT we now know about australopithecines and their ancestors was yet to be discovered in 1997 when I was pondering whether I should next search for the ancestors or the descendants of *anamensis*. After much thought, the australopithecines won the day. As a group, they show a remarkable longevity for any hominin lineage (some three million years), taken in unison from the first australopithecines at Kanapoi, 4.2 million years ago, to the most recent robust *boisei* at Peninj, a full three million years later. Late nonrobust forms show some new trends in their hands and possible behaviours that suggest meat consumption. And most critically at this time, our own genus *Homo* turns up between 2.5 million and two million years ago. *Homo* evolved from a completely unknown ancestor among that elusive diversity of hominin species and gained all the right adaptations to thrive in the changing conditions. Sorting that out would be my next target.

11

‖‖‖

A FRIEND FOR LUCY?

IMPATIENT TO GET TO WORK, WE HAD BROKEN OUR CARDINAL rule of beginning our field season in June—after the rainy season is truly over—and we found ourselves halted at the banks of the Kalakol River on the west side of the lake. The normally dry sand river was sodden. It would take some careful negotiating for our vehicles to cross, especially for the heavily burdened expedition lorry. We camped that night on the riverbank to the sounds of an unusual and loud chorus of frogs occupied in their frenzied search for a mate before all the water was gone. The familiar surge of excitement and anticipation that accompanies the approach to a new site was even sweeter than usual. It was May 1998, and for the first time, Louise would join me as coleader of the expedition. We each had our own plans for the fieldwork.

At Kanapoi, we had found evidence of an early biped—but not of diversity at this stage of human evolution. Yet I still did not believe that the *anamensis-afarensis* lineage was the only species of hominins alive between 4.2 and three million years ago. It couldn't be! A few years earlier, Michel Brunet had announced a new species of this age, *A. bahrelghazali,* found in Chad's Djurab Desert. But Brunet had only discovered one specimen, and it was the front part a mandible. This was not nearly diagnostic enough to persuade cautious scientists—especially the "lumpers" who steadfastly insist on the minimum number of species possible—that this was anything other than an *A. afarensis*.

So we bent our steps to Lomekwi on the other side of the soft sands of the Kalakol River. Lomekwi's sediments are of the same age as *A. afarensis,* and I had high hopes of finding a completely different-looking creature.

During this time, the australopithecines were undeniably the most common and successful hominin line, cropping up across a wide geographical area—Ethiopia, Kenya, Tanzania, Chad, and South Africa—a clear signature of evolutionary success. But at 2.33 million years, a completely different creature suddenly appeared on the scene as evidenced by the shape of the small teeth and rounded palate of the first unambiguously *Homo* maxilla found by Johanson's team in the younger beds at Hadar in Ethiopia in 1994. This upper jaw shows clear evidence of an alternative dietary strategy to that of the megadonts we saw in the last chapter as it does not have the same thick enamel and hefty chewing molars. And after two million years, *Homo* became more common. The antecedent to this new type of hominin could have evolved elsewhere before migrating into the Omo-Turkana Basin. A mandible dated at 2.8 million years was published in 2015 from a site called Ledi-Geraru in the Afar region of Ethiopia. This older specimen shows a mix of characters intermediate between those typical of the australopithecines and *Homo.* But in 1998, nothing like this had been found yet. By looking in sediments around the three million-year mark, I hoped I might find evidence of diversity within the basin at the time Lucy lived. I certainly meant to try.

We had settled on Lomekwi because of its richly fossiliferous sediments and its geologic time frame. It was not that far from Nariokotome, where Kamoya discovered the first fragments of the Turkana Boy fifteen years before. My decision to open a new field site here was based on discoveries from those earlier expeditions. When the field crew had surveyed the Lomekwi sites in the early 1980s, they found two intriguing mandibles, which coincided with the time when *A. afarensis* lived at Hadar. These two curious finds had remained undescribed largely because no one understood what they represented. I hoped that if I spent more time at Lomekwi, I might find clues as to what these mandibles were. For her part, Louise was studying for her PhD at the University of London and planned to survey the younger sediments represented at Lomekwi and other sites to the north. She wanted to investigate whether there was any evidence to support Elisabeth Vrba's provocative turnover-pulse hypothesis. Vrba had predicted that there was a major change in the faunal assemblages in Eastern and Southern Africa 2.5 million years ago and that the emergence of *Homo* was part of this turnover event. In addition to studying the large faunal collection housed in the Nairobi National Museum, Louise needed to collect evidence of the fauna from horizons close in age to 2.5 million years ago.

Having successfully traversed the wet sands of the Kalakol River the following morning, we drove north, exploring many sand rivers to locate a suitable site for our camp. We finally settled on a campsite right on the lakeshore not far from the Kangatukuseo River and conveniently situated near the small village of Kataboi, which had a borehole installed by missionaries that produced delicious drinking water. We would also have the untold luxury of plenty of washing water from the lake, so we would have to carry only drinking water to the camp. But as soon as the lorry turned off the road to drive down the sand river to the shore, it became hopelessly stuck. We spent the afternoon and the whole of the following day unloading the lorry and ferrying the equipment and supplies to the campsite by Land Rover. We set up a lovely camp along the lakeshore, with our tents comfortably nestled among doum palms. But as the afternoon progressed, the sky turned ominous, with huge clouds piling up in impressive columns on the horizon. The waves beat against the beach and the cooling breeze became a gale. The following morning, we awakened to a vastly different scene from the picturesque camp of the previous evening. All the tents were standing in inches of water, and the store tent had been blown over and was lying upside down beside our drenched three-month supply of food!

Somewhere under the tangle of canvas and guy ropes, two tethered goats bleated pitifully. Everyone was shivering with cold. Somehow Louise managed to light a small fire to supply us all with tea while we set about repairing the devastated camp. But we say a bad start means a good season.

The 1998 field season was indeed productive. Our new trainees included Justus Erus and Stephen Muge, who had first worked with us during their school holidays at Nariokotome excavating the Turkana Boy. There was certainly a little adjustment required with the injection of young new blood in a team that had worked together for decades. Nzube rather famously remarked on this once when a new recruit took time off for a shoulder stiff from carrying loads of dirt from a sieve. "The new model," he succinctly stated, "is not the same as the old one!" So perhaps the division of labour that unfolded was a natural one. Louise worked with the young team exploring the west side sites with exposures dated between three and 1.3 million years. Going quite far afield in the younger sediments that extended far to the north, they took with them a minimum of equipment each week and slept wherever they stopped for the day. Meanwhile, my generation of the field crew and I surveyed the exposures dated between four and three million years. We spent our weeks happily camped close to a waterhole in the dry sandy bed of the beautiful, shady Lomekwi River that was filled with huge acacia trees abounding with birds. Our new arrangements were significantly less luxurious than in Richard's day: by substituting metal camp beds with bedrolls that we laid out at night on a tarpaulin and replacing a full gourmet kitchen with a trunk of canned necessities, both teams travelled remarkably light. With radios in the cars, we could communicate regularly with one another provided that we remained in a line of sight.

On weekends, we all returned to our much more luxurious and delightfully scenic base camp. We updated our records, packed the fossils for the long and bumpy journey to the museum in Nairobi, and caught up with camp chores. Our lakeshore camp provided a truly special setting, affording us the opportunity for early morning swims and spectacular walks along the beach. After working inland at Lothagam and Kanapoi, I found this a welcome change.

Having Louise with me was a joy. I once again had someone with whom to share the great excitements and trials of leading the expedition. And once again I had a pilot and the untold convenience that this brings in such remote areas. After his accident Richard had bought a Cessna 210. This model had the same six-seat configuration as the faithful 206 it replaced, but in place of a big underbelly to stow luggage, it had retractable wheels. The result was a faster trip, but the new plane carried less cargo and required a longer strip to land on. Although Louise had obtained her pilot's license several years earlier, it took a while before she could build up her flying hours to the point where her father felt she was competent enough to use his plane. While I was working at Kanapoi, Louise had persuaded Richard's brother Jonathan to lend her his two-seater plane, and we had several eventful flights when I needed to get to Nairobi. In 1998, while at Lomekwi, Richard finally allowed Louise to fly his plane but only if she had an experienced pilot with her. This meant that we could fly faster and with more luggage, but it also meant that the copilot had to return to Nairobi, which left us without the aircraft.

Frank Brown, who had done so much of the geology for Richard in the early days, had joined us once again to study the geology at Lomekwi. He was assisted by a young Kenyan geologist, Patrick Gathogo. Patrick had first joined us in the museum as a volunteer and had proved himself to be one of the hardest-working people in the department. Frank was so impressed with his commitment and skills in the field that the following autumn he arranged for Patrick to enroll as a student at the University of Utah. This was the beginning of a great partnership between them. Frank passed on to Patrick as much as possible of his incredible knowledge of the Turkana Basin geology—the cumulation of decades of research.

Patrick would later build on this for his graduate studies and develop a long-term study of the geology to try to piece together and integrate all the separate chronologies from different parts of the Turkana Basin. Frank and Patrick first concentrated on the Lomekwi sites, and their detailed interpretation of the geology gave us a fine resolution for the relative placement of the Lomekwi fossils. At the same time, Ian McDougall, who had given us those precious pumice dates at Lothagam, collected samples for additional dating.

The distinctive green and brown claystones of the great ancient Lonyumun Lake are visible in sediments to the south of the Lomekwi site. This is the vast lake that we saw at Lothagam with the squashed snails below the basalt and briefly in the middle of the Kanapoi sequence far to the south. Just under four million years ago, this lake retreated from Lomekwi, and the Omo River took its place as the permanent water coursing through the Omo-Turkana Basin from the Ethiopian Highlands far to the north.

The Lomekwi sediments are sandwiched between the Murua Rith Hills that run north-south slightly inland from the lake. The steep, craggy slopes of the ancient lava that makes up these hills have heavily influenced the geology of the site because of a long series of ephemeral rivers that cut down the hills in dramatic gorges and drain across Lomekwi into the lake to the east. The geological footprint left behind by these ancient rivers is a distinctive alluvial fan: tumbled rocks and boulders dumped rapidly out of fast-flowing water in a characteristic fan shape. This pattern is repeatedly overlain by a second competing system of deposition that has its source in the much larger and perennial Omo River. Deposition from this river system was much more sedate, and the fluvial deposits are characterized by a series of upward-fining layers typical of the floodplains of a slow, meandering river system like we saw in the Upper Nawata at Lothagam.

In Lomekwi, these two opposing patterns of deposition replace each other through the sequence as the course of the meandering river changed through time. We were fortunate that there are some very clear markers to help orient ourselves in this repeating sequence.

There are two beds of volcanic ash that were brought downstream by the Omo River after volcanic eruptions in the Ethiopian Highlands. The bottom-most and the oldest of the two at about 3.6 million years, the Lokochot Tuff is harder to find. Far more common and impossible to miss is the Tulu Bor Tuff at 3.4 million years. This tuff layer represents a huge ashfall from a vast volcanic eruption, and the chemical signature of this eruption has been picked up in ashes as far afield as the Mediterranean, the Red Sea, and the Baringo basin. Luckily for me, most of the fossils at Lomekwi are either just above or just below the distinctive and well-dated Tulu Bor marker. Just below the Tulu Bor Tuff is an unmissable third marker known as the Burrowed Bed. This bed is packed with the burrows and tunnels formed by worms and other invertebrates that lived on the shores of a shallow lake that briefly replaced the Omo River slightly less than 3.6 million years ago.

Modern seasonal sand rivers still drain from the hills, and we frequently found ourselves well and truly stuck as a result. That first year, the rains persisted later than usual, and a fine layer of wet clay was often concealed beneath the deceptively firm-looking sand, and our vehicles often landed up to their axles in the mud. Because of the rain, there were many frogs and toads, which were probably the attraction for the inordinate number of snakes. Hardly a day went past without a snake of some kind being spotted in the camp, and on many days, we encountered at least three deadly carpet vipers.

During that 1998 field season, as well as recovering many beautifully preserved fossils of a large variety of mammals, we discovered a number of isolated hominin teeth, but, disappointingly, nothing very significant or definitive. There were indications that the isolated teeth from Lomekwi differed from those of the contemporary Australopithecus, but the evidence was tantalizingly limited. We returned to Nairobi at the end of the three months feeling that there was surely an interesting hominin to be found at Lomekwi but that we would have to work very hard to find it. And it was only a year later that we struck gold when we visited a set of new exposures that Frank Brown and Patrick Gathogo had spotted and urged us to survey.

The whole crew was reunited in one place, in preparation for leaving, when we finally got around to visiting the new exposures in the final week of the 1999 field season. At the end of a long, hot, and unrewarding morning on the second day at this site, Justus Erus asked me to look at something he had found. A few broken and weathered fragments of what appeared to be a hominin face were scattered on the rocky surface.

At last, here was a new fossil that might answer some of my questions about the diversity of hominins at the time of the early australopithecines. But it was in terrible shape and covered in matrix, so there was no way to know how significant this might be. On the positive side, it was just beginning to erode out of the soil. Perhaps if we were lucky, more of the skull would be hidden in the ground and in better condition. I gently removed some of the soil to see if there might be more underneath, and it looked promising. But it was too late in the day to begin the delicate task of excavating such a fragile and potentially important find. I painted the bone fragments with Bedacryl before carefully replacing the soil and covering the area with a protective rock. I was worried about hyaenas: particularly after the near tragedy of the pig skull at Kanapoi and a subsequent occurrence when they seriously damaged a young elephant skull. Apparently attracted by the strong smell of the Bedacryl, hyaenas had dug up the fossils hoping to find food.

On my return early the next morning, I was appalled to find that the stone covering the spot had been moved and the ground disturbed. Had a hyaena really been able to move this stone? My heart sank as I realised that an inquisitive Turkana herdsman, who must have been watching us from a nearby hill, had come down to see what had so interested us and had curiously poked and prodded the ground with his spear. Never before had something like this happened, and I was dismayed by the damage. I grimly and painstakingly collected all the bone fragments, and the field crew carefully sieved all the surrounding soil.

I then began to excavate the remaining bone buried in the ground. I was amazed to find the rest of the skull that went with our damaged face. But this too was in very bad shape. Roots had penetrated the bone, and large cracks ran through the specimen. Added to this, the bone was fragmented into millions of little pieces that were cemented together with a hard matrix. I could tell that I had found something potentially significant, and I desperately hoped that the damage the fossil had sustained from the spear and the roots could be repaired.

We extracted the skull and carefully packed it into a box for the journey back to camp. The following days were spent sieving the slope below and digging a small excavation to make sure there were no more pieces still hidden in the ground. We then took several sacks of the sieved sediment back to camp, and this was washed with lake water and sieved again. Meanwhile, I reconstructed as many of the broken pieces as I could to make the specimen stronger before taking it back to Nairobi. This initial reconstruction revealed an intriguing face that seemed unusually long and flat with rather small teeth. Was this really A. afarensis? And if it wasn't, could the damaged bones tell us what we needed to know? I couldn't wait to find out.

It was a long time before we could resolve the question as to what this skull represented. After our return to Nairobi, Christopher Kiarie, who is an excellent preparator, spent nine painstaking months removing the sandstone rock grain by grain from the bone. As more of the pieces were cleaned, we were able to start reconstructing the skull. The more we put together, the more unusual the face became. It didn't look like we had ever encountered one of its kind before, and we were impatient to begin the study and make comparisons with other known specimens. We enthusiastically began a scientific study of the skull as well as the other hominins we had found at Lomekwi. A new long-time collaboration also began with Fred Spoor, an excellent Dutch anatomist and palaeontologist, who was at the time working at University

College London before moving to the Natural History Museum. Fred had first worked with us in the field at Kanapoi, and his cautious and meticulous approach was perfectly suited to the controversial task ahead of us. What was this flat-faced hominin, and how did it fit into the evolutionary picture that had been built up gradually over many decades of discovery?

When the Lomekwi skull was finally prepared, Louise, Fred, and I arranged to visit Addis Ababa to make comparisons with similar-age fossils that had been found in Ethiopia. The first order of business was to ascertain if our flat-faced specimen could be a variant of A. *afarensis,* since the sediments at Lomekwi covered the same time interval as those from which A. *afarensis* derived. Tim White and Don Johanson were generous in allowing us to look at their specimens, and Berhane Asfaw and Bill Kimbel, who were both in Addis at the time, willingly shared their fossils and ideas with us. Berhane and Bill are both old friends whom we have known for many years. Bill has an extensive knowledge of both A. *afarensis* and early *Homo,* and we frequently bounced ideas off him. We greatly valued his insights and the scientific discussions with Berhane about our Lomekwi skull that made our visit to Addis all the more fruitful. We took numerous measurements and observations of all the *afarensis* specimens in the Ethiopian collection. We became increasingly excited, as there were apparent differences even before we could fully analyse the data.

If our flat-faced friend was not A. *afarensis,* we needed to rule out affinities with any of the other known species of the same time interval. Later that year, the three of us travelled to South Africa and studied the specimens of A. *africanus* and *Paranthropus robustus* housed at Witwatersrand University and Pretoria's Museum of Natural History. Back in Nairobi, we included detailed comparisons with the Kenyan collections in the Nairobi National Museum.

Although Kiarie had done a superb job of cleaning the fossil, there was nothing he could do about the fact that the skull was distorted. The entire skull was slightly off-centre, and numerous hairline cracks crisscrossed its surface. These cracks had filled with matrix as the bone became fossilised and had distorted the specimen. This made comparisons much more difficult and interpretations more tentative. But whilst the distortion of the cranial vault was so extreme as to conceal its real shape, the face did not appear to have been as seriously affected. Nevertheless, if we were going to name a new type of hominin, we needed to be sure that our measurements were not off because of these distortions.

We were lucky that technology invented for medical purposes works perfectly well for fossils too and that we could use CT scans to look at the inside of the cranium to determine the distortion and then use computer-imaging technology to reconstruct it with a far greater degree of accuracy than would otherwise be possible. We could also take scans of the tooth roots normally hidden from view. The CT results confirmed our initial measurements, and the most outstanding observation was that *A. afarensis* was very different from our Lomekwi skull.

We immediately started thinking about what this skull meant. Having spent the last decade trying to find diversity among the hominins between four and three million years ago, we were now faced with the usual problem of how much variation should reasonably be accommodated in a single species. If Lomekwi represented something different, we would have to be very sure that our analysis was faultless before we redrew the family tree. We were fortunate that both *A. afarensis* and *A. africanus* have very large known samples. It would be relatively straightforward to take a series of measurements, analyse these statistically, and see if the Lomekwi skull fell within the range of variation seen in either one of these species.

Fred Spoor compiled a large database of all the relevant characters and measurements of each specimen that we had examined in Ethiopia, South Africa, and Kenya. Along with the digital images we had taken of many of the relevant fossils, this database enabled us to assess the morphology of the new fossil skull character by character. To our delight, the data showed the Lomekwi skull to be significantly different from all australopithecines in a number of ways.

Most incongruous was the flat face of our new Lomekwi skull, and we soon found ourselves calling it Flat Face. In particular, Flat Face had big deep cheeks extending from below the eye socket towards the teeth, which created a tall malar region. The snout was also less primitive looking in that it didn't protrude nearly as much as that of the more apelike *afarensis* (in other words, its face was less prognathic). When we took a line from just below the lower rim of the nasal opening directly down to the central incisors and measured the angle between this line and the one formed by the sockets of the upper teeth, we found that Flat Face was less apelike and more humanlike (it showed less subnasal prognathism). In later hominins, including humans, this angle is much greater.

But what did the differences in facial shape mean? Evolutionary changes in facial shape are driven largely by dietary requirements, and the unusually small teeth of our new fossil supported such a hypothesis. Our work at Lothagam indicated that new dietary opportunities had opened at the end of the Miocene. Although many herbivores then became grazers (needing bigger, thicker, taller, or serially developed teeth), others had evolved mixed feeding strategies, and some had remained browsers—this herbivore diversity in size and ecology was mirrored by the large number of different carnivores. How had early hominins responded to these new opportunities? If these new feeding niches could also be exploited by early hominins, we should not be surprised to find differences in the dentition and shape of the face among different lineages of hominins. The distinctive facial features of Flat Face and its small teeth could indicate a different diet than that of *A. afarensis*.

There were additional differences between Flat Face and its contemporaries. Flat Face had terribly damaged teeth that had broken off and left only a few fragments in place, and these could tell us very little. However, there was one upper second molar that was complete enough to be measured although it had a large matrix-filled crack. After correcting for this crack, we found that this tooth was smaller than any other upper second molar we had measured for our database. To be sure, we rechecked our measurements on this tooth through a microscope and on computer images.

The size of the opening of the ear was also very small, similar to what we had seen in the Kanapoi temporal bone of *A. anamensis* and Yohannes Haile-Selassie's new *anamensis* male skull from Ethiopia. So besides being different from *A. afarensis* in the shape of the face and the size of its teeth, Flat Face also retained some primitive traits that had disappeared in *A. afarensis*.

As we completed our comparisons, we had to decide where Flat Face belonged. The more we looked at the evidence, the more difficult it became to incorporate this specimen in any of the known species. *A. afarensis* was the most likely given its contemporaneous age, but Flat Face was clearly different. We felt that this was not a good fit, and we began to think of our skull as a new species. But how did Flat Face fit into the human family tree?

If we were to name a new species, we would also have to decide if Flat Face belonged to a known genus. Naming a new genus would shake up the family tree far more than the announcement of a new species, and our working assumption had always been that Flat Face belonged to one of the four known genera: *Ardipithecus, Paranthropus, Australopithecus,* or *Homo*. The problem was that the data did not sit comfortably with this assumption, since, with only one exception, none of these genera included species with a similar long flat face. Flat Face did not show any of the strongly primitive characters of *Ardipithecus*, so it was easy to exclude it from that genus. And Flat Face had almost none of the specialised features of the robust australopithecines grouped together under *Paranthropus*. *Paranthropus* does have a flat face, but it is different: it looks as though the nose had been punched hard and pushed in so the face is "dished." Plus, *Paranthropus* has huge megadont cheek teeth that could not be more different from the small teeth of the new fossil from Lomekwi. We could find no grounds for including it in this genus unless it could perhaps be an ancestral species for this group, which we considered unlikely because of its completely different facial morphology. The only possible place within an existing genus was either in *Australopithecus* or *Homo*.

With one exception, Flat Face also lacked the specialised features associated with our own genus. Of all the skulls in the collections, Flat Face seemed to most resemble a specimen we had found in 1972, the skull called 1470 that I had the pleasure of putting together fragment by fragment at Koobi Fora with an infant Louise. This extraordinary fossil still remains very much an enigma. It had a very similar face with little facial or subnasal prognathism. At 1.9 million years (almost 1.5 million years younger than Flat Face), 1470 is considered to be an early example of *Homo,* and the remarkable likeness gave us great pause. Was the Lomekwi skull ancestral to 1470 and the *Homo* line? As we shall shortly see, the assigning of 1470 to *Homo habilis* is clouded in controversy as well.

The only other available genus was *Australopithecus.* Should we then attribute Flat Face to *Australopithecus* since that was the only genus believed to have been living at that time, or should we emphasize the magnitude of the differences by naming a new genus? The problem with this was the absence of megadont cheek teeth that characterize the australopithecines. Unable to decide, Fred and I wrote two papers that argued for each of these solutions. Put this way, the answer was clear. The paper attempting to incorporate Flat Face as a species of *Australopithecus* was unconvincing whereas the alternative paper describing a new genus provided a logical argument. So we took the plunge and named both a new species and a new genus. We gave our new skull the name *Kenyanthropus platyops:* the flat-faced man from Kenya.

Flat Face was splashed across the front pages of the world's leading newspapers, and in spite of our trepidations, the paper was generally well received by our colleagues. The significance of the find was far-reaching. Just as Louis had predicted all those years ago and as I had long believed, diversity in the fossil record did indeed extend to at least 3.3 million years ago. Moreover, A. *afarensis* was not necessarily the ancestor of all later hominins. Later hominins may have had their ancestry in *Kenyanthropus* or *Australopithecus* or in some other as yet undiscovered genus or species. The few sceptics who disagreed with us were most concerned about the possible effects of the distortion of the skull on our assessment of its shape. The strongest disagreement came from colleagues such as Tim White, who firmly believes that human evolution was a single evolving lineage with no diversity at this point in time.

Dissent in science is often a positive force that stimulates further research and new discoveries. Our discovery had upturned conventional wisdom and led to new questions. But in order to prove our conclusions and convince the sceptics, we needed additional—and undistorted—specimens. This, of course, is easier said than done and remains a challenge for the future.

My family
in 1958.
*Courtesy of the
Epps family*

Me with Richard's dog Ben and
our two pet otters awaiting a lake
crossing to Koobi Fora, ca. 1970.
*Bob Campbell, courtesy of
the Leakey family*

Counterclockwise from right: Nzube, Richard, Kamoya, and a heavily armed and curious
local herdsman stare at the complete skull KNM-ER 406 exactly as we discovered it on our
memorable camel trip in 1969. *Bob Campbell, courtesy of the Leakey family*

Richard and I inspect some of the many hundreds of fragments of 1470 in 1972. I spent hours gluing this specimen together, usually with Louise keeping cool in a basin of water at my feet. *Bob Campbell, courtesy of the Leakey family*

Not wanting to miss out on the excitement in the field, I took Louise (left) and Samira with me to Koobi Fora within weeks of their births, and throughout their childhoods they regularly joined us in the field. *Bob Campbell, courtesy of the Leakey family*

The camels proved to be highly contrary and single-minded, requiring a considerable amount of coaxing and coercing in 1969. Much to Nzube's relief (and ours) we soon substituted them for a more reliable, mechanical conveyance.

Bob Campbell, courtesy of the Leakey family

The enigmatic Flat Face, at last cleaned of matrix, presented us with a set of completely new features at an early stage of human evolution, which led us to announce a new genus and species, *Kenyanthropus platyops*, in 2001. *Bob Campbell, courtesy of the Leakey family.*

Before satellite technology transformed our means of communication, Kamoya needed all the equanimity he could muster as he spent many frustrating hours trying to make radio contact with Nairobi through the control tower. *Bob Campbell, courtesy of the Leakey family*

Before the advent of GPS, I spent an inordinate amount of time on elevated plateaus trying to orient myself against our aerial photos. *Courtesy of Louise Leakey*

At Kanapoi, and later at Lomekwi, we used "hill crawls" on particularly prolific areas to localise our sieves and excavations around points where we found hominins. We recovered both a deciduous molar and a capitate this way—the very bones we needed the most. *Courtesy of Louise Leakey*

Fossil hunting is an acquired skill. In addition to perseverance and a large element of luck, it is essential to have a mental template of the various bones. We found a far greater variety of fossils after I taught the team about the morphology of modern mammals, which expanded their search-image library. *Courtesy of Louise Leakey*

Our scenic lakeshore campsite at Lomekwi came to an ignominious end on the very first night in 1998 when a heavy storm upended all our tents, including the all-important store tent. *Courtesy of Louise Leakey*

At Kanapoi, we conducted some of our most extensive and time-consuming sieves of my career, moving mountains of dirt by dental pick and paintbrush, and checking all of the sieved soil for fragments of fossil bone. *Courtesy of Louise Leakey*

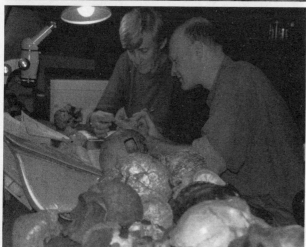

Back in the lab in 2000, Fred Spoor and I compare our specimens to the casts of other hominid finds to ascertain what we had found. *Courtesy of Louise Leakey*

Louise refuels the plane at Ileret with help from our camp manager, Mutuku, ca. 2011. *Mike Hettwer / hettwer.com*

The field crew relax in whatever shade they can find during the hottest part of the day before prospecting again in the late afternoon. Because ambient temperatures fall so little at night, Turkana is one of the hottest places on earth. *Mike Hettwer / hettwer.com*

An example of early aerial photographs with fossil finds marked on them. Before GPS, a pinprick with an associated number written as small as possible on the back was the only way to map the location of our discoveries. All of this data has now been digitised, which makes it considerably easier to analyse.

Courtesy of Meave Leakey

In 1951, Mary Leakey painstakingly traced the evocative rock paintings from 186 sites in Tanzania, and thirty years later, I was privileged to help her compile a book of the lifelike images and scenes depicted in this ancient art. In this image, figures excitedly dance and cavort around an elephant they have caught in a trap: a poignant image of daily life that gives a unique window into the behaviour of our ancestors. *Courtesy of Meave Leakey*

The construction of permanent facilities by the Turkana Basin Institute dramatically altered our ability to conduct long-term fieldwork. The Turkwell campus, pictured here, is situated on the banks of the Turkwell River. *Courtesy of Louise Leakey*

Once Louise completed her studies, she joined me in leading the Koobi Fora Research Project. She proved an invaluable addition as well as a welcome companion, as shown here in 2002. *Courtesy of Josephine Dandrieux*

Samira's daughter, Kika, together with her cousins Seiyia and Alexia, illuminated for me the vital (and highly enjoyable!) role played by grandmothers in human evolution. A cornerstone of our modern reproductive strategy is to have grandmothers help rear the next generation, which allows for a shorter interval before the next child. *Courtesy of Samira Leakey*

The Koobi Fora spit, where we had our base camp for many years, proved a magical setting with abundant birds and wildlife. It was also a strategic location to defend against roaming bandits as it was protected by the lake on three sides. *Courtesy of Fraser Smith*

Preparing large fossil specimens for safe transportation back to Nairobi was an elaborate process that involved reinforcing them with branches and encasing them in protective hessian sacking covered with plaster of paris. *Courtesy of Fraser Smith*

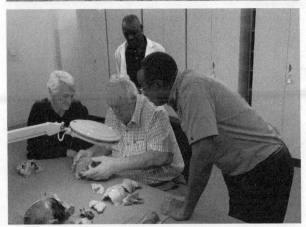

Richard and I revisit some of our early finds at the National Museums of Kenya in 2020. We enjoy an intellectual debate about human evolution to this day. *Courtesy of Marta Lahr*

EARLY *HOMO:* A HORRIBLE MUDDLE

FLAT FACE OPENED PANDORA'S BOX BY DISPROVING THAT THE single ancestor for the human line began with Lucy at slightly more than three million years. Johanson and White's simple family tree—with a single forked branch leading from the *A. afarensis* stem to the robust australopithecines on one twig and *Homo* on the other—was in tatters. *K. platyops* has to fit in there somewhere. But where? And the incontrovertible evidence of our having more than one possible ancestor begs the bigger question: did we evolve from an early megadont australopithecine or from something else?

Part of the problem is that the minute we start talking about our own, human, line, emotions start to fly. Much stronger sentiments accompany any decision about designations to the genus *Homo* than any other. Perhaps this is why Darwin picked the humble barnacle as his study subject when he set out to find evidence to support his theory of evolution. The very mention of our own genus resonates with a deep and compelling desire to know where we come from, and opinions have been governed more closely by emotions than science. People all too frequently have unstinting confidence in the truth of their convictions, however weak or contradictory the evidence may be. Fuelling this cocktail is the frustrating paucity of new specimens to resolve old disputes.

The first sensational discovery of a fossil human ancestor in Asia by

Eugène Dubois in 1891 in Java is the earliest example of this. This discovery, which came to be known informally as Java Man, owes a great deal more to serendipity and dogged determination than any well-founded expectation of success. Dubois was Dutch and completely fixated with the notion of finding the "missing link," then known purely hypothetically as *Pithecanthropus alalus*. (Although no such creature had been found at the time, a biologist named Ernst Haeckel had postulated that such an "ape man without speech" was only awaiting discovery.) Moreover, Haeckel wrongly believed that humans were more closely related to Asian orangutans and gibbons than African gorillas and chimpanzees. Since Dubois's opportunities were tied to the location of Dutch colonies, some of which were in Asia, he decided quite logically that Indonesia was the only suitable option available for his search. A doctor by training, Dubois secured a post with the Royal Netherlands East Indies Army as a medical officer in 1887 and staked everything he had on the chance of finding his beloved missing link in what was then the Dutch East Indies. Prevailing against bouts of high fever from malaria, hellish tropical heat, and difficult conflicts with his field crew, Dubois struck gold relatively quickly, finding first a molar and then a skull cap with unmistakeably human attributes at a site called Trinil. Hot on the heels of this momentous discovery, Dubois's team returned to work in 1892 after the rainy season and unearthed a complete thigh bone that showed uncontrovertibly that his hominin walked upright, which led Dubois to give it the appellation *Homo erectus*.

Dubois paid a heavy price for the audacity of his discoveries, which went against the conventional wisdom of his time. Darwin's idea that humans evolved from apes was preposterous enough. But Dubois's fossils hailed not from Europe but from the Far East. Worse still, the association of such a modern-looking leg bone with such a primitive skull cap was cause not only for great consternation but derision and ostracism from the scientific community. Dubois was devastated and became increasingly reclusive and possessive of his fossils, refusing to allow any researchers to see them for many years.

Over the decades since Dubois's first discoveries, there have been a great number of other important and exciting finds in Asia. Initially, all the evidence of *H. erectus* was in Indonesia, where Dubois was based. Several decades after Java Man was unearthed, a site in China would make headlines. At this time, in the early 1920s, a great deal of confusion was caused by the fraudulent Piltdown Man, which was constructed from the skull of a modern human and an adulterated jaw of an orangutan with the teeth carefully filed down to present the comforting impression that our human ancestor was large-brained and, most important, British. Nevertheless, scientists had by then accepted Dubois's finds as genuine hominins, and Asia was considered a legitimate place to search for human origins. Davidson Black, a young Canadian physician, was one of the scientists who had collaborated on the interpretation and study of the Piltdown fake fossils and, like Dubois, was obsessed with finding the missing link. He secured himself a post at the Peking Union Medical College in 1919, and in between his teaching duties, he made forays to various caves, most notably to Dragon Bone Hill in the Zhoukoudian cave system. Following the discovery of an initial hominin tooth, extensive excavations ultimately yielded a skull cap in 1929 that was even more complete than Dubois's Java Man. This came to be known as Peking Man (*Sinanthropus pekinensis*).

In retrospect, this fabulous fossil seems to have been ill-fated from the outset. Excavated by dim candlelight in the cramped recesses of a cave, it was extracted dripping wet and rendered dangerously soft and fragile from the heavy clays. The leader of the excavations, Pei Wenzhong, arranged a continuous shift to carefully turn the skull over a fire to painstakingly dry it out. Then it was wrapped in glue-soaked gauze before being bundled up in numerous protective layers and dispatched by bus to what was then Peking. In spite of the prodigious care taken to protect this valuable fossil, geopolitics would ultimately seal its fate —along with numerous other irreplaceable finds representing at least fifteen hominin individuals recovered from the Zhoukoudian excavation over the following decade. In 1938, the excavation was abandoned because of the increasing interest in the site shown by the Japanese who had already overrun large parts of China by then.

By this time, Black had died and been succeeded by a German Jew, Franz Weidenreich, who had fled Europe for a life of exile in America before agreeing to replace Black in China. Fearing the degree of interest shown by the Japanese in the fossils and concerned that they would be looted, Weidenreich took meticulous notes, photographs, drawings, and measurements of all the hominin specimens and cast replicas of every bone. The director of the Chinese Geological Survey then decided that the fossils should be smuggled out of the country, and a careful plan was secretly hatched. The fossils were to be transported to the port of Qinhuangdao by US marines and picked up by the SS *President Harrison*. But in the ensuing chaos after the Japanese bombed Pearl Harbor on December 7, 1941, the fossils vanished and were never seen again. Thank goodness for Weidenreich's foresight and his meticulous records and casts; for without these, Peking Man and his kin would have been lost to science forever.

Meanwhile, another intrepid and daring individual was flying in the face of conventional wisdom by looking for human ancestors in East Africa—my future father-in-law. A German butterfly collector, Wilhelm Kattwinkel, was the first to notice the wealth of fossils at Olduvai in 1911. Louis first visited the site in 1931 after seeing some of the fossils collected by Kattwinkel and a subsequent German expedition to Olduvai in 1913. Louis straightaway found many stone tools and fossils of other animals that persuaded him of the huge potential of the site. It was this conviction that made him return to Olduvai again and again over the following decades, adamant that humans had their origins in Africa and that further searching would eventually lead to the discovery of the maker of the stone tools. But Louis had so many ideas, projects, and potential projects that there was never enough money or time for him to sit for very long in one place. And finding hominins usually takes a great deal of patient searching. It is not surprising that from the time that Louis first started to work at the site until 1955, the only ancient hominin bones to emerge from the gorge were two pieces of a hominin skull. These were found by Mary on her first visit in 1935 when she was twenty-two and very much in love with both Louis and the combined allure of the spectacular landscape and the intriguing artefacts that littered the steep Olduvai slopes. Two teeth, a canine and a molar, are the only other hominin bones they ever found there until 1959, the year that Mary made her momentous discovery of Dear Boy, the first complete *Paranthropus boisei* ever found.

Mary writes with characteristic wry humour that Louis "was sad that the skull was not of an early *Homo,* but he concealed his feelings well and expressed only mild disappointment." They had been looking for close to thirty years, and Mary had produced a find that was nothing short of spectacular. One can only wonder at the mildness of disappointment she must have felt at his lukewarm response! But even if the skull was not all that Louis had hoped for, it did completely change the nature of their research and many aspects of their lives by attracting funding from the National Geographic Society. So ended the snatched snippets of time they could spend at Olduvai after scrimping and saving to get there. And because they were spending so much more time at Olduvai, a glut of finds soon followed. In a letter dated December 8, 1960, Louis wrote to his close confidante, famed anatomist Wilfred Le Gros Clark, "You may be wondering as others have done, why we are finding so much hominin material now . . . The answer is simple. We have been working *continuously now since February, with adequate funds* and a huge labour staff, and have already put in some 72,000 man hours this year. Had we been able to do this sort of thing before, we would have had the results before."

For Mary, the reward of finding this skull was nothing compared to the joys that lay in the excavation that followed. Mary was an archaeologist, and until she started working with Louis, her only experience was excavating archaeological sites and making exquisite line drawings of much more modern flint artefacts in Europe. She wasn't particularly interested in the fossils as her great love was stone tools. But other

than Mary and Louis, who had seen with their own eyes the vast numbers of tools scattered about the gorge, few people expected that there
would be any tools at 1.75 million years. With the new funding, Louis
and Mary at last had the money to recruit and train staff to help with
the excavations, and this task fell to Mary. She was ferocious in establishing her trademark exacting standards—her meticulous excavation
techniques are emulated to this day. Kamoya and Nzube, who formed
part of Mary's team of sixteen men recruited to excavate the skull site,
still get exercised today when they recount the rewards and travails
of working for the fierce and fastidious matron they came to greatly
respect.

Mary's new team removed the overburden from an area that covered about 925 square metres (equivalent to a very comfortable house)
to reach the level from which the skull had been excavated. Due to the
steepness of the slope, this entailed digging down a staggering seven
metres at the back of the excavation and removing several hundred
tons of earth and rock by wheelbarrow. This enormous pile of dirt still
stands today where it was dumped in front of the site, and having conducted so many excavations of my own, I never fail to be impressed by
the sheer effort this must have required. Mary's just reward was 2,470
artefacts. What made the find so significant was that only sixty of these
were stone tools. The rest were flakes and debris left over from making
the tools. Mary and Louis had long believed that hominins would have
been returning regularly to favoured spots to eat and sleep, but they
were alone in believing that evidence could be found in support of this.
The excavation showed for the first time that the stones had been left
there and used deliberately for specific purposes.

In addition to an assortment of fossils of animals ranging from antelopes to rodents, birds, and even chameleons and lizards, this giant
excavation had one further gem to offer: some bones of a second hominin that didn't seem to square with the robustness of *P. boisei*. Although
Louis initially described these smaller fossils—two lower leg bones,
some skull fragments, and two teeth—as Dear Boy's "wife, mistress,

mother or sister," he and Mary soon suspected that they represented a second species, and that this, not *P. boisei,* was the maker of the stone tools. Mary writes of this second smaller-toothed hominin that "he must be considered as the tool-maker of Bed I, in preference to *[Paranthropus] boisei,* who seems unlikely to have progressed beyond a tool using stage of development." That more than one species of early human ancestors lived contemporaneously at that time was almost as controversial as the notion that they had been living in cooperative groups with home bases. It was only after their eldest son, Jonny, then just nineteen years old, found much more complete skull fragments, a mandible, some hand bones, and a complete foot nearby that they were able to announce their phenomenal discovery to the world.

This second breakthrough find happened one day in 1960 only a hundred feet away from the enormous ongoing excavation of the *P. boisei* site. Jonny had noticed a strange-looking lower jaw that his father immediately recognised as the first sabre-toothed cat to be found at Olduvai. Because this was such a rare animal, Louis suggested that Jonny sieve the site to try to recover further pieces. Jonny didn't find any more of the cat—the sieving team recovered a hominin tooth and toe bone instead. Before long, a handful of foot bones emerged from the excavation. These clearly belonged to one individual because Mary carefully extricated them from a small area of about one square foot. This momentous find was followed by several finger bones, a collar bone, and thin and fragile pieces of skull. Finally, the excavation yielded two rather large parietal bones (from the top of the skull) that lacked the sagittal crest that was such a prominent feature of *P. boisei.* Mary and Louis's suspicions from the first excavation were confirmed —this was a much more humanlike creature than Dear Boy. Because the mandible, hand, and skull fragments belonged to a juvenile while the foot bones, clavicle, and radius were those of an adult, they were accorded two accession numbers—Olduvai Hominins (OH) 7 and 8.

It is almost impossible to imagine how exciting this must have been for Louis and Mary, and also how difficult it would have been to put the new fossils they were finding in a proper context. The australopithecines from South Africa and the assortment of finds from Asia had not yet been grouped together as the single species *H. erectus* that we now believe they represent, and Mary and Louis had nothing local to compare their bones with. But these spectacular discoveries would completely change the picture of human evolution as it was then understood. East Africa went from being a complete nonentity palaeontologically speaking to becoming accepted as the very cradle of mankind just as Louis had always stubbornly insisted.

In 1964, Louis and his colleagues Phillip Tobias and John Napier proposed that this new hominin from Olduvai was actually an early *Homo,* and the more complete juvenile mandible and the two pieces of skull from OH 7 were selected as the type specimen. This was hugely controversial because it was thought at that time that the first "human" attribute to evolve was the large brain. An obvious criterion to admit a fossil skull to the *Homo* genus was therefore a minimum brain capacity. In 1948, Sir Arthur Keith, one of the early giants in the field and a prominent anatomist of the British palaeoanthropological establishment, set this "cerebral rubicon" for the *Homo* line at an arbitrary 750 cubic centimetres. He arrived at this figure largely on the basis that it would exclude the South African australopithecines while including the material from Asia. This criterion is far too simplistic. For example, a brain size in and of itself tells us little; it is the size of the human brain relative to our bodies that makes us such a brainy species. We have since learned that brain enlargement was an attribute acquired quite late in human evolution as a result of other adaptations made possible by the move to bipedality. But this knowledge would come from discoveries that would follow much later.

Louis and his colleagues blasted the credibility of Keith's definition of a cerebral rubicon for *Homo.* They were obliged to lower the cerebral rubicon to 600 cc to accommodate the new fossils within the *Homo* genus. And it is not surprising that this caused an uproar, particularly because Tobias estimated the cranial capacity of the juvenile skull to be 680 cc by reconstructing the skull from the two fragmentary parietal bones. (Tobias later revised this to 647 cc for the juvenile but extrapolated that the brain would have been 674 cc in adulthood). Never one to skirt controversy, Louis also tickled tempers with the choice of species name, *habilis,* which was suggested by Raymond Dart at Louis's invitation. Louis, Phillip, and John expanded the definition of *Homo* to include bipedalism, a precision grip, and the lower brain-capacity threshold of 600 cc. *Habilis* ("handy man") references the inference Louis made about his hominin's ability to make stone tools based on the fact that the relatively complete hand bones showed a precision grip. The very real problem with this is that you cannot link bones and behaviour that we are no longer witness to—we can only *infer* who the maker of the tools was. They could have been made by other contemporary hominins. Nevertheless, the name *H. habilis* stuck, and OH 7, or Jonny's Child, became the type specimen for the earliest species of *Homo* known at the time.

H. habilis SEEMED destined to continue on the controversial path that began with its naming. In addition to the disagreements on the basic attributes of *Homo,* there was a second level of dispute. During the rest of the 1960s and the 1970s, many additional specimens were found at both Olduvai Gorge and East Turkana that showed a large variation for a single species. This was especially contentious for the earliest specimens thought to date from around the two-million-year mark. At the centre of the dispute were two quite different-looking and well-preserved adult skulls from East Turkana found by members of Richard's legendary field crew. These two fabulous fossils came in our windfall of finds that we enjoyed in the 1970s at Koobi Fora that mirrored the spate of good luck that followed Louis and Mary through the 1960s. Both have been attributed to *H. habilis,* but there were some disquieting differences between them. Everything was small about one of these skulls, KNM-ER 1813. It was lightly built, small brained, small toothed, and small faced, and had a very generalized morphology. The other, KNM-ER 1470, was larger brained and, like *Kenyanthropus platyops*, is remarkable for its flat face.

KNM-ER 1470 was discovered on July 27, 1972, along the Koobi Fora ridge by Bernard Ngeneo. It remains one of my favourite fossils ever because of the happy memories I have of reconstructing it with hippos playing in the lake and baby Louise playing at my feet in a cool basin of water. Against this backdrop and out of thousands of nonde-scriptive bone fragments emerged one of the most exceptional, complete, and enigmatic finds our team had ever made.

It was evident that 1470 was not an australopithecine, and it had all the indications of being far more human than anything else we'd found at Koobi Fora because its brain was much bigger. Desperate to know just how much bigger, Richard, Bernard Wood, and Bob Campbell resorted to pouring sand into the skull cavity and transferring the sand into a rain gauge, the only implement we had on hand for measuring volume. Using this ingenious solution, we learned that the brain capacity was equivalent to eight inches of rain, and we then began the business of converting this to something meaningful. Richard was so excited that he proudly wrote the camp diary entry for September 9 himself: "Worked on the first brain size estimate for the new skull. Plasticene (play dough), masking tape and sand in conjunction with a rain gauge (inches) and a 5 cc syringe resulted in an 800 cc minimum estimate. Historic moment!" For Richard, this was incontrovertible proof that he had a very early *Homo,* and he rushed back to Nairobi as soon as he could to share his great excitement with his father. He was just in time as Louis would be alive for only a few days more, and 1470 brought father and son together in a way that they had not enjoyed in decades.

One of the reasons that Richard and Louis were so excited about this large brain-capacity estimate was because the layers where 1470 was found were below the now-infamous KBS tuff, which we believed at the time to be more than 2.6 million years old. Such a date associated with such a large brain was what Louis had long wished to discover, and for that brief evening they shared together, Richard and Louis basked in the belief that this was indeed so. In fact, the KBS tuff was much younger, and 1470 would later be pegged to a date of 1.9 million years.

One of Richard's rare diary entries in the camp log is for June 21, 1974. I was in Nairobi, having just given birth to our second daughter, Samira, and I was desperately anxious for any news of new finds. Richard writes dismissively and notes almost with disgust the possible hominin finds that had been collected just in one week: "Very poor specimens as follows. FS 100- 2 pieces cranial vault from 6B above middle tuff. FS 101- scrap of hominin molar from 6B above middle tuff. FS 102- ? Tibia pieces close to CHARI bone. FS 103- Cranial vault fragments above middle tuff #1." Sure enough, every one of these four was indeed hominin — but nobody saw fit to celebrate!

After discovering something as spectacular as 1470, it was difficult to get excited about little fragments here and there — we really needed another skull. That happened at the end of July in 1973. Kamoya turned up in camp in the early afternoon to tell us he had found a piece of another skull south of the Koobi Fora peninsula in sediments close in age to those where 1470 was found. We immediately set off in Richard's plane to have a look. All we could see was a beautiful set of teeth embedded in an upper jaw that was barely protruding from the soil next to some skull fragments lying nearby. This was the only hominin we have ever found that was so easy to reassemble — a sharp contrast to 1470. As Richard excavated each fragment over the next two days, I was able to reassemble the whole skull in the field beside him. As the skull gradually took shape, we found that we had a beautiful specimen. It was a small skull that appeared to have come from a diminutive individual. This was the skull that was given the accession number KNM-ER 1813.

KNM-ER 1813 looks quite different from 1470. It has a smaller brain capacity (510 cc) than 1470 (752 cc). It is not only smaller, but it also lacks 1470's flat face, and the cranium has a different profile. It is unfortunate that 1470 has no teeth preserved, but from the appearance of the tooth roots, it seems to have had larger teeth than 1813. However, we can't actually tell where the crowns of the teeth broke off from the roots. Because upper molar roots are always splayed at the tips and taper towards the crown, this makes a big difference. If the crowns broke off partway up the roots, this would give the false appearance of much larger teeth. Until we find another specimen with teeth, we cannot be sure about how big 1470's teeth were.

KNM-ER 1813 closely resembles another important find from Olduvai, OH 13, nicknamed Cindy after Cinderella because of its delightful daintiness. Louis grouped Cindy in *H. habilis* along with Jonny's Child. His decision must be seen in its historical context: at the time, only australopithecines had been recognised from Africa, so he was focusing intently on how different his new fossils were from the robustness of *P. boisei* rather than on differences among his gracile small-toothed fossils. Moreover, both Cindy and Jonny's Child were quite incomplete. It probably didn't occur to Louis that he might in fact have two rather than one additional species. Given the evidence at the time and prevailing opinions, such an assertion would never have been accepted. As more complete fossils emerged over the following decades and Africa became an acceptable cradle for mankind in public opinion, the differences among different specimens assigned to *H. habilis* became more apparent.

A further problem with inferring species from these skulls is that without any reasonably complete postcranial fossils associated with the skulls, we cannot tell how large brained the individuals actually were since what we are really interested in is the relationship between brain and body size. For instance, 1813 could have been a small (probably female) individual, and if we knew what her body dimensions were, her relative brain size might not have been particularly small. Rather than a cerebral rubicon dividing true humans from other hominins, a brain-to-body-size ratio is much more informative. However, since we so rarely find postcranial and cranial fossils that clearly belong to the same individual, this is not easy to estimate.

Most scientists prefer to err on the side of caution and therefore tend to lump everything into the minimum (preferably one) species for any age. *H. habilis* has therefore become a catch-all name for a collection of rather different-looking specimens. It is almost certain that they don't all belong together, though few agree on which fossils can be grouped, which of these should belong to *H. habilis,* and what to do with the others that do not. One of the main problems is that the *H. habilis* type specimen, OH 7, is not only a juvenile but also an incomplete mandible. Most of this mandible is in beautiful condition, but the two sides have been crushed together during fossilisation, so it is impossible to see exactly what the shape of the mandible would have been. On top of these problems, mandibles can be highly variable among individuals of the same species, so they make a notoriously bad choice for a type specimen. As a result, there has been much disagreement as to whether this specimen should be grouped with the smaller hominins or with 1470. And because, like it or not, Jonny's Child is the type specimen, this is central to the question of how researchers understand *H. habilis.*

There has, therefore, been a long-standing muddle about early *Homo.* The controversy settled for a time when scientists largely agreed that two species are represented in the collection of fossils grouped today as *H. habilis.* One of these is smaller and more gracile (represented by 1813 from Koobi Fora and OH 13 from Olduvai) than the other (represented by 1470). But I and many others found it hard to place OH 7 in either of these two groupings with any conviction. At our own dinner table, this was not a subject on which agreement was ever possible, and it was considered taboo by mutual consent!

To make things even more confusing, the Russian anthropologist Valery Alexeev muddied the waters further in 1976 by assigning 1470 to its own species, *rudolfensis.* When Richard and Alan Walker first described 1470, they deliberately avoided assigning it to any species because they recognized the lack of evidence to relate it properly. And for many years, the consensus among palaeoanthropologists was to live with the imperfectly understood grouping. But Alexeev changed all that, and 1470, along with one or two mandibles, were assigned to *Homo rudolfensis.* Other than the *H. erectus* fossils, all the other *Homo* specimens from the earliest Olduvai sediments as well as all those from similar-age deposits at Turkana are lumped together as *H. habilis.*

But until the placement of OH 7 could be firmly established in either of the two groupings, what should take the name *habilis* remained disputed. But new technology allowed us to revisit this question decades after these controversial fossils were discovered. Performed correctly by a qualified anatomist, digital reconstruction allows for far more precision in a virtual construction than our early attempts with modeling clay ever could. In 2015, Fred Spoor created a digital model using CT scanning of the OH 7 mandible in order to correct the distortions caused by the sides of the mandible being crushed together. When Fred and his colleagues compared this reconstruction to other specimens of early *Homo,* their results showed that the variability across the sample is inconsistent with their all belonging to a single species. The reconstruction shows that the mandible is remarkably primitive, bearing a closer resemblance to the morphology of *Australopithecus afarensis* than to *Homo erectus.* Their reconstruction of the cranial capacity of OH 7 using the parietal bones yielded a much larger brain (between 729 cc and 824 cc) compared to the 680 cc arrived at manually in the 1960s, and they concluded that 1470 could not belong to the same species as OH 7.

But both species are very different from the lineage of robust australopithecines. The robust australopithecines very likely occupied a feeding niche of tough, fibrous vegetation (using their megadont strategy of big cheek teeth, relatively small incisors, thick enamel, and little brains with big crests for muscle attachments). In contrast, the more gracile hominins probably ate softer foods such as fruits, insects, and small mammals (using their much smaller molars, relatively big incisors, thinner enamel, small muscle-attachment areas, and small faces relative to the size of their brains). If *Homo* evolved from the australopithecine line, it made a radical departure from the megadont trend in an as yet undiscovered intermediate species somewhere between three and two and half million years ago. Or perhaps its origins lie in an omnivorous dietary strategy that evolved in parallel much earlier —*Kenyanthropus* shows that this scenario is entirely plausible.

Since *H. habilis* appears rather suddenly in the East African record—fully formed with a morphology that would not evolve much over the subsequent half a million years—it is quite probable that it evolved elsewhere and then moved into the drying woodland habitats alongside *H. erectus.* Of course, given that there is such a huge gap in the sedimentary record in East Turkana, it is equally possible that all this evolution happened right there and we have absolutely no way of finding the evidence for it. Other sites might yield the answer to this, and perhaps one already has. There is a 2.33-million-year-old maxilla missing its third molar that was discovered by Don Johanson's annual field expedition to Hadar in 1994. This delicate, well-preserved fossil is indisputably early *Homo.* It has the classic U-shaped *Homo* palate. Don Johanson and Bill Kimbel at the Institute of Human Origins in Tempe, Arizona, together with Yoel Rak, have extensively studied the fossil, A.L.666-1, and concluded that its closest affinities lie with *H. habilis.* But it is too incomplete to say any more than this. An even older mandible, at 2.8 million years from Ledi-Geraru, is currently the oldest known precursor to *Homo.* This specimen corresponds in age to some isolated teeth found in Koobi Fora in 1978 that share some of the same transitional traits that suggest the teeth could be from the same ancestor to early *Homo.* These finds are important for they represent some of the earliest evidence we have of our own genus.

While *H. habilis* is mysterious because of its great antiquity and its presumed status as one of the earliest *Homo,* even more intriguing is the origin of *H. erectus.* Louis assumed that *H. habilis* was the direct ancestor of our own species. However, this is another example of a scientist thinking with more heart than evidence. Indeed, Louis was to miss the import of his own momentous *H. erectus* discovery altogether. In 1960, he came across a pile of bone fragments that looked suspiciously like a tortoise. On closer inspection, Louis elatedly realised they were the pieces of a hominin skull without the face. Glued together, the pieces of OH 9 presented a nicely preserved, thick-boned skull cap with prominent brow ridges and all the trademark features of *H. erectus*. But Louis had been heavily influenced by Sir Arthur Keith, who had been a victim of the Piltdown hoax. Louis was looking for a large-brained ancestor so he was completely thrown by the small size of *erectus*'s brain. Ironically, OH 9 is among the largest brained of all the East African *erectus* specimens that have subsequently been found. But other fossils that would hint at the pivotal importance of this species would be discovered only after his death. Louis had very little to compare his Olduvai specimen to, which made it all the easier to be swayed by his strong preconceptions.

Louis and Mary never found evidence of *H. habilis* and *H. erectus* in exactly the same level. Believing that *H. habilis* was the maker of the numerous tools scattered about Olduvai Gorge, Louis always regarded *H. erectus* as a dead-end branch of the family tree and therefore not particularly interesting. We now know that Louis was wrong. *H. erectus* is indeed extremely interesting. As we shall see, there is currently broad consensus that *H. erectus* was the first species to move out of Africa—and was the ancestor to *Homo sapiens*.

One thing is fairly certain about the assortment of fossils variously classified as *Homo habilis* and *Homo rudolfensis:* they can all be clearly distinguished from *Homo erectus*. But what distinguishes them are features they lack rather than features that are in and of themselves characteristic of this hodgepodge group. The origin of the *Homo* lineage remains obscure, and it continues to perplex me a great deal. We can solve this enigma only with new fossil evidence. I have long hoped to find another 1470 to try to get to the bottom of this puzzle. Indeed, the focus of the Koobi Fora Research Project today, after more than three decades of searching, is to solve this ongoing conundrum. The ultimate discovery would be a complete skeleton, which would tell us so much more than we can learn from a skull alone. Until someone can unearth such a miracle, our story continues with *H. erectus*.

PART III

13

BECOMING GRANDMAS

MY WHOLE PERSPECTIVE ON JUST HOW DIFFERENT WE ARE from all other mammals stopped being mere academic curiosity when my first granddaughter arrived in August of 2004. Here before me was a totally helpless and, it has to be said, screaming and colicky infant, who turned more than just her mother Louise's life upside down. And even though she cried almost all the time, how I loved her! In between burping and nappy changes, baby Seiyia got me thinking in a whole new light about the significance of the different stages of life history and why the differences between humans and other animals are so important in explaining how we have become what we are. I had wondered with trepidation what my new role as a grandmother would be. But Seiyia's arrival also highlighted two unique features of the human life history. The first of these was just how dependent and undeveloped a human baby is at birth. The second was more of a surprise: that my role as a grandmother is important in the grand scheme of human evolution.

Women are reproductively active for about as long as a chimpanzee female is—approximately thirty years. But our life histories depart in all other respects, especially in infancy and postmenopause, and these differences relate directly to our large brains.

Strangely enough, the gestation periods for a chimpanzee and a human are quite similar—eight months for a chimp compared to nine for a human. We would thus expect the babies to have similar-sized brains. However, the average size of a newborn chimpanzee's brain is 137 grams and a human baby's is 364 grams. These differences increase as the infants grow, and the average volume of an adult human's brain is 1350 cc compared to a chimpanzee's 400 cc—more than triple. But at birth, a chimpanzee, like all other primates, is already grown enough to function independently of its mother. It can cling tightly to her as the troop moves and is quickly adept in locomotion and basic socializing. In contrast, a human baby needs its mother for absolutely everything for many months. In general, mammals' life histories are closely related to the size of their brains. The problem for a human baby, however, is that if it stayed in the womb for the length of time we'd expect for its brain size, it would never make it out into the world past its mother's hips!

Standing and walking upright imposes all sorts of constraints relating to balance and locomotion, which resulted in significant changes to the shape of the pelvis. To accommodate an efficient striding gait, the shape of the human pelvis differs significantly from that of a quadrupedal ape. Mammalian birth canals are encircled by the bones of the pelvis (the ilium, ischium, and acetabulum) and the sacrum, so changing the shape of these bones alters the shape and size of the birth canal. Our pelvis has developed big "wings" (the ilia) to attach the abductor muscles that propel our legs. The edges of the ilia also attach to muscles from our torsos that keep us upright. The angle of the ilium is much more vertical compared to the elongated, diagonally positioned ilium in a chimp. But the shape of our pelvis is constrained by the need to keep our centre of gravity as close to a vertical axis as possible. If the heads of each of our femora were farther apart, we would waddle rather than stride (which is, incidentally, why men can usually run faster than women—because women have wider hips). As a result, the acetabula, which accommodate the heads of each femur, must remain as close together as possible. This in turn confines the width of the birth canal at the outlet, and the result is that the inlet and outlet of the human birth canal are quite different in size and shape.

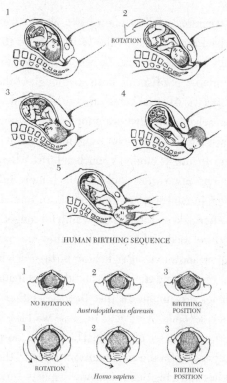

HUMAN BIRTHING SEQUENCE

Birthing human babies is a challenging compromise between the size and shape of the brain and the pelvis, and our babies are much less developed at birth than those of other primates.

If we model the length of gestation for a human foetus to be developmentally equivalent to that of a chimpanzee at birth, the gestation would be twenty-one months. Nature's solution is for the human foetus to gestate for as long as possible in the womb and then continue to develop on the other side of the birth canal in a "secondary altricial" state. This is all about our big brains: at birth, a human baby has a brain that is 25 percent of the size of an adult's, but this doubles a year later to 50 percent. But in all other primates, the growth of the brain slows at birth. In other words, the brain of the human baby continues to grow at the fast foetal rate for twelve months after birth.

In general, mammals follow one of two alternative reproductive strategies. Altricial birthing allows the birth of more than one infant at a time: usually a whole litter can be accommodated in the womb to maximise the number of offspring born so enough of them survive to propagate the species. Altricial babies are born blind, deaf, and unable to fend for themselves, like newborn puppies. An extreme example of an altricial baby is a joey—the infant kangaroo that, still in foetal form, struggles its way up into its mother's pouch at birth, where it latches on to a teat and keeps on growing until it is much more fully developed. Precocial babies, in contrast, are generally born one at a time. This enables the mother to keep them in the womb for longer. Because they are born fully developed and independent, they are more capable of escaping predators and surviving right after birth, much like a newborn gazelle is able to get up and run after its mother in a matter of minutes.

Human babies are unique in that they fit neither pattern completely. They are secondarily altricial because we have reverted from the common primate pattern of precocial births to a more altricial state in order to compensate for the constraints imposed by the combination of a bipedal pelvis and a big brain. In some respects, human babies are still precocial—they soon open their eyes and they can hear and respond to stimuli—but they are altricial in their complete dependence on others and their undeveloped motor coordination.

Another consequence of the unique shape of our pelvis is that, unlike all other primates, humans find it next to impossible to have unassisted births. While a baby's head is narrowest between the ears and widest from the forehead to the back of the head, the human birth canal begins at the inlet with its widest point from side to side but shifts midway by 90 degrees so the long axis is front to back at the outlet rather than side to side. Consequently, the baby must contort and twist around during its passage through the birth canal so its widest dimensions (the shoulders and head) are always aligned with those of the changing shape of the birth canal. The baby, which enters the birth canal facing sideways, exits facing backwards. A chimpanzee or monkey mother can easily reach down to guide her forward-facing baby out of her own birth canal, clear her infant's breathing passage, and begin to nurse. Because of the precocial development of the foetus, it can grab hold of its mother and assist in its own delivery once its hands are free. For a human baby, this much exertion so early on is impossible. Moreover, because the baby is born facing backwards, the mother risks injury to its spinal cord, nerves, and muscles by pulling against the natural curvature of its spine. Complications to delivery, such as the umbilical cord being wound around the baby's neck, can likewise be solved only by a helping third party, so humans in all but exceptional circumstances rely on assistance at birthing. For millennia, experienced grandmothers or elder women have acted as midwives to assist with the risky business of birthing human babies.

The abbreviated gestation period for a human baby can be explained by problems arising from the combination of two key human traits—bipedalism and a large brain. Extended childhood, as well as an extended life history long after menopause, are likewise unique to our species, and both are related to our unusually large brain. The extended childhood serves several functions. Delaying the growth to a full-sized adult body permits more energy to be spent on the calorically costly growth of the brain before puberty. The body then catches up at the end of adolescence with a final growth spurt. During childhood and adolescence, the smaller body size ensures that the young are nonthreatening to adults and that they are the recipients of social goodwill and teaching. Our large brain allows us to be an unusually innovative species, and our ability to speak enables us to form a collective body of ever-increasing knowledge that far exceeds that which could be achieved in the span of a lifetime. This body of knowledge is passed from generation to generation, and speech is a pivotal force in making us who we are today. Needless to say, modern humans therefore need an extended childhood for all this learning, and grandparents, with the accumulated wisdom of age and a postmenopausal period free from the constraints of childbearing, are traditionally the best-positioned members of a social group to make this contribution.

Having adult females around for such a long time who are not reproductively active may seem at first glance to be a poor strategy for a species' survival. But if we look more closely, we find that this is a very successful strategy as it reduces the interbirth interval for reproductive females. On average, chimp mothers are able to bear young only every five and a half years (although this can vary between four and six and a half years) because each infant must be fully weaned and independent before the mother can turn her attention to a newborn. Human mothers are usually able to bear young at twice this pace. The two-year-old is nowhere near as developmentally advanced when the new sibling arrives, but the grandmother can help. With their long postmenopausal period that allows for collective nurturing, human grandmothers significantly increase the fertility of the group.

Looking at the life history of ancient hominins is an illuminating way to explore how human our ancestors were and when they became more like us. But if we are to learn anything about brain size relative to body size and how the mechanics of birthing were accomplished, it is essential to have skeletons associated with skulls. We know from Lucy's skeleton and other fossils from South Africa that the australopithecines, while bipedal in stature, retained small brains and did not suffer the complications of birthing faced by modern humans. But what about our direct ancestors, *Homo erectus*?

The jewel in the crown of the *H. erectus* sample is the Turkana Boy. This is the remarkably complete skeleton that began as an unpromising matchbox-sized piece of skull found by Kamoya on a short Sunday stroll from his Nariokotome camp in 1984. Because the excavation extended into a high bank, tons of soil that covered the horizon in which the bones were buried had to be removed, but the dividends were unimaginable as bone after bone emerged. Five field seasons later, however, the cost of continuing the enormous excavation simply outweighed the remote chance that we might find the missing hand and foot bones, which are particularly informative and which we desperately hoped for. But even without these missing appendages, we had an amazingly complete skeleton, skull, and mandible of a single individual. This is what makes this find so valuable: having all these elusive bones from a single individual permits many deductions that are impossible to make from single limb bones or unassociated skulls. Frank Brown reconstructed the geology for us, and the samples of tuff that he sent to Ian McDougall came back with a date of between 1.51 and 1.56 million years. This was later narrowed down to a most likely age of 1.53 million years.

One of our first questions was how old the Turkana Boy was when he died. We could quickly see that he was not yet fully grown because all of his lower teeth had erupted except for the wisdom teeth (third molars) and the milk deciduous canines were still in place in the upper jaw—although we could see the adult canines just peeping out of the sockets. Alan Walker was soon asking my children and the local youths helping out on the site to open wide. But Louise and Samira were both dentally younger than the Turkana Boy—and none of the other children could tell us how old they were in years because they were only able to say that they were "born the year of the bad drought" or some other seminal event that their parents remembered. Alan then turned to the mouths of some of Louise's friends and provisionally settled on a working age of eleven to twelve based on the sequence and timing of human dental eruption today.

This date differed somewhat from our initial ballpark estimate based on the long bones. In juvenile mammals, it is the shafts on our bones that grow the most while the articular ends (epiphyses) do not grow at the same rate. The two parts join together only when the growth process is nearly complete. In humans, the lower epiphysis on the upper arm bone (the humerus) is the first of the long bones in our limbs to fuse. In the Turkana Boy, this bone was almost completely fused whereas none of the other epiphyses had begun to fuse. The Turkana Boy was therefore a subadult who had not yet reached puberty and probably still had to undergo the final growth spurt. This timing, based on comparisons with modern humans, suggested a skeletal age for the Turkana Boy of eleven to fifteen years at the time he died.

The question was: did the Turkana Boy grow along a human growth trajectory or not? This would make a difference in how old he was when he died. Irrespective of his absolute age, his relative age would remain the same whatever growth trajectory he was on — he was about 65 to 75 percent of the way towards being fully grown. But a chimpanzee needs fewer years to arrive at the same point of maturation than a human. Had Alan looked into the mouths of chimpanzees rather than children, he would have decided that the boy was approximately seven and a half.

Once the Turkana Boy's bones had been cleaned of matrix and the fragments reassembled, Alan enlisted the expertise of several colleagues to help him get to the bottom of this conundrum. The first order of business was to figure out the stage of foetal development at birth. Would a *H. erectus* baby be precocial, like a chimp, or secondarily altricial, like us? This would depend on the size and shape of the birth canal, and on the development of the brain from birth through adolescence. Remember that the human brain is 25 percent of the adult size at birth, doubles during the first year, and doubles again to reach its adult size at the end of adolescence. In contrast, a chimpanzee's brain is already 40 to 50 percent of its adult size at birth.

The bones of the Boy's skull were not yet fused together and were so superbly complete that Alan was able to make a remarkably accurate reproduction of the inner surface of the skull and mould an endocast. Because the brain fits snugly into the skull, this endocast gave a very precise size for the Turkana Boy's brain. When he died, his brain was 880 cc, and regardless of whether he followed an ape or human growth trajectory, his adult brain would have grown to only about 909 cc. This is two-thirds the size of an average modern human brain (1,350 cc) but more than double that of a chimp (400 cc). We might then expect that for *H. erectus* females, birthing had already begun to present difficulties.

Alan faced a more formidable challenge next. From the partially complete subadult male pelvis, he needed to estimate the size and shape of an adult female birth canal. Alan reconstructed the pelvic bones as best he could and arrived at a pelvis shape much like ours only narrower. By his calculations, the baby's head at its widest point could have been no larger than 110 mm, which would accommodate a brain size no larger than 231 cc. To start life at birth with a brain size of 231 cc and arrive at an adult brain size of 909 cc, the brain would have to quadruple in size. In short, it would be impossible for the growth to follow an ape trajectory and end up with a brain this large. Alan concluded that the female birth canal would have been too small to have allowed the passage of a *H. erectus* baby that was born precocially —with its brain size already nearly half grown as happens in a modern-day African ape. *H. erectus* would have already had to adopt the secondarily altricial strategy of later humans.

Alan's conclusions lean heavily on his pelvic reconstruction, which suggests a slender-hipped physique that could not admit a very large-brained baby. But because of the fragmentary nature of the Turkana Boy's pelvis, Alan's reconstruction was part science, part art—a fact he admits to in his gripping account of his study of the skeleton, *The Wisdom of the Bones*. Since this groundbreaking study by Alan and his colleagues in 1993, additional evidence has come to light from new scientific technology that reopens the question of whether *H. erectus* was already following a life history pattern like ours or whether it grew up more like an ape.

It turns out that a critical piece of evidence already existed—found in 1936 at Mojokerto in Java. This specimen is a well-preserved cranium of a young *H. erectus* that likely lived at approximately 1.5 million years ago (the exact location of this early find is unclear, constraining attempts to date it through modern techniques). Because this specimen has no teeth, it was hard to estimate its age, and for a long time, it was speculated that it died when it was between five and six years old. Only recently, CT scanning techniques have shown this to be far too old an estimate and suggest that the child was probably only about twelve months old when it died. The giveaway was two small bones on the base of the skull right in front of the foramen magnum that fuse together right around puberty. The suture between these bones (the spheno-occipital synchondrosis) gives a good estimate of the child's developmental age, but it is often difficult to see this clearly in a fossil. Using CT scanning, however, it became clear that this suture in the Mojokerto skull was still unfused. Suddenly, we had a one-year-old baby with a brain size we could reliably estimate—and this allowed us to learn much more about infant brain development.

The Mojokerto baby had a brain capacity of 663 cc—which, for a one-year-old *H. erectus* on a human growth trajectory, seems implausibly large. If *H. erectus* grew on a human trajectory, the brain at age one would still have to double before puberty, bringing it to more than 1,300 cc, which is the size of an average human brain. But if it grew like an ape, it would already be 70 percent of an adult's size. When we compare the baby's brain size to an average size for all the adult *H. erectus* brains from Asia, the Mojokerto baby has a brain that is 72 percent of an adult's. This strongly implies a trajectory that is as close to that of a chimp's as to a human's and that the Mojokerto baby was more likely born as a precocial infant or, at most, partially secondarily altricial. It is probable that the fragmentary nature of the Turkana Boy's pelvis —combined with the extrapolations that Alan had to make about its maturation into adulthood as well as what the female *H. erectus* pelvis would have looked like— introduced a considerable degree of error and explains Alan's contrary result.

Dental development can also tell us a great deal about the childhood phase of life history. Alan enlisted the expertise of Holly Smith, who specialises in life history patterns and determining the age of a fossil individual from the development of its teeth. She began by mapping out the sequence of tooth development in a monkey, a human, and an ape. Because the Turkana Boy's pattern of tooth development in no way resembled the macaque monkey, she discarded this from her comparison. Every human's teeth erupt in roughly the same order at roughly the same age. Instead of only looking at which teeth had already erupted, Holly could get a far more detailed sequencing by using X-rays to look at the development of the roots and crowns of the teeth within the jaw. She was able to plot a chart that showed what stage of development each tooth would be at for every year. At age five, the human adult front teeth and the first molars have fully formed crowns but only partially formed roots. On the other hand, the premolars and second molar have incompletely formed crowns, and the roots have not even started to develop, and the third molar has neither crown nor root formed. In contrast, a chimpanzee grows its teeth more quickly and not in the same order. Because it is much more prominent, the canine starts to grow a little later and takes longer to fully develop.

Holly then superimposed the sequence and timing of the Turkana Boy's tooth development over her charts of what age the corresponding state of development would be in a chimp and a human. Neither one was a perfect fit, but the majority of the Turkana Boy's teeth had developed approximately as much as those of a nine-to-ten-year-old boy or a seven-year-old ape. His canine developed more like ours than a chimp's, which makes sense because it was relatively much smaller than a chimp's. Holly concluded two things: first, that the Turkana Boy would have been about nine and a half years old when he died; second, that he grew up on a time scale intermediate between chimps and humans so he would not have been comparable to a nine-year-old human in terms of his social development. In Alan's words, the Turkana Boy had "the height of a fifteen-year-old, with the brain of a one-year-old . . . a staggeringly different view of our ancestors."

New technology using sophisticated microscopes has also yielded important information about age and dental development. If you look at a tooth in cross section under a powerful microscope, you can actually count the daily increments of enamel secretion. Much like rings formed in a tree trunk at regular intervals, these daily growth increments form clear lines known as the striae of Retzius. On the enamel surface, these Retzius lines manifest themselves as perikymata: horizontal lines and grooves running across the circumference of the tooth that are clearly visible under the microscope. They give us a chronometer that is both accurate and perfectly preserved, and allows us to compare the rate of accumulation of enamel on the teeth of humans, apes, and fossil hominins. Because we can count the daily increments, we can count the number of days it takes for each perikymata to develop and estimate the length of time it takes for each tooth to grow. This gives us an accurate means of determining the actual age of an individual until the teeth are fully formed.

Application of this technique allows us to go back to the Turkana Boy and compare the new methodology with Holly's results and see how his true age matches the estimated skeletal age that Alan originally based on a human growth trajectory. Since Alan's seminal work on the Turkana Boy, another colleague who specialises in the microstructure of teeth, Chris Dean, has also studied the Turkana Boy's teeth. Chris found that the rate of enamel accumulation in *H. erectus* was faster than in modern humans. He estimated that the true age of death of the Turkana Boy was approximately nine years — reassuringly close to the dental age that Holly estimated using a different methodology. By comparing the rate of enamel formation of tooth samples of fossil hominins and apes and modern humans and apes, Chris found that although the *H. erectus* enamel is thick like that of modern humans, it grows fast like modern and fossil apes. Plotting all of his samples on a scatter plot, Chris found that the trajectory of enamel growth of *H. erectus* is intermediate between ape and human but more similar to that of apes. So it seems that *H. erectus* grew neither like a modern human nor a modern African ape but like something in between, and this is what we would expect for an animal with a brain size that was also halfway in between.

Dental records can offer insight into the final part of the life history cycle as well—where humans again depart from other apes. We find a surprising number of old individuals with very worn teeth throughout the hominin record. Wambua's maxilla of *Australopithecus anamensis* found in our very first week at Kanapoi is one example. In 1970, I also found an *H. erectus* mandible with very worn teeth at Koobi Fora. This mandible is what first set me to thinking about how old this individual was when it died and what the life expectancy of *H. erectus* would have been. In 2004, two researchers from the University of Michigan decided to look at this in detail. Rachel Caspari and Sang-Hee Lee looked at the fossil record to see if there was any evidence of increased adult survivorship over time. By looking at the degree of wear on the teeth of 768 hominin fossils from the past three million years, they were able to obtain a rough measure of the ratio of old to young individuals in the sample. Their results show that this ratio has indeed changed from australopithecines to *H. sapiens*. The data support the conclusions from our study of the gestation and early childhood patterns: while *H. erectus* had begun to enjoy greater adult survivorship, this is not very pronounced compared to earlier hominins. The marked increase in longevity that characterizes modern humans is unique to our species and has come into play only relatively recently.

Nevertheless, the life history pattern of *H. erectus* does seem to have begun to depart from the norm for primates. In human terms, the Turkana Boy's development corresponded to eleven years of growth instead of a chimpanzee's seven—but the Turkana Boy was actually nine. This slightly extended childhood would have been matched by the expectation of slightly increased longevity, which in females probably extended beyond menopause. The most fascinating implication of this is that such a departure from the normal primate mode must have gone hand in hand with heightened social cooperation. Higher primates are intensely social creatures, but the bonds in primate societies are in general based on mutual self-advancement. The exception is the mother-child bond, which is so strong that the mother will put herself in extreme danger to save her child, and she may even carry a dead baby for several days before finally abandoning it: behaviour that can surely be described as grieving. But the success of the human strategy of having postreproductive females to allow increased fertility in the actively reproductive group only works if the females can spend less time mothering each child before the next one comes along. This in turn depends on having social bonds that extend beyond the mother-child relationship. Somebody has to teach the youngsters during their extended childhood when Mum is busy.

It is very difficult to infer social behaviour from ancient bones, of course. Yet there is compelling fossil evidence that social bonding had already begun to strengthen in *H. erectus*. The first time Richard and I were alerted to this intriguing possibility was back in 1970. One of the most talented members of our field crew earned his merit badge by sheer audacity propelled by his conviction that fossil hunting was his calling. Bernard Ngeneo turned up one day and began washing dishes in our Nairobi home. Richard and I were completely bewildered when we discovered after a few days that neither one of us had employed him! However, his affable manner, helpfulness, and initiative soon had him in our formal employ. Not very long after he turned up in our kitchen, we found Ngeneo at the end of our driveway, bag in hand, as we were leaving for the field. "May we give you a lift somewhere," Richard enquired curiously, to which Ngeneo replied, "Yes, please—to Koobi Fora!" Ngeneo manoeuvred himself out of our Nairobi kitchen into the Koobi Fora kitchen and soon afterwards into the camp kitchen of an archaeological dig being conducted by one of our dearest friends and closest colleagues, Glynn Isaac. From there, it was but a small step to find himself on the path between the camp and the dig that was walked along countless times each day by every trained member of Glynn's archaeology field crew. It is testimony to Ngeneo's innate talent that it was he who discovered beside this well-trodden path a hominin femur with a startling story.

This femur, like the Turkana Boy's, was about 1.6 million years old. On the shaft, thickened bone showed that this individual had suffered, and recovered from, a severe break to its leg. Anyone who has broken a leg bone knows that a biped cannot get along unassisted until it heals. For the bone to have healed as it did, somebody had to have cared for this individual, bringing them food and water, and protecting them from predators. Motives of self-interest cannot explain such behaviour, so the degree of social bonding must have been considerable. It is unfortunate that we were unable to tell from this femur how old the individual was when the break occurred, although we do know that it had reached puberty so it is possible that its mother took care of it.

Even more compelling evidence of extraordinary caring behaviour was unearthed by Kamoya in 1973. A decade before he would find the Turkana Boy, Kamoya had already discovered a partial 1.7-million-year-old *H. erectus* skeleton, KNM-ER 1808. We could tell it was *H. erectus* from its teeth and the partial skull. However, this female skeleton is much less well-known than the Turkana Boy because so little can be said about its morphology. She died from a very severe pathology that is manifested by an extra layer of diseased bone laid down on the surface of her limb bones. Alan Walker was also tasked with studying 1808, and it took him a long time to identify what caused the abnormal calcification of her bones. Eventually, he took thin sections across one of the broken ends of bone and brought the slides to a group of medical professionals at Johns Hopkins University School of Medicine, where he was teaching at the time. Drawing up a list of possible causes, the doctors eliminated those that didn't fit until a consensus was reached.

Barring an unknown disease that has since disappeared, they settled on hypervitaminosis A, a massive overdose of vitamin A. Knowledge of just how lethal too much vitamin A can be is common to the Inuit and other polar communities who hunt carnivores. European explorers of the polar regions learnt the hard way when their rations expired and they were forced to eat their huskies for survival. An excess of the vitamin causes the fibrous tissue attachments connecting the muscles to the bone to rip free, and the blood vessels haemorrhage, forming huge clots and further separating the muscle and bone. The intervening space full of blood clots then becomes calcified, leading to an extra layer of bone. It is an excruciatingly painful and debilitating condition accompanied in extreme cases by delirium, disorientation, and dementia as well as severe pain in the joints and stomach cramps. The extent of the blood clotting and the thickness of this calcified layer visible in 1808's long bones show not only the severity of her case but also indicate that she must have lived for weeks or months after ingesting the toxic dose. The social implications are staggering—there is no way she could have survived that long without round-the-clock nursing, and as an adult female, her bond with her nursemaid(s) cannot have been the exclusive one she had as an infant with her mother, who would have long since passed on. There is only one way to explain how far her disease had progressed: she lived in a social group in which members took compassionate care of the infirm to an extraordinary degree.

Another interesting aspect of her diagnosis is the question of how she came to overdose on vitamin A. Vitamin A accumulates in the liver and is especially concentrated in carnivore livers where it cumulates through successive consumption of herbivore livers, which is the most likely source for 1808. We don't know if *H. erectus* was able to hunt carnivores or if 1808 came across a sickly or dead predator. Either way, the hominin would have had to be able to successfully fend off other scavengers that had a far more deadly arsenal of teeth and fangs in order to obtain this food.

The Turkana Boy might also have needed some assistance, and a disability might explain the circumstances of his death. Why a youth entering the prime of his life should come to such a precipitous end facedown in a swamp is hard to fathom. But his vertebrae show a marked scoliosis, an asymmetry resulting from an S-shaped side-to-side curvature to his spine. In 2006, flash flooding in the Omo River produced a twenty-foot-high wall of water that swept away countless livestock and homesteads. Such a circumstance could easily have trapped an individual with scoliosis, who may have been impeded from moving fast enough to get out of the way of the impending torrent. The massive sedimentation accompanying such flash floods would also account for the extraordinary preservation of the Turkana Boy's bones. Because he was swept into a swamp and submerged in shallow water, scavengers did not find his carcass, and it was buried and fossilised. As the water receded and the swamp dried, hippos walked all over the mud-covered boy. We found imprints of hippo feet in the excavation, and the weight of the hippos snapped some of the boy's ribs, causing them to lie vertically in the sediment. As we shall see, mobility is one of the key advantages that *H. erectus* capitalised on as it evolved. His scoliosis was a disability that would have probably impeded his survival to adolescence without the support of a social group.

We can say with some certainty that *H. erectus* was on the threshold of becoming human when the Turkana Boy was alive. His tall slender frame already closely resembled that of any modern desert-dwelling human—but his brain had an awfully long way to go. Although the Turkana Boy's remains answered many questions about his kin, I remained mystified. What were the selective pressures that drove the evolutionary changes for *H. erectus* to realise this remarkable potential?

14

GROWING BRAINS

OUR HOME ON THE EDGE OF THE RIFT VALLEY IS GRACED WITH fabulous views and large open spaces. Thin fingerlike ridges dotted with whistling thorn acacias roll south away from our house along the edge of the escarpment, and vegetated valleys run down to the plains on the floor of the rift below. The alkaline Lake Magadi peeps out from behind the great hulk of Mount Olorgesailie, and the opposite flank of the Rift Valley, the Nguruman escarpment, rises majestically up from the plains, occasionally merging with the hazy sky. On clear days, you can see Lake Natron, which marks Kenya's border with Tanzania, and the distinctive shapes of the ancient volcanoes in Tanzania that stud the skyline—the still active Ol Doinyo Lengai; Mount Kilimanjaro, with its flat cap of snow; the broad-based and towering Gelai; and countless other dormant volcanoes. It is a breathtaking panorama—one that changes endlessly with different light conditions.

I take our dogs for long walks here as often as I can. Our most recent dog was typical of the dogs from Turkana and a far cry from the pedigree species bred for specific traits. She was lean, independent, and catlike, and had the unenviable name of Fuzzy (bestowed on her by Seiyia despite the dog's minimalist sleek fur adapted to the dry heat of the desert). Fuzzy, like the dogs who came before her, would invariably catch the scent of a reedbuck, hare, or some other delectable treat. She would take off in high pursuit, leaving me in the dust pondering the improbability that our ancestors could have ever caught anything without the benefits of a bow and arrow, catapult, or a gun. Most mammals compete for survival using their strength, agility, and speed. In contrast, humans are weak, slow, and awkward. There has been a persistent bias in our thinking that humans must have therefore won out by using their superior cognitive abilities rather than any athletic competency (brains over brawn). But none of this has ever explained how the brains grew so big in the first place, and ever since it was established that bipedalism and manual dexterity preceded brain expansion, a convincing theory of what really happened has been elusive.

Although we don't know nearly enough about the earliest *Homo,* including *Homo habilis,* it seems that by the time *H. erectus* enters the picture, a strategy for securing meat with relative ease and safety had been adopted. The modern human brain, which is only 2 percent of the body's weight, consumes a greedy 20 percent of the body's metabolic resources. However, the total basal metabolic rate of modern humans is no more than the average for a mammal of similar size. Something had to give, and indeed, the human gastrointestinal tract is unusually small for a primate of similar body size. It is our consumption of meat that allows us to obtain enough nutrients to feed our great brains in spite of this drastic reduction in the gut, thus opening a whole new realm of possibility for an enlarged, calorically expensive brain to evolve with all the resulting benefits.

Homo erectus is at the centre of this enigma. Despite its living in harsher and more uncertain times than our earlier ancestors, its evolving progressively bigger brains over the course of a million years, and its successfully conquering new territories, the awkward tools *erectus* uses get no more advanced. However, did *H. erectus* kill anything with these unwieldy tools? Or, if it was scavenging instead of hunting, how could *H. erectus* compete successfully with other scavengers and ferocious carnivores?

It had always been Louis's opinion that early *Homo* obtained meat by scavenging, not hunting. To prove this, he tried to reconstruct events "exactly" as they would have been. This is vintage Louis, and imagining the scene never fails to make me giggle. One day at dawn, Louis and Richard removed every single article of clothing from their persons and strode off onto the Serengeti plains in search of some meat. They were armed with the sort of weaponry that Louis imagined *H. habilis* to have used: easily available natural objects, in this case a long bone from a giraffe to serve as a club, as well as its jaw bone, which would presumably have been a lethal-enough blunt object when applied with force. They soon spotted the telltale circling of a large number of vultures that signified a recent kill. It was a fresh zebra, and they settled down to wait for the pride of lions to finish breakfasting before making their move. Then, as Louis expected they would, the lions went off to drink, and Louis and Richard rushed in ahead of the hyaenas, vultures, and jackals. Louis had used some of the time while they waited for the lions to leave by fashioning a stone tool, which he proudly boasted took him just thirty-five seconds. While the elder Leakey kept the other scavengers at bay by energetically brandishing the long bone club, the younger rushed in and hacked off a leg using the stone tool. At this point, the hyaenas and vultures got rather angry, and father and son had to run for it—but they did so in triumph, with the joint of zebra.

For a long time, I and many others also assumed that *H. erectus* was predominantly scavenging. But what aroused my curiosity the most was that the specialist carnivores become extinct in Africa after 1.9 million years: the sabre-toothed cats, the false sabre-tooths, and many of the hyaenas all die out as *H. erectus* multiplies and spreads across the planet. Something obviously happened to push these successful predators to extinction while the less specialised lion, leopard, and cheetah remained. I didn't think that it could be coincidence that the spectacular and rapid rise of *H. erectus*, the superflexible generalist, coincided so precisely with this phenomenon because having *erectus* scavenge their kills would not have been sufficient to wipe these specialist carnivores out. I continued to puzzle over what adaptation made our ancestor so astoundingly successful.

But when I first heard Daniel Lieberman give a lecture, something clicked. Dan is a brilliant scientist with an approachable manner, curly hair, and a perpetually amused look on his face that matches his great sense of humour. He also has an amazing lab full of specialised treadmills and other exercise equipment hooked up to computers that he uses to look at the ways muscles and bones interact. Dan, who is passionate about human evolution, is a committed marathon runner. All those grueling long runs led Dan to think long and hard about the limits of physical endurance, and his obsession is what led to his collaboration with another functional and evolutionary morphologist, Dennis Bramble, who is based at the University of Utah. The pair came up with an inspirational explanation for how the comparatively puny *H. erectus* may have overcome relatively large and fast herbivores with sharp horns and deadly hooves without being injured.

What I completely failed to comprehend as my dogs raced happily away after the fitter antelopes who always escaped unharmed is that I actually stood a better chance of catching the beast than they did. The piece of the puzzle that I overlooked is that I was comparing human athletic prowess with that of sprinters. It is true that we are poor competitors in the fields of strength and speed—but in persistence running, we are in fact the stars of the whole animal kingdom. If I (before I became a grandmother anyway!) had kept after the antelopes, they would have tired rapidly. If I chased them at midday, they would have tired even more quickly and collapsed of heat exhaustion. Poaching restrictions aside, I could have bounded up to them and clobbered them over the head with a nearby rock were I so inclined.

If you don't believe me, consider this. An amateur runner with little training can sustain an average speed of five metres per second and can easily cover ten kilometres on a daily basis without tiring overmuch. With more fitness training and more innate talent, a world-record-holding Kenyan can run at 6.5 metres per second and sustain this speed over a marathon distance for two to three hours (although not every day). In October 2019, Eliud Kipchoge achieved what had always been considered beyond the realm of human possibility by running a full marathon of 42.195 kilometres at a blistering pace of 1 hour, 59 minutes, and 40.2 seconds. In contrast, a dog of similar body mass can sustain a gallop at 7.8 metres a second but can only do this for ten to fifteen minutes. Dogs such as huskies are the elite athletes of their species. These specially bred dogs can run up to fifty kilometres. However, there are two provisos: they can only do so at a trot, and they need frigid conditions to keep them cool.

The way that nonhuman mammals keep cool is by panting, which is rapid shallow breathing. Although an animal takes ten times as many breaths panting as they would breathing normally, the exchange of air is limited to a space in the upper pharynx, so there is little or no intake of new oxygen into the lungs and bloodstream to fuel the aerobic exercise of the muscles. This makes extended galloping impossible while panting. The world's premier sprinter, the cheetah, gets so hot that it has to stop after just one kilometre. It can reach a staggering speed of 110 kilometres per hour (70 miles) in seconds, but a chase usually lasts only 20 seconds and rarely exceeds a minute. A human generates ten times more heat while running than walking, so you can imagine what hopeless athletes we would be if we relied on panting to keep us cool. Instead, humans have evolved an efficient alternative cooling system that allows us, in Dan's words, to "dissipate copious quantities of heat while running in hot, arid conditions." A key way we do this, of course, is by sweating profusely when we run. We also shed our fur sometime in the past, which makes sweating still more efficient because fur traps air and moisture in a layer above the skin and reduces convection.

This combination of efficient sweating and no fur is what singles out humans. Horses are the mammal that we immediately think of as endurance runners that sweat. And Dan and Dennis talk about horses. Horses can outrun humans with a maximum galloping speed of 8.9 metres per second for a 10 km race. But for longer distances, a horse's pace must slow to about 5.8 metres per second. At this speed, a horse can run for about 20 km a day—anything more, and the horse will suffer irreparable damage to its muscles and skeleton. It turns out that Dan and Dennis were not the first to question a man's prowess compared to that of a horse. Two inebriated men in a Welsh pub were overheard by the landlord debating this very point in 1980, and a challenge was born, lubricated by several additional pints. The landlord turned the private argument into a public challenge, and ever since, man and horse race in an annual marathon event over the hilly moors of Wales from that very pub. Dan points out that, on occasion, the man outruns the horse in this unlikely race. In 1999, the fastest horse beat the runner by only eighty seconds, and in 2004, the horse was beaten by a runner by two minutes. But horses have to carry riders in this race, which unfairly handicaps them for our comparison. But these particular horses have also been carefully bred over generations to improve their endurance and speed, and the race is in cold, wet, windy Wales, which tips the advantage away from humans.

Other long-distance runners that come to mind are wild dogs and wolves. The wild dog, which hunts in the hot African savannah, is perhaps the most apt comparison for the purposes of looking at *H. erectus*. These dogs are impressive. They hunt in packs that range in size from two to as many as sixty individuals and reach speeds of up to a staggering 18.3 metres per second (64 kph or 40 mph), and they can sustain a slightly slower hunt speed of about 50 kph for at least 5 kilometres in hunts that can last up to an hour. Their prey includes gazelle, wildebeest, and zebra, and range in size between 20 and 126 kilos (44 to 277 lbs.) compared to the wild dogs' body weight of 17 to 30 kilos (37 to 66 lbs.). One pack that scientists followed in the Ngorongoro Crater in Tanzania made an average of two kills per day with a hunting success rate of 85 percent, and another pack in the Serengeti achieved success 70 percent of the time. These figures far exceed those of other predators. These dogs hunt only in the cool of early morning and late evening to avoid getting heat exhaustion, so their high success rate during hunting is all the more impressive compared to those of other African plains predators.

We alone are capable of endurance running at midday in hot climates, and it turns out that traditional human societies in tropical arid habitats still practice persistence hunting. Societies where this technique has been documented include the Bushmen of the Kalahari, the Tarahumara of northern Mexico, the Navajo and Paiutes of the American Southwest, and the Australian aborigines. In all these cases, the hunters prefer the hottest possible time of day and year. The favoured speed in all these groups is between the prey's preferred trotting and galloping speeds—an awkward speed for prey that cannot gallop for sustained periods and don't have long enough to cool down in between galloping bursts before the hunters catch up again.

What is remarkable about this hunting strategy is that it unites the benefits of low risk, low metabolic cost, no need for sophisticated technology, and a relatively high success rate. Of the hunts documented by one researcher in the Kalahari, 50 percent were successful, which is higher than the rate achieved using a bow and arrow. Two further elements contribute to the high success of this type of hunting. First, tracking is a critical and difficult skill that has to be learnt. The better the tracker—which entails both reading the tracks on the ground and anticipating the animal's movements—the faster the hunter can track the prey, and the sooner it will tire and collapse from overheating.

Second, it requires a high degree of cooperation and a cohesive social group with strong bonds. When a hunter returns to camp empty-handed, he or she can depend on other members of the group to help replenish the additional calories expended in the hunt as well as their normal metabolic requirement. When the hunt is successful, the highly caloric meat is sufficient to feed the whole group. Like modern humans who use this strategy, the wild dogs that have been studied in the Serengeti are notable for the extreme amity that exists between members. Even with a whole pack crowding around a single kill, there is little strife. And even more remarkable, dominant adult males will wait while the pups take their fill.

For modern humans—with their big brains and capacity for speech

—planning and executing a successful hunting strategy and passing complicated tracking skills down from one generation to the next are relatively straightforward to master. The crux of Dan and Dennis's argument is that *H. erectus* could have done this too. They point to its physique, its brain size, and its ability to make symmetrical tools that require the maker to work from a preformed mental template.

There are two possible holes in their argument that need to be dealt with, however. First, could early hominins have run over rocky, thicketed, and uneven ground without sustaining serious injuries? When I confronted Dennis with this objection, he provided a surprising and compelling response. He pointed out that in the United States "very long endurance races (50 to 100+ miles) are popular and are generally run over rough, hilly and mountainous terrain with plenty of rocks, thickets, and the like. There is a surprising lack of serious injury among the participants—exhaustion and dehydration are the actual killers." Dennis also mentioned another fascinating fact about these cross-country ultramarathons: at this distance over this terrain, middle-aged women are much more competitive with men. So presumably for *H. erectus*, every member of the group could participate in long-distance hunting with equal success—grandmothers included. Dennis's answer satisfactorily dispenses the first problem but leads us directly to the second, potentially bigger flaw in the argument. How did *H. erectus* avoid death from dehydration? The only solution will never be provable—but surely a manually dexterous creature capable of fashioning stone tools and carrying these from the factory site would have been capable of making a receptacle from a gourd or the stomach of prey to carry some water on the marathon hunt. It is also worth noting that modern desert-dwelling people are notable for how much less water they need to survive than we soft town-dwellers do. Quite probably, *H. erectus* was also efficient with its water metabolism.

Alan's calculation of what the Turkana Boy's brain would have reached at adulthood (909 cc) is not far off from that of a one-year-old human baby. One of the interesting developments in early childhood learning that I experienced after I had my own children is the use of signing to help babies to communicate before they can speak. If a mother is diligent at teaching her baby the signs, and begins between the ages of six and eight months, the child will be using the basic signs —I am hungry, I am sleepy, my tummy hurts, more, and enough— within only six to eight weeks. By age one, the child will be mimicking and learning new signs almost as fast as the mother can teach them and will have quite a big repertoire of signing vocabulary. As any mother or grandmother will testify, babies understand a great deal even though they cannot speak.

My point is that one-year-olds are really smart. At age one, my granddaughter, confronted with the sight of a carcass of a lamb hanging in a cool room off our kitchen, pointed and said, "Baaa!" At aged two, Seiyia could identify the birds in the bushland around the house. Once, when we were on a walk together, she pointed at a lamb frolicking playfully in the grass, and said, "Babu [her grandfather] cooks really tasty chops!" Not long after this, while watching the DVD of *Finding Nemo*, she announced how much she liked eating fish. From a very tender age, Seiyia has always had a clear appreciation of where her dinner comes from. I don't think it is a big stretch to hypothesize that *H. erectus* would have already developed some effective form of communication that it used to hunt and gather food cooperatively even though it only had the brain of a one-year-old.

In any case, because of our secondarily altricial development, human babies are not as good an example as other members of the primate family. Other primates give a better minimum expectation of what *H. erectus* would have been capable of. During the dry season, the local troop of baboons that frequents the valley beneath our house stealthily approaches the vegetable garden every Sunday at one p.m. I don't know how they know it is Sunday lunchtime, but they do—and they also know that this is the time when they are least likely to be caught when they breach the fence and start ripping up our precious vegetables. It is really remarkable and truly inconvenient at times! But more formal observations confirm how intelligent other primates are and how highly evolved primate social systems can be.

The celebrity of the chimpanzee world, Kanzi, and his younger sister, Panbanisha, offer a fascinating glimpse into how versatile the cognitive skills of chimpanzees are. Kanzi is a bonobo chimpanzee orphan that was an infant in the 1980s when scientist Sue Savage-Rumbaugh was engaged in trying to teach his adopted mother, Matata, basic words and symbols at the Great Ape Trust near Des Moines, Iowa. Like any child, Kanzi loved to mess up the teaching aids that Sue was working with or jump on the keyboard of lexical symbols or partake in other amusing games. Sue had little inkling of how much Kanzi was absorbing passively as he played until Matata was taken away for breeding when Kanzi was just two and a half years old. Kanzi, completely bereft and devastated at the loss of his second mother, turned in desperation to the closest social bond he had left—Sue. To Sue's astonishment, Kanzi turned to the keyboard more than three hundred times the very day his mother was taken away and asked Sue for food, affection, and help finding his mother. Kanzi had learnt these symbols the way human babies learn language—by being around those who were using them. But Kanzi had never before had the motivation to use them, and Sue realised the implications of this serendipitous revelation. Chimps, like humans, are sponges for information as infants. But she was barking up the wrong tree trying to teach words and sentences out of context to adult chimpanzees.

The research program was completely overhauled. Before long, Kanzi was defying all preconceived notions of what linguists believed to be possible for a nonhuman. He was talking about places and objects that were out of sight, referring to the past and the future, and he was able to understand new sentences made up of familiar words. Linguists were in an uproar and required still more convincing. They asserted that Kanzi could have been responding to body language and facial expressions rather than words and that language included the ability to use metaphors and figures of speech. So Sue donned a welder's mask to completely obscure her face and stood stock still while she asked Kanzi the unlikeliest of questions. He passed with flying colours. And Panbanisha, on one occasion when she disliked the behaviour of a visitor, used the symbol for "monster" to refer to the offender. Even more astounding is another story that Great Ape Trust researcher Bill Fields tells. Kanzi wanted to refer to a Swedish scientist named Pär Segerdahl, who Kanzi knew would be bringing bread with him. Not having a symbol to refer to the Swede, Kanzi pointed to the symbols for bread and the pear fruit. When asked, "Kanzi, are you talking about pears to eat or Pär?" Kanzi replied by pointing at the man.

The study with the bonobos is part of a growing body of work that has upturned a basic tenet of linguistic theory held since the 1950s: that the key to language is a uniquely innate human understanding of the grammar or the rules of language. But there is increasing evidence that language also depends on an ability to imagine the world from another's perspective. This "theory of mind" holds that the meaning of words depends on the social context that produced them. For this reason, autistic people struggle enormously with social interaction although they have a perfect grasp of the mechanics and rules of language. But to understand the true meaning of what is being said, they also have to consciously learn a whole set of signals and nonverbal cues — such as eye contact and body posture — to help them figure out what people mean: skills that nonautistic people don't even know they are using. Neuroscientists have a biological explanation for this. We have specialised brain cells called mirror neurons that reflect physical actions as well as emotions. These neurons fire when you stick your tongue out and when you watch somebody else stick their tongue out. But they also fire when you feel pain or when you see another person in pain, so they are essential for empathy. It is the ability to empathise — to put ourselves in another's shoes — that allows us to understand the social context in which the words are being used. Without this, a lot of what can be accomplished with language simply disappears. It turns out that people with autism have mirror neurons that don't always fire correctly.

Not only could Kanzi and Panbanisha comprehend spoken language that included metaphor (the monster and the pear) and the past and future tenses, but the evidence that they also showed empathy is very convincing. Kanzi's friend Bill Fields is missing a finger on one of his hands. On one occasion, Kanzi was grooming Fields, and when he got to the hand with the missing finger, he used the keyboard to point to the lexicon symbol for "Hurt?" In much the same way, Richard had a bizarre bonding experience with an untrained adult chimpanzee. He was in a room with many others including the chimpanzee's adopted human family. The chimpanzee entered the room, looked around, and made a beeline for Richard, who was sitting with his legs extended in front of him. Richard was rather alarmed to be singled out by this rather large strong adult chimp. But the chimpanzee merely settled itself at Richard's feet, rolled up the leg of his trousers, and experimentally tapped his artificial leg. The chimpanzee then looked at him directly in the eye before exiting the room without further ado. How the chimpanzee divined that Richard had no legs I simply don't know —but it clearly did, and it cared enough to investigate.

Empathy is the ingredient that allows us to use language for the purposes of social bonding. Obviously, it is impossible to see from the fossils whether *H. erectus* would have been already using language for both practical communications and social bonding. But based on the evidence of modern bonobos and very young children, it seems likely that at least rudimentary bonding by language had begun. *H. erectus* probably had the brains to track and hunt as well as the social bonds to cooperate in the hunt and support the hunters when they were not successful. But did they also have the physique to persistence hunt using endurance running?

The focus on early bipedalism was previously centred on the ability to walk, not run, and there are more than a dozen skeletal features found in *H. erectus* that improve both running and walking. Dan and Dennis had another look at the skeletal features found in *H. erectus* to see if there are any characteristics that specifically pertain to running.

The biomechanical differences between running and walking were the obvious place to start. At this point, it all gets horribly technical. One of the differences between running and walking is the way that the kinetic and potential energy are stored during each new step. When we walk, as we extend each leg, the body's centre of mass rises as potential energy for half the stride and is then released as kinetic energy during the second half much like the way a pendulum keeps swinging. During running, the model is like a coiled spring rather than a pendulum. Elastic energy is stored in collagen-rich tendons and ligaments in the leg in the first half of the stance and then released in the second half of the stance that propels the body into an aerial phase. This means that any derived features pertaining to spring-mass mechanics are direct evidence for improving running capabilities.

Since walking is inherently more stable than running, especially in bipeds, a second aspect of biomechanics significant to our enquiry is stabilization, particularly that of the head and trunk. Trunk stabilization, which stops bipeds from toppling over, is primarily accomplished by contracting the gluteus maximus (the buttock muscles). Head stabilization is more complex but no less important. Everyone has experienced the disorienting sensation of being stationary beside a moving train or other vehicle and believing that the vehicle you are sitting in is the one moving. In the same way, spinning rapidly in a circle makes us dizzy. Our balance relies on the interplay between the balancing structure in the inner ear (the labyrinth) and the eye. When we move, the labyrinth senses the movement and signals to the brain, which then corrects for the moving image on the retina so our gaze seems stable to us. When quadrupeds run, their heads are counterbalanced by the weight of their bodies and their somewhat horizontal necks. As bipedal humans thrust the weight of their bodies forward, there is nothing to counterbalance the head, which is thus subject to rapid pitching. If we didn't feel this pitching movement through heightened sensitivity in the inner ear, we would not be able to stabilize our head and we would fall over in a dizzy heap.

Dan and Dennis found that there are indeed some indications of specialization for running in *H. erectus*. The shape of the pelvis indicates that the areas for the attachment of two muscles critical in trunk stabilization are much larger in early *Homo* compared to *Australopithecus*. And complementary research by Fred Spoor, Bernard Wood, and Frans Zonneveld shows even more concrete evidence of features that relate to head stabilization. High-resolution CT scanning techniques show that the semicircular canals in the labyrinth of the inner ear are larger relative to body mass in *H. erectus* than in *Australopithecus* and chimps. The presence of such heightened sensitivity is what gives the individual greater balance during running, which involves more pitching than walking or climbing through trees. Taking all this evidence together, Dan and Dennis conclude that it is difficult to think of any human activity other than running that would have selected for increased sensitivity to head pitching.

We can make an educated guess and say that *H. erectus* had both the cognition and the physique for endurance running. This could be the solution to the conundrum of how *H. erectus* could survive and thrive in the opening habitats. It didn't need sophisticated tools nor did it need to scavenge to avail itself of a whole new feeding niche—a niche that was its ticket to a bigger-brained future.

15

||||||||||||||||||||||||||||||||||||||||||||||||||||||

THE ICEHOUSE

WHEN SAMIRA WAS AT SCHOOL LEARNING GEOGRAPHY, OUR fruit basket was filled for a few weeks with oranges with all the earth's continents drawn lovingly but crudely around their smooth skins — a line circling the girth of the orange to represent the equator, the stem representing the North Pole and an abstract and misshapen Antarctica scrawled around the bottom over the South Pole. "You be the sun, Mummy, stay still!" she would cry as she walked wide circles around me until she got dizzy. But she always maintained the orange stem at the same tilted angle as she orbited me, her sun.

If the earth were oriented "upright" on a true vertical axis, the equator would always be the closest part of the earth to the sun because of the earth's curvature, and the north and south poles would be permanent twilight zones. Yet the equator is the closest point to the sun only on days in March and September (the equinoxes). As Samira earnestly explained to me using her orange, there would be no distinct seasons anywhere on earth without this tilted axis.

Samira's geography teacher naturally did not tell her that it's a lot more complicated than this. The earth's seasons, in fact, depend on three different cycles of long-term changes in our orbital pattern around the sun that last different intervals of time and affect the amount of solar radiation beamed to earth. This heavenly variation is what has shaped the massive climatic shifts that have continually reconfigured the landscape over millions of years—and the evolution of all animals, including our ancestors, was shaped ineluctably by this process.

At the time when *Homo erectus* appeared in Africa, climatic variations were particularly extreme. Tim Flannery puts it most succinctly in his gripping yet depressing account of our role as perpetrators of climate change, *The Weather Makers*.

> The environment of these distant ancestors was very different from the one that we inhabit today, for their world was dominated by an icehouse climate, in which the fate of all living things was determined by Milankovitch's cycles. Whenever they conspired to expand the frozen world of the Poles, all over the planet chill winds blew and temperatures plummeted, lakes shrank or filled, bountiful sea currents flowed or slackened, and vegetation and animals alike undertook continent-long migrations.

Homo erectus is remarkable for how it was able to spread across the globe and flourish for nearly two million years through many intense climatic upheavals. It is all the more notable that its contemporaries (the other *Homo* species and *Paranthropus*) eventually died out between two and one million years ago. We cannot fully comprehend *Homo erectus*'s dispersal to the far reaches of our planet and its longevity as a successful species without first trying to fathom the world it inhabited—and this entails an abbreviated side trip into the fascinating and immensely complex global climate system. It is a story replete with unexpected twists and turns, and seemingly insurmountable obstacles faced by colourful characters. Each new startling idea was met with a frosty reception, and many different branches of science were involved in eventually finding the evidence to prove as fact what was initially dismissed as fantastic fiction.

• • •

THE FIRST CYCLICAL SHIFT that affects the earth's seasons relates to the degree to which the earth is tilted. The 23.5-degree angle that Samira was taught at school actually represents the current position, which is midway in a range of 22.1 to 24.5 degrees that cycles approximately every 41,000 years. This three-degree difference in the angle of tilt in the earth's axis obviously accentuates or moderates the severity of seasonal differences in climate and makes summers and winters much harsher than when the earth's tilt is at its least severe. Climatologists call this shift in the earth's tilt "obliquity." And as we shall see, the dates of the equinoxes are not immutable either.

There is a second major cyclical change in the earth's orbit called "eccentricity." This is determined by the celestial movements of other planets. If the sun and the earth were alone in the universe, the earth would make an even circle around the sun each year. In reality, however, the earth's orbit is also shaped by gravitational forces exerted by other planets, mainly the heavyweights Jupiter and Saturn and occasionally Venus. The changing proximity of the other orbiting planets affects the strength of gravitational forces pulling on the earth's orbit and alters the shape of our path around the sun from a rounded to an elongated ellipse in a cycle spanning roughly 100,000 years.

Since its orbit is elliptical rather than round, the earth is not always equidistant from the sun in its annual journey. It reaches its closest point (the perihelion) around January 3 each year, which is currently a difference of three million miles from its farthest point (the aphelion). But based on the shape of the ellipse at different phases of the eccentricity cycle, the changing distance of the earth from the sun dramatically affects the amount of solar energy reaching earth. During the most elliptical phase of the 100,000 year eccentricity cycle, the earth receives some 30 percent less sunlight at the aphelion than it does at the perihelion, introducing yet another cyclical impact on the degree of the earth's seasonality.

There is yet a third cyclical pattern, which is due to a wobble in the earth's orbit caused by the slight equatorial "bulge" in the girth of the planet and gravitational forces exerted by the moon and the sun. For me, the easiest analogy is a figure skater whose spin is off balance. While her skates are spinning from a fixed position on the ice, her body is not rotating symmetrically around this fixed point, and her extended arms and second leg appear to seesaw as she spins. The earth's wobble is just one rotation of the figure skater's spin, but it occurs in super slow motion over some 26,000 years. The earth's wobble also has an important effect on the earth's seasons because, combined with the elliptical shape of our orbit around the sun, it affects the position of the four cardinal points—the summer and winter solstices and the two equinoxes—in a slightly shorter cycle lasting around 23,000 years. Today, the North Pole is tilted most directly towards the sun on June 21. But in another half cycle 11,500 years from now, the South Pole will be tilted most directly towards the sun on June 21. This gradual progression of the cardinal points along the orbital path is called the "precession" of the equinoxes.

This 23,000-year precession cycle has left its indelible mark in the sediments at Turkana. When we first began working at Koobi Fora, we constructed our camp out of thatch *bandas* using the tough reeds that grew along the lakeshore. Our *banda* met its unfortunate demise when the paraffin freezer started a fire that reduced it to a small pile of ashes. Building a replacement stone structure took some time because we carried a few flat slabs of sandstone back to camp at the end of each working day rather than stopping our fieldwork. Each slab was unique and packed with interesting fossils, mostly fish and molluscs. One particularly attractive piece received pride of place by our front door, and it boasted a large and perfectly preserved hippo anklebone.

For years, I had pondered these distinctive bands of hard sandstone packed full of snails that we consistently found sandwiched between finer and softer layers. In its undisturbed state, the banding looked to me like a repeating sequence that cropped up all over the Turkana Basin. Frank Brown was the first to look into the reason behind this banding that is most clearly shown in the sediments in the Omo Valley

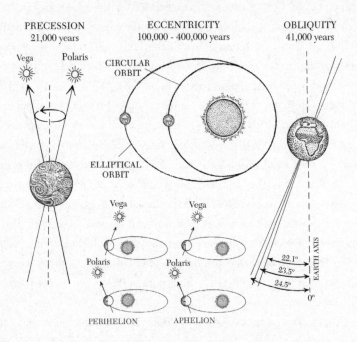

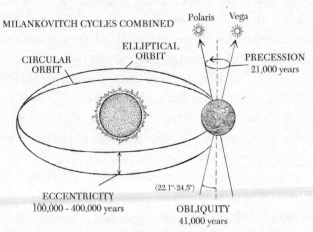

The cyclical patterns of orbital eccentricity, obliquity, and precession influence global climate, which in turn drives environmental changes that shape evolution. These processes have been particularly important for the human story in the last million years.

in Ethiopia. With accurate dates from volcanic layers above and below the bands, Frank and his colleagues were able to work out the length of time represented by each repeating layer in the Kibish Formation of the Omo.

Sure enough, the stacked sequence of floodplains seems to have been laid down with astounding regularity as if the swing of a pendulum on an old grandfather clock marked a time period of about 20,000 years to the tune of the precessional cycle. These regular bands of molluscs were laid down close to the shore of a lake that was shrinking and expanding with the variations in climate. In between the mollusc bands and the layers of lacustrine sediments and occasional beds of algal stromatolites marking wet spells, there are softer fluvial deposits (the fine sandy layers laid down on the banks of rivers). These are distinctive for their fossil root casts and other organic matter derived from the ancient riverbanks and floodplains that advanced as the lake receded. Craig Feibel and his students also found the exact same signals at Koobi Fora even though they are less obvious than in the Omo Valley.

There is another link that ties neatly into the same story. Some of the rainwater feeding the lake and river system in the Turkana Basin drains from the Ethiopian Highlands to the north. These highlands are also the source of tributaries that feed into the Mediterranean and the Gulf of Aden to the east where scientists have extracted long sea cores from the ocean floor. When the fluvial sediments in Turkana mark a dryer spell, these deep-sea cores tell a story of thick clouds of dust that blew off the Sahara and the Arabian deserts, settled over the water, and percolated gently down to the sea floor. Conversely, when the lake in Turkana swelled during the wet part of the precession cycle, the inflow of vast volumes of fresh nutrient-rich water from the Ethiopian Highlands into the Mediterranean was a boon for all manner of tiny organisms. These are preserved in the sea core as thick layers of organic matter known as "sapropels." Because the sea cores also meticulously preserve layers of volcanic ashes that are accurately dated, we can ascertain that the Mediterranean dust and sapropel layers are the flip side of the same story unfolding in the sediments of Turkana. Of ten sapropels known from the Mediterranean during the Late Pliocene and Early Pleistocene, nine have probable known correlates in the Koobi Fora sediments.

These three cyclical patterns—obliquity, eccentricity, and precession—are the Milankovitch cycles that drove the wild fluctuations and extremes of the "icehouse" climate of *Homo erectus*. They are named after the brilliant Serbian scientist who hypothesised in the 1920s that the celestial movements of planets influenced the wax and wane of ice ages. Others before him had also puzzled over the existence of past ice ages. Bernard Friederich Kuhn attributed the erratic placement of boulders in his native Switzerland to ancient and extensive glaciation. The idea that the earth's surface—and its climate—could change this much was highly controversial for the prevailing worldview in all Christendom was that God had made the world and all its creatures in the form they are found today. Consequently, the subject of past ice ages was long and fiercely debated among geologists before it gained broader acceptance.

It was James Croll who first correctly deduced the effects of our changing orbit around the sun on global climate. Impoverished and almost entirely self-educated, Croll's humble and poorly paid position as a janitor at what is now the University of Strathclyde in Glasgow gave him unfettered access to all the great books in the library. Here, in 1864, he turned his attention to finding an explanation for the waxing and waning of ice ages.

Croll asserted that the changes in eccentricity acting in concert with the precessional cycle would be capable of setting off a response in the global climate system that could result in an ice age. He termed the long interval when eccentricity is large enough to induce glaciation as a "glacial epoch" and the intervals separating them as "interglacial epochs." The next part of Croll's argument is key: that the astronomical changes are merely a triggering mechanism whereby a decrease in the amount of winter sunlight favours the accumulation of more snow. Ice is white, of course, and its greater reflective properties are a great coolant because much of the solar energy beamed to the earth is reflected back out to space. Thus the reflective properties of ice and snow magnify and augment the orbital effect. This is what climatologists today call "positive feedback."

Croll published *Climate and Time in Their Geological Relations* in 1875 that summarized his findings and earned him huge recognition. The only problem was that while the length of the precessional cycle had been calculated, methods to determine the exact ages of glacial deposits had not yet been invented. Croll's ingenious theory could not be tested, and after much intensive debate, it almost fell by the wayside in the absence of conclusive proof. It was fortunate that his ideas were retrieved from obscurity by Milutin Milankovitch many years later.

Milankovitch, blessed with a formal education, a considerable intellect, and the self-confidence of relative youth, ambitiously set out to "develop a mathematical theory capable of describing the climates of the Earth, Mars and Venus, today and in the past." This endeavour would take him thirty years to complete, and his labourious manual calculations are the bedrock of modern climate science. When Milankovitch's epic work was at last complete, geologists worked in earnest to try to find proof of actual glacial and interglacial periods that corresponded to or disproved his mathematical predictions, but limitations in dating techniques hindered the debate raging among the geologists in the first half of the twentieth century. And it was James Croll who had correctly predicted many decades earlier where such evidence might be found: on the sea bottom, where he thought "skeletons, shells and exuviates of creatures that flourished in the seas of these periods" would be preserved. This indeed proved to be true. An 1872 British expedition found that the sea floor was blanketed with sediments with the evocative name of "fine-textured oozes" that were composed almost entirely of the fossilised skeletons and shells of sea-dwelling creatures as predicted by Croll.

Tiny floating planktonic and bottom-dwelling organisms called foraminifera live within small shells made up of calcium carbonate. Radiolarians, another group of plankton, have skeletons made of more durable silica rather than calcium carbonate. Whenever either group of plankton outgrows these shells and sheds them for the next size up —or when they die—it is their discarded shells that sink to form the ocean floor. In parts of the ocean exceeding four thousand metres in depth, the skeletons disintegrate under the sheer pressure as they rain down through the depths. However, vast areas of the sea floor, especially in temperate and tropical seas, do carry a fossil record of their remains that goes back hundreds of thousands and even millions of years.

Croll correctly intuited that some planktonic organisms live only in warm water while others survive in cold. As the global climate swings between one extreme and the other, the geographic range of these temperature-sensitive species shifts. But this evidence remained hidden for decades. A way had to be found to extract a core of the sludgy ocean sediments without interrupting the sequencing of the layers.

The first attempts to get a sea core at the beginning of the twentieth century were based on the rudimentary principle of ramming a hollow pipe into the seabed. The longest cores that could be achieved with this type of technique reached three metres. But these cores showed three distinct bands of ooze, and the layers did not all bear the remains of the same plankton. One species of foraminifera played a key role at lower latitudes: *Globorotalia menardii,* which is abundant in warm water today. This was present in the top and bottom layer but not in the middle layer. The presence of *G. menardii* was a clear indication that the sediments were deposited during a warm interglacial period. And the change between the upper layer and the second layer without any *G. menardii* happened very abruptly an estimated 11,000 years ago.

These exciting initial results soon prompted a more determined effort to develop technology that could retrieve deeper cores. In the late 1940s, a Swedish expedition sailing around the world on the *Albatross* was able to retrieve cores between ten and fifteen metres long that represented a time interval of up to one million years. Their results recorded at least nine glacial periods during the Pleistocene. The cores also showed that layers of high concentrations of fossils alternated with layers of low concentrations, which reflected the glacial-interglacial conditions.

A second way that ooze can tell us about global climate was developed in the 1950s. Oxygen atoms in water, like the carbon atoms in grasses, are not all the same. In its stable form, oxygen can have either sixteen or eighteen neutrons. The much rarer ^{18}O isotope is heavier than ^{16}O. Because ^{16}O evaporates more easily than ^{18}O, ice sheets are almost entirely composed of this lighter isotope (ice sheets build up when snow precipitation does not melt from one season to the next). This means that during periods of glaciation, when much of the evaporated water is locked away in miles-thick ice sheets, there will be a higher ratio of ^{18}O to ^{16}O isotopes in the remaining seawater than during hotter, wetter intervals. This isotope ratio is captured by the foraminifera, which use carbon dioxide dissolved in the water to build their calcium carbonate shells. In the same way that fossil tooth enamel preserved the ratio between C_3 and C_4 plants in the diet and documented the climate change we recorded at Lothagam, foraminifera shells preserved in the ocean sediments bear a signature of the mixture of oxygen atoms in the seawater in which they lived. As foraminifera skeletons settle on the sea bottom, changes in the ratio of ^{18}O to ^{16}O are recorded layer by layer in the ocean sediments and bear witness to glacial and interglacial epochs through the ages.

Engineers have now perfected sea-core drilling techniques that enable them to penetrate deep into the vertical miles of fossil-bearing sediments on the ocean floor to extract cores several kilometres long that represent millions and millions of years. But in the meantime, a second obstacle had presented itself: how could the various sea cores taken from distant points around the vast oceans be correlated with one another to build a global, composite climate picture?

This is where Giuseppe Folgheraiter enters the story. He was studying ancient pottery and noticed that the clay somehow took up the magnetic signal from the earth. In 1906, Bernard Brunhes expanded on Folgheraiter's 1894 treatise on the magnetic alignment in baked pottery and bricks when he deduced that ancient lavas were likely to behave the same way. When Brunhes set out to measure the magnetism of ancient lavas, the result was surprising: some of the lava flows had inverted magnetic fields. His radical conclusion was that at some points in time the earth's magnetic field had reversed (so a compass would point south, not north). But this novel and unprecedented idea went down among the scientific community like a lead balloon.

Some twenty years later, the Japanese scientist Motonori Matuyama found that ancient lava flows in Japan and Korea also showed at least one magnetic reversal in the Pleistocene. Nobody believed him either. Then, in the 1950s and 1960s, scientists found that ancient lava flows in Russia, Iceland, and Hawaii also exhibited the same phenomenon, and a critical mass of evidence seemed to persuade the sceptics. By this time, new potassium-argon dating techniques had been invented using the rate of isotopic decay of an unstable potassium isotope into an unstable isotope of argon—the dating method we used to date ash layers at Turkana. This potassium isotope has a much longer half-life than carbon, and this allowed us to obtain much older dates than using the radiocarbon dating technique (which can be used accurately only on the first 50,000 years). Since volcanoes are chockablock with easily datable materials, the timing of the magnetic reversals could now be ascertained by using the new method to date the ancient magma flows on land.

We are now in the last epoch of "normal" polarity that started in the Pleistocene, which was named the Brunhes Normal Epoch in 1963 in honour of the French geophysicist. The last reversal took place 772,000 years ago, and the preceding period of reverse magnetism is called the Matuyama Reversed Epoch. Two short periods of "normal" magnetism were also discovered that fall within the Matuyama epoch. One of these short normals, which lasted approximately 40,000 years, is called the Jaramillo normal event after a small river in Mexico. The second and older normal was first found at Olduvai Gorge, and it is called the Olduvai event. It lasted for 30,000 years around 1.8 million years ago and is a very useful marker for palaeontologists.

The work on palaeomagnetism marked a true breakthrough. For lo and behold, the sea cores also contained evidence of the same magnetic reversals. By stacking the data from cores drilled from a variety of sites in the world's oceans and correlating the magnetic reversals, trends in the global palaeoclimate have now been reconstructed that extend all the way back to sixty-five million years ago (although the record does become less detailed in the older sediments). Scientists are now able to read the wealth of information preserved in the cores — not just from the oceans but also from the ice of the Arctic and Antarctic — to an astonishing degree. The sea cores are startling in the extent that they support Croll and Milankovitch's theory that changes in the earth's orbit and tilt are primary drivers of climate change.

THE PRESENT-DAY (Late Cenozoic) glacial age has borne witness to the whole history of human evolution and long predates our initial split from the ape lineage. As the polar ice has accumulated, the tropics have seen a trend in increasing aridity marked by ever more violent fluctuations in climate. The shift towards a drier climate and the consequent opening of forests and woodlands that we documented at Lothagam between seven and five million years ago — and was already evident at Kanapoi between four and three million years ago — would continue in response to the accumulation of ice in Antarctica propelled by different Milankovitch cycles. The current glacial epoch got going in earnest 2.6 million years ago and marks the beginning of the Pleistocene. At first, the ice sheets were relatively small and changed little. Between 2.58 and 0.78 million years ago, the world's climate got cooler, ruled largely by cycles of obliquity over 41,000 years that at particular points in time generated continental ice. However, glacials and interglacials only became clear at the onset of the Middle Pleistocene when eccentricity forces shift the earth's climate to an approximate 100,000-year cycle of much greater temperature amplitude.

These warm intermissions between mini ice ages can make the terminology horribly confusing: when people refer to the last ice age, they often do not mean the last fifty-five million years, which encompasses the present glacial age. They are usually referring to only the last glacial epoch, which ended some 10,000 years ago before the current warm interglacial period that we are enjoying today began. From 2.5 million years onwards, once both poles had iced over, a marked global cooling trend sets in, leading to increased general aridity in the tropics punctuated by ever more violent swings and upheavals in the global climate. In the last 800,000 years, there have been at least eighteen separate advances and retreats in the ice that fall in approximately 100,000-year intervals under the telling influence of the eccentricity cycle.

On the grand scale, this is the icehouse world that *Homo erectus* evolved in, and it must surely have played a huge role in shaping our ancestor's evolution. Many different strands of evidence in the burgeoning field of climate science all point to this time interval as being particularly variable. Dramatic climate change also led to a corresponding sweeping turnover of flora and fauna, opening new dietary niches and closing old ones. For her PhD, which she completed in 2001, Louise looked at the fossil record of bovids in the Turkana Basin to see if she could detect any significant changes in fauna in this very time period. The idea that changes in fossil biodiversity could be correlated with climate swings was first mooted by Elisabeth Vrba in the 1980s. Vrba studied African bovids from the Miocene to the end of the Pleistocene, and she brought together palaeontologists, climatologists, oceanographers, and geologists to look at the evidence from many different angles. It was Vrba's hypothesis that Louise tested for her PhD thesis.

But Louise had found that it was very hard to correlate changes in fauna with specific climatic events. For starters, there are large gaps in the sedimentary record. Added to these challenges, our dating instruments are still quite crude even with all the advances that have occurred since we started working at Turkana. Unless we find a way to date the fossil-bearing sediments themselves, our techniques depend on finding a datable volcanic layer above or below the sediments in which we find the fossils. All told, there is still a significant margin of error to our calculations of any particular geological slice of time that makes it impossible to precisely correlate local habitats to specific climate cycles—for there are simply too many climate shifts. Frank Brown, who knew the geology of the Turkana Basin better than anyone, pointed out that in addition to annual seasonal cycles and Milankovitch's three megacycles spanning thousands of years, there are also decadal (ten year) and multidecadal cycles. So changes by a mere five hundred years could mean the difference between a maximum and minimum in a millennium (thousand-year) cycle. With our blunt dating instruments, all these climatic cycles on multiple levels mean that there is always a degree of shooting in the dark involved in our reconstructions.

Moreover, whenever we try to reconstruct ancient environments, we can't be sure if the specific area being described is representative of a larger locale or if it is highly individual. This is clearly seen around the modern Kerio River in Turkana: the vast expanse of arid, barren landscape is abruptly replaced with a lush and dense riverine forest flanking the river. This forms a wildlife corridor for creatures large and small, including migrating elephants. Since deposition only occurs where there is water, a future palaeontologist sampling the fossil record along the modern Kerio could mistakenly arrive at the conclusion that in our time the entire area was covered in enough vegetation to support these large pachyderms and a host of other wildlife when in reality the narrow green belt was flanked by barren desert.

But while Louise could detect no clear evidence that confirmed Vrba's 2.5-million-year fauna turnover event, there seemed to be more evolutionary action going on 1.8 million years ago. Her thesis results are corroborated by subsequent studies of the fossil fauna from the Omo Valley and East and West Turkana by scientists René Bobe, Gerald Eck, and Kay Behrensmeyer (the very same Kay who found those artefacts at Koobi Fora in 1969 and after whom the famous KBS Tuff is named). They found a definite trend over time towards increasingly arid-adapted bovids that favour open habitat. Furthermore, they found three peaks in species richness—3.8 to 3.3 million years ago, 2.8 to 2.4 million years ago, and 2 to 1.4 million years ago. And they saw four episodes of relatively high faunal turnover—the biggest and most significant of which took place between 2 to 1.8 million years ago, which also coincides with the time *H. erectus* evolved.

This evidence of increased faunal diversity associated with shifts in the global climate corresponds with the remains that the hominins left behind at sites known as living floors. These are particular stratigraphic layers where the bones, stone tools, and other remains that indicate hominin occupation and activities have been preserved, such as the large living floor Mary excavated at the Dear Boy site at Olduvai. Although the living floors do not reveal whether the butchered remains of animals were scavenged from other predators or obtained by hunting, they do give us rare insights into the hominins' varied diet. At Turkana, archaeological sites dated between 1.8 to 1.4 million years show that the hominins—that we assume to be *H. erectus*—ate a huge variety of different species of different shapes and sizes. They include one type of elephant, one species of giraffe, and several species each of rodents, monkeys, small carnivores, zebras, rhinos, hippos, pigs, and antelopes. The hominins also ate turtles, crocodiles, fish, snakes, and birds. The living floors found at Olduvai Gorge show a similar diversity. In addition to these animals, whose fossils have been preserved, the hominins would almost certainly have also been eating birds' eggs, grubs, grasshoppers, and crustaceans. All told, as a result of the changing habitat, there were lots of delectable choices on the *H. erectus* menu!

Yet, in spite of the massive new feeding opportunities presented by the many flourishing species of herbivores, previously successful specialist predators—sabre-toothed cats, false sabre-tooths, and hyaenas —were pushed to extinction by a competitor also at the 1.9-million-year mark. There is interesting research into the evolution of tapeworms that suggests these hugely successful parasites decided to jump ship—or at least broaden their options—as early *Homo* evolved and shifted from hunted to hunter.

Tapeworms have two different hosts during their life cycle—a "definitive" host that bears the tapeworm during its reproductive phase and an "intermediate" host that acts as a vector for the parasite. I, along with others far more knowledgeable about parasitology, had always assumed that people picked up tapeworms when they began to domesticate livestock. But molecular biologist Eric Hoberg and his colleagues have used divergence data analyses to identify two separate occasions when tapeworms switched definitive hosts and evolved into new species. Both times predate animal domestication, both occurred in sub-Saharan Africa, and both times the original host was from the carnivore guild (hyaenas, jackals, wild dogs, and big cats) and the new host was a hominin. Most intriguing of all, they estimate that one of these host switches occurred between 1.71 and 0.78 million years ago. This means that *H. erectus* must have taken the tapeworms with them when they migrated to Asia.

Like the specialist carnivores, many of the monkeys also largely disappear, particularly the bigger, more specialised ground-dwelling colobines. *H. erectus* was probably eating them too. There is a fossil site called Olorgesailie, not far from Nairobi, where I sometimes took the children for weekends when we didn't have time to go to Koobi Fora. The site is relatively young, less than one million years old, and is famous for its impressive accumulation of Acheulian hand axes, which were made by *H. erectus*. Not far from the vast jumble of stone tools, there is a veritable graveyard of broken bones of a species of monkey called *Theropithecus*, and stone tools are scattered among the broken bones. When we compared the dietary adaptations of *Theropithecus* and *Paranthropus* in chapter 10, we found that some of these monkeys became really large. Those found at Olorgesailie show that at this point in time the largest *Theropithecus* were the size of a female gorilla and must have been truly impressive animals. I had long ago noticed that the monkey bones all shared extremely unusual and repeated breakage patterns. These breaks are not at all typical of carnivores or natural processes. For example, one of the arm bones, the ulna, was consistently split lengthwise rather than snapped into pieces by the teeth of a carnivore, and one of the foot bones called the calcaneum was always split longitudinally. The evidence strongly suggested that hominins frequently made meals of *Theropithecus,* and Alan Walker's wife, Pat Shipman, subsequently studied the break patterns and found this to be the case.

Climbing up the food chain from a predominantly herbivorous animal to a meat-favouring omnivore would have had profound implications for the home-range requirements of *H. erectus.* Like all predators, individual groups of hominins would have required bigger hunting ranges in their search for meat. Combined with their newfound success as hunters, the increased aridity during glacial periods would have further increased the size of territory that individual bands of hominins needed. Their ability to move into these grasslands would have been uninhibited initially. But as the *Homo erectus* population grew with its successful new dietary strategy, this expansion across Africa could not continue ad infinitum.

Could erratic climate and a corresponding massive faunal turnover have served as a trigger for *H. erectus* to move out of Africa? During the more "normal" periods of glaciations that have characterized the climate with few intermissions for some three million years, the great impenetrable dryness of the Sahara rendered much of Africa inhospitable. But the Sahara also prevented easy access to the land bridges out of Africa that would have been exposed when the great ice sheets lowered sea levels. But during the few windows of wetter, warmer, and more hospitable interglacial periods, a corridor through the Sahara would have opened—but rising sea levels would have cut off the land bridges out of Africa. Either way, *H. erectus* had to find its way through the formidable physical barriers that the weather extremes presented to dispersal before it could flourish across the globe.

16

||||||||||||||||||||||||||||||||||||||||||||||||||||||||||||||||||||||||||||||||

THE FIRST EXPLORERS

I SOMETIMES IDLY WONDER WHERE I WOULD GO IF I COULD CON-
jure a time machine for just one trip. Backwards, almost certainly, after
a lifetime of trying to fathom our past—but how far back? The world
of *Homo erectus* is one I would be sorely tempted to pick above all the
others. I would want to see for myself how this ancestor actually dealt
with the ferocious carnivores, how they really hunted and scavenged
with their rudimentary stone tools, and how far ahead they thought to
store food for lean times. I'd like to know how far the bonds of social
cooperation had grown and whether grandmothers enjoyed a brief period after menopause when they taught the youngsters how to survive
the hardships of life in the Icehouse.

From the cozy warmth of my time bubble, I'd love to see the frozen
spectacle of the earth gripped in the vise of vast sheets of ice during
glacial periods. With the greatly lowered sea levels during these times,
I would want to chart the altered shapes of continents and see land
bridges where today there is endless deep blue ocean—for the intrepid
hominins would not have travelled very far without those periodic land
bridges that opened up migration routes. My time travel would also
offer me a glimpse of the other animals inhabiting the world at that
time; just as we have posited that the evolution of our own species was
driven by the massive upheaval and alteration of the global climate, the
same forces opened up new dietary opportunities for a plethora of new

species across the animal kingdom. I would want to plug all the gaps in this story and see how much of it we've got right. Perhaps I could even sneak a peek at the other mysterious hominins, such as *Homo habilis*, who for a brief time were sharing the world with *Homo erectus*.

JUST HOW did the first intrepid *H. erectus* get out of Africa? We know that they had accomplished this amazing feat by 1.8 million years ago when we find the first evidence of Eurasian hominins in Dmanisi near Georgia's modern capital of Tbilisi. This site has revealed a veritable mine of beautifully preserved skulls, mandibles, postcranial fossils, and artefacts of *H. erectus*. These were discovered completely by accident plumb in the middle of the ruins of a medieval castle where archaeologists were engaged in the pursuit of knickknacks and artefacts left behind by far more recent ancestors. But the archaeological record yields frustratingly few clues of the path *H. erectus* took to get to Dmanisi. The most obvious route would be to follow the Nile—the only river that is known to have traversed the Sahara in the last two million years. There is scant if any evidence that the hominins followed the Nile through the desert, but this certainly does not mean they never did. The hominins presumably might also have taken a northern passage by crossing the Sahara during one of the wet periods that would have shrunk the desert considerably and provided life-sustaining water in the small lakes and wadis that briefly materialized at that time. Nevertheless, the few archaeological sites that have been unearthed in North Africa, although poorly dated, show that one way or another *H. erectus* did get across the Sahara.

Having reached the North African coast, *H. erectus* would have found only two possible ways to cross the expanse of water in the Mediterranean. The shortest route would be across the Strait of Gibraltar, which is currently almost fifteen kilometres wide. The only other alternative would be via the island of Sicily from what is now Tunisia. But the Mediterranean in both these places is really deep. To completely close the gap across the Strait of Gibraltar, sea levels would have had to drop by more than 300 metres, and a drop of 200 metres would narrow it to 6.5 kilometres, still a formidable distance. With a drop of 150 metres, the distance between Africa and Europe narrows to between two and five kilometres according to a team of scientists from the Catalan Institute of Palaeontology Miguel Crusafont i Pairó in Sabadell, Spain. Such a drop would have made for a few shorter crossings between newly exposed islands and doing this is theoretically possible. The longer crossing to Sicily would have required a dramatic drop in sea level to overcome the current two-hundred-metre-deep divide, and the complete absence of any African fauna in the Sicilian palaeontological record at this time does not support this scenario.

One way or another, *H. erectus* was able to breach the barrier of the Mediterranean at some point. An archaeological site in Italy called Pirro Nord has interesting artefacts and African-looking species of fauna. Sites in Spain from 1.4 to 1.2 million years ago also show that these hominins had made it across to Southern Europe. These explorers probably became refuge populations of *H. erectus* that were prevented from moving farther north by colder climate and the impenetrable ice sheets.

Two alternative routes circumvent the Sahara altogether by hugging the Ethiopian Highlands. The first one follows the western shore of the Red Sea northwards to cross out of Africa at what is now the Suez Canal, and the second crosses the Red Sea directly at the western tip of the Arabian Peninsula across the Bab-el-Mandeb Strait. Many archaeological sites scattered along the eastern shore of the Red Sea and on the tip of the Arabian Peninsula suggest that it was one or the other of these routes that was ultimately successful rather than the North African route. This route would also have worked for any hominins who found themselves on the east bank of the Nile in what is now Egypt.

Yet further formidable geographical barriers awaited the new emigrants—first the Arabian Desert and then the high Zagros mountain range that flanks the desert's northwestern boundary. But we know that some intrepid *erectus* were successful because of the *H. erectus* found at Dmanisi at 1.8 million years. Still more surprising is the rapidity with which they spread across Asia after forging a route out of Africa. Trinil is the site in Indonesia where Dubois discovered the very first specimen of *H. erectus,* Java Man, and it has also yielded additional specimens, including the Mojokerto child that has been dated to approximately 1.5 million years. At nearby Sangiran, a number of sites of varying ages yielded evidence of *H. erectus,* the oldest of which is dated at slightly more than 1.6 million years.

There are pitifully few clues in the archaeological and palaeontological record to tell us what path these first explorers followed, and the clues that do exist have generated doubt and controversy because of unreliable provenance and dating. Much of the evidence that these early *erectus* left behind is probably now buried under the sea on the previously exposed continental shelf. Nevertheless, get there they did. As the crow flies, ignoring the vast mountains, rivers, and oceans, the distance to Java from Dmanisi is nearly nine thousand kilometres. The actual path must have been much longer, most probably following the contours of continents and tracing coastal routes in order to avoid the world's highest mountain range, the Himalayas. Archaeological evidence indicates that *H. erectus* reached the far east of China at a place called Majuangou in the Nihewan basin some 1.66 million years ago. In North China, there are well-dated archaeological sites that show a continuous presence from the first known *erectus* at 1.66 million years through the next 340,000 years. In other words, these explorers were able to survive at the relatively high latitude of thirty-five degrees north (think Beijing, Tbilisi, and Chicago) through the climatic upheavals that surely followed the species' expansion. And perhaps the greatest testimony to its success is the fact that *erectus* flourished in eastern Asia for more than a million years.

It is most likely that *H. erectus* left Africa in successive waves, climate permitting. They wouldn't all have made it successfully, and their distribution would have waxed and waned with the vagaries of the weather. Some groups of *H. erectus* almost certainly would have become isolated by the sudden onset of a new climate cycle and become refuge populations cut off by the returning ice and desert during glacial periods or by rising sea levels during interglacials. These populations might have persisted for a while before either dying out or merging back with the larger population once the geographic barrier retreated again. This could help to explain the extraordinary degree of variation that is witnessed across the whole sample of *H. erectus* as well as the perceived differences between the specimens found in Asia and Africa.

Refuge populations of *erectus* might also have remained isolated for far longer periods and continued along a separate evolutionary trajectory. Exciting evidence of an extreme example of this has recently been unearthed on the Indonesian island of Flores in one of the most significant and intriguing finds in decades. This discovery ranks among the most controversial finds of all time and has generated an inordinate amount of discussion and divergent opinion. Flores is one of a series of islands that stretch in a lazy arc like a string of beads from the mainland peninsula of Malaysia towards the northern tip of Australia. It is easy to see how in ancient times when sea levels were much lower that many of these islands would have formed a long peninsula. But one of the things that makes the discovery so interesting is that even if sea levels had fallen dramatically, any creature wanting to reach Flores would have had to cross three deep stretches of water that separate the Sunda Shelf from Flores and span at least nineteen kilometres each.

The fossil evidence on Flores supports the idea that deep waterways isolated the island because all the remains are from species that could have reached the island by swimming (crocodiles and a type of elephant), by flying (birds, along with the plant seeds and spores lodged on their feet), or by accidentally hitching a ride on natural rafts of floating vegetation that were flushed out to sea by flooding rivers (tortoises, Komodo dragons, rats, and molluscs). But there is one very interesting exception. In 1998, a team of scientists led by the Australian archaeologist Michael Morwood discovered some crudely flaked pebbles at a site called Mata Menge. Not only were these artefacts similar to those found with *H. erectus* in Java, but they were found to be approximately 700,000 years old—far too old to have been made by *Homo sapiens*.

It is unlikely that hominins could have survived such a long sea passage by unintentionally drifting on a natural raft. And humans are not strong enough swimmers to undertake such a journey without a raft. The irrefutable conclusion is really startling: that the maker of these tools must have used a rudimentary type of watercraft to reach the island deliberately—even if it was just a few logs lashed together with lianas and powered by some kind of paddle. Nobody had thought that human ancestors were capable of seafaring voyages until much later —between 60,000 and 40,000 years ago. Plus, everybody had assumed that it was *H. sapiens*, not *H. erectus*, who first deliberately set sail for other shores. In the absence of fossil evidence, some scientists were sceptical that these old stones were artefacts at all. If you think about it, however, it isn't that much of a stretch to believe that by 700,000 years ago *H. erectus* had already learned to cross stretches of water without swimming—for its path this far had already taken in many great rivers —the Euphrates, the Indus, and the Ganges, for starters. And since *H. erectus* first left Africa some 1.8 million years ago, water levels had been rising and falling like a yoyo thanks to the Milankovitch cycles.

This first find on Flores was then followed in 2004 by an even more astonishing discovery when Morwood and his team of Indonesian and Australian researchers discovered a hominin skull and partial skeleton in a cave called Liang Bua. The stratigraphy of Liang Bua is complex, but more than a decade's work has established that the hominins discovered by Morwood lived there between 100,000 and 60,000 years ago. The specimen was surprising not just for its recent age but also because of its extremely small stature. The bones were of an adult female, and they revealed a creature that would have stood only one metre tall —slightly smaller than even Lucy, the smallest known australopithecine. If you estimate the Flores hominin's weight by the cross-sectional area of her femur—the method used to calculate that of the Turkana

Boy—she would have weighed about thirty-six kilograms (close to eighty pounds).

But perhaps most surprising of all was how primitive some features of this new hominin were. It had quite a broad pelvis and a femur with a long neck that looked quite different from ours. To match its tiny frame, the brain capacity, between 380 and 417 cc, was also smaller than any other known hominin's. Then there were unmistakably modern characters—the small teeth and the shape of its little face. The scientists concluded that it most closely resembled *H. erectus,* albeit in miniature. They named it *Homo floresiensis* and nicknamed it the Hobbit. Since the first skull and skeleton were discovered, remains of other individuals have also been recovered.

But what could explain why the Hobbit was so small? The explanation that makes the most sense is the one offered by the researchers when they published their spectacular find. They argued that *H. floresiensis* had been subject to island dwarfism: a phenomenon that takes place when animals get isolated on an island without any predators. Animals tend to either get smaller to escape the notice of predators or bigger to intimidate them. But both strategies come at the expense of expending energy less efficiently. And since the carrying capacity of islands is constrained by their size, islands can sustain bigger populations of smaller animals. In the absence of predators, every animal will tend to move towards the calorically efficient size of a rabbit: bigger animals get smaller, and smaller ones get bigger. The selective pressures to change size once the threat of predation has been removed are remarkably strong. In Sicily, there are fossils of dwarf elephants that suggest the process of downsizing took less than five thousand years. On Flores, there is evidence of island dwarfism among other mammals —the fossils include a giant rat and a pygmy elephant. What makes the scientists' claim so controversial is that it shows that hominins responded to the evolutionary pressures on islands in the same way as other mammals did in spite of having bigger brains and stone tools, which might arguably have buffered them somewhat from evolutionary pressures.

Many believe this claim to be so unlikely as to be impossible. Some critics asserted that the bones were the remains of a *H. sapiens* with a genetic disorder known as microcephaly, which causes a very small brain. Morwood and his colleagues had tried to figure out how brainy the little Floresian actually was by comparing estimates of her ratio of brain to body weight with that of other hominins. This ratio is called an encephalization quotient or EQ. The Hobbit's EQ is very small and falls within the range for an australopithecine. But when compared in shape to those of australopithecines, *H. erectus,* different modern apes, and *H. sapiens,* all the principal brain components on the Flores endocast most resemble *H. erectus* and differ markedly from a microcephalic human. In other words, the new hominin wasn't at all brainy compared to modern humans. Although its brain is surprisingly small, *H. floresiensis* probably matched the cognitive powers of *H. erectus.*

Others have recently put forward the idea that the little people were *H. sapiens* who were the victim of a dietary deficiency that leads to a genetic deformity affecting the function of the thyroid. This "congenital hypothyroidism" results in severe dwarfism and reduced brain size, and victims of this disorder are called "myxoedematous endemic cretins." The proponents of this theory claim that there is nothing about the Hobbit's morphology that rules this explanation out, but they have never examined the fossils themselves, and several scientists who have done so have strongly disputed their findings.

Moreover, the postcranial remains would seem to suggest that the island dwarfism explanation is the correct one. *Floresiensis* had a wrist and shoulder morphology not found in either modern humans or Neanderthals but found in older ancestors. Wristbones are very hard to describe to the nonanatomist because there are a lot of them, so their scientific names are hard to keep straight, and they are all arranged to fit together like a complicated 3-D jigsaw puzzle of movable parts.

They have many different surfaces that articulate with other adjacent bones, and all these surfaces have changed shape during the evolution of our flexible, dexterous hand. To further complicate matters, we don't actually know what the *H. erectus* wrist looks like because hand and foot bones are delicate and rarely preserved—and are even harder to find associated with skulls, which is necessary to correctly identify which species they belong to. We continued the excavation of the Turkana Boy for five years in the hope of finding the hands and feet, but we were ultimately unsuccessful. So, for many measurements, the Hobbit could only be compared to the modern apes, *Australopithecus afarensis,* the Neanderthals, and modern humans, but not to early *Homo.*

The capitate, the bone we searched long and hard for at Kanapoi, is one of the three telltale wristbones that was recovered from the Hobbit (the other two are the trapezium and the trapezoid). Matthew Tocheri, an expert in hominin wrist morphology at the Smithsonian Institution, took a number of measurements of these three bones and used multivariate analysis to compare those of *H. floresiensis* to the bones of modern apes, baboons, Neanderthals, and both archaic and modern *H. sapiens.* The results are startlingly consistent: the Hobbit wrist comes out time and again much closer to that of modern apes and baboons than to those of Neanderthals or any modern human. Our wrist morphology is formed after only eleven weeks of embryonic development. This is before the genes that would cause abnormal development, such as the microcephalic condition, have kicked in. So the differences in our wrist from the Hobbit's are strong evidence that the Hobbit descended from an earlier ancestor—one that left Africa before the wrist evolved to our modern form.

The partial skeleton from Flores also includes much of the shoulder: a partial collarbone (the clavicle), a partial shoulder blade (the scapula), and an almost complete upper long bone from the arm (the humerus). The clavicle and scapula form the socket that the top of the humerus fits into, so the shape of these three bones is very important in determining the mobility of the arm. Morwood teamed up with anthropologists Susan Larson and Bill Jungers at Stony Brook University to study the Flores shoulder. They found that the Hobbit's shoulder morphology is more primitive than that of *H. sapiens* and is more similar to that of the Nariokotome Boy than to modern humans—a form that is intermediate between human and ape.

Humans have a long collarbone whereas Flores has a relatively short one. And the Flores shoulder blade is lengthened compared to ours. Another key feature that they studied is called "humeral torsion"—the angle of the head of the humerus compared to the angle of the base of the bone at the elbow joint (the lower articular surface), which originally in mammals was 90 degrees. We, in contrast, have a much bigger angle between the two. For the Hobbit, the angle is about 120 degrees, and although the Turkana Boy is missing the top of the humerus, estimates suggest 111.5 degrees of torsion. Similarly, the Dmanisi hominins exhibit low torsion at 110 and 104 degrees for the two different humeri. Thus the torsion in the Hobbit appears to be much closer to that of *H. erectus* than to that of modern humans, which can be as high as 180 degrees. Because of this high torsion, our arms fall naturally from the body with the palms facing inwards to our thighs. In the Hobbit, the hands would have faced forwards.

We don't have any *H. erectus* feet to compare to those of the Hobbit, but here too its morphology is surprisingly primitive. Compared to all the other small features of this diminutive creature, it had surprisingly big flat feet. While human feet are about 55 percent of the length of the femur, the Hobbit's feet were 70 percent the length of this long bone. The Hobbit's foot most resembles early hominin feet such as *H. habilis*'s. It had a big toe like ours for "toeing off" while taking a step, but its gait would have been much more awkward than ours because of its very large feet.

The complex mixture of features in the skeleton of the Hobbit can't be explained by pathological conditions and, incredible as it might seem, are most consistent with a process of island dwarfism among an isolate *Homo erectus* population. The recent discovery of hominin fossils of a similar small size at the 700,000-year-old Mata Menge site have given us a potential timeline for the evolution of *Homo floresiensis*.

It is mind-boggling to think that more than 140,000 years after *H. sapiens* evolved, an island dwarf that derived from an *H. erectus* ancestor still persisted in an isolated pocket. Until only relatively recently, *H. sapiens, H. neanderthalensis,* the elusive Denisovans, and *H. floresiensis* were all sharing the planet, a remarkable diversity at a relatively late stage of our evolution. The Hobbit and the pygmy stegodon both apparently became extinct after a volcanic eruption on Flores rendered it uninhabitable for a while. But in the plethora of islands that make up the Indonesian archipelago, Flores is not alone for its legends of the "little people" who live in the forests. Might there be still more island dwarfs hiding in the remote forests or only recently becoming extinct? It is an alluring idea.

17

A VERY GOOD HOMININ

IN 1971, AT THE SHARP CRACK OF A GUNSHOT AND THE SENSA-
tion of a bullet whizzing past his ear, Bernard Ngeneo dove beneath the
dubious protection of a small wait-a-bit thornbush. He was prospecting
for fossils with the rest of the team at Ileret near Koobi Fora. In those
days, *shifta* bandits roamed far and wide in the northern reaches of
Kenya so run-ins with small groups of armed men were not uncommon.
With a friendly gift of some tea or tobacco, fresh water, and perhaps a
meal, these meetings usually passed without live gunfire. But this time,
as bullets ricocheted about, Ngeneo was doubly lucky. Not only did the
bullets narrowly miss their mark, but beneath the bush, to his delight
and great surprise, Ngeneo found a beautifully preserved complete
hominin mandible! The bandits were soon completely forgotten in the
excitement and subsequent controversy around KNM-ER 992. This
fossil would end up being erroneously classified as a new species, and
it would be almost three decades before we could sort out the ensuing
tangle with another spectacular fossil find.

The year 2000 at last brought me back to Ileret and the very sediments where Ngeneo had made his headlong dive under a bush. I had
long planned to return to the east side of the lake, but with one thing or
another, my fieldwork on the west side had taken more than a decade
to complete. Every site we explored had proved to be so much more
exciting and productive than we could have hoped for. But all the work

I had done and the research by others in numerous related fields now seemed to be beckoning me inexorably to the 1.8-million-year mark.

We knew from the geological record and climate science that by this time the accumulation of ice sheets over Antarctica had flung the world fully back into an ice epoch. The palaeontological record showed that the increased extremes in climate had in turn caused a huge turnover in the fauna as old dietary niches literally dried up and new species evolved to meet new opportunities. Emerging out of all this change was our direct ancestor—the spectacularly successful *H. erectus*. I was convinced that I could no longer delay, and I resolved to look more closely at this pivotal age in our prehistory in East Africa.

Louise was also anxious to return to the east side of the lake. She was concerned about the increasing incursion of livestock in Sibiloi National Park and the desperate plight of young people at Ileret who had no prospects for employment. All this was contributing to threaten the fossils and the wildlife that the park had been established to protect. In the intervening years since we last scoured the East Turkana sediments, the squat, square, and orderly police post of Ileret, which used to comprise a few buildings arranged with military precision, had mushroomed into a sprawling town of thousands. Still leading a partially nomadic lifestyle, its inhabitants brought untold numbers of goats with them, and the formerly well-vegetated landscape once filled with abundant wildlife now more closely resembled the denuded west side.

The Koobi Fora base camp, built up with such care by Richard so researchers could continue to work the rich deposits, had fallen into a state of terrible disrepair under the management of the National Museums of Kenya, with the museum staff often enduring many months without resupply of food and water   or even salaries—because they had neither spares nor fuel for their vehicles. Still, it was exciting to be back in this magical part of Kenya, and we soon had a basic operation outfitted and running again.

We moved into the old sandstone *banda* that Richard and I had occupied in the 1970s, which still had our old furniture including a battered chest of drawers full of a very colourful collection of my vintage '70s outfits. Now almost my exact same size, Louise spent an enjoyable afternoon trying these retro outfits on with much enjoyment and mirth before we discarded them to make room for new field clothes. Not long after this, we discovered that our new wardrobe had invaded the home of a small African rock python that had taken up long-term residence in the chest of drawers. Each morning, Louise and I would investigate whether the python was in Louise's drawer or mine and wear each other's clothes to avoid disturbing our handsome tenant!

As usual, the highlight of that first season back on the east side came towards the very end. Fredrick Kyalo Manthi, who had sampled the microfauna at Kanapoi, arrived for a rushed thirteen-day visit stolen out of his time with his family in between semesters at the University of Cape Town where he was in his third year of undergraduate studies. Kyalo announced without preamble that he had come to find a hominin for me — and not just "any old hominin, but a very good one." Perhaps I rather dampened his enthusiasm with my very sceptical rejoinder that he'd better hurry as he hadn't much time! But precisely thirteen days into his visit, following the Nzube method, Kyalo waited until the very hottest part of the day just before lunch to tell me that he had something "rather interesting" to show Louise and me. "Rather interesting" is another way of saying, "I have a monkey to show you," which usually means "I think I have a hominin, but I don't want to be disappointed if it isn't." Enjoying the thrill of anticipation that is always heightened at moments such as these, we willingly accepted the detour and hastened after him across the oven-hot exposures.

To this day, I have no idea how he spotted it. Not far from a scattering of scrappy bovid bones, Kyalo pointed to a very unprepossessing layer of cementlike rock in a rather flat area of land denuded of all but the toughest tussocks of camel grass by Ileret's voracious goats. As I stared past Kyalo's finger through the blinding glare, my eyes at last made out a slightly protruding bump in the rock that could have been anything. But all at once, the large distinctive circular earhole that can only mean a hominin skull leapt into focus. And right where it should have been, I noticed what looked like a thin brow ridge barely peeping out of the calcrete matrix that completely obscured everything else. Kyalo's really good hominin was way beyond our wildest dreams—but we wouldn't know that for some time because we couldn't see how much more of the skull would be concealed in the rock.

Kyalo went back to the car to retrieve some picks, hammers, chisels, and some water, and sent the rest of the crew back to camp for lunch. Louise ran back to collect the camera, and she managed to drive the car farther up the soft sand river so we wouldn't have to carry the rock-encased skull such a long distance. Close to four hot and sweaty hours later, we had at last managed to extract a solid lump of rock encasing the skull. Kyalo led us triumphantly back to camp and on to Koobi Fora, where the rest of the crew excitedly awaited us with a celebratory meal of delicious fresh tilapia plucked out of the lake specially for the occasion and washed down with the rare treat of icy cold beer.

Kyalo's spectacular discovery was the beginning of a spate of really good luck. Hot on its heels, two Turkana members of our field crew, John Kaatho and Robert Moru, produced beautiful and important finds—an upper jaw that would prove immensely significant because it was not *H. erectus* and a stunningly well-preserved lower leg bone (tibia). Four isolated teeth in various degrees of preservation rounded out the year's haul of hominins.

It took Christopher Kiarie almost a whole year to meticulously remove all the grains of sand firmly cemented onto the skull—with all of us periodically peering impatiently over his shoulder and through his microscope to watch the bone emerge at a glacial pace. He did an incredible job, and his reward was an exquisitely preserved skull. It was complete but for the face, and apart from one small broken area on the top of the skull, it was undistorted. How different from the sorry state of Flat Face when its battered skull first emerged from the rock! Best of all was the incredible degree of detail on the base that showed every orifice where blood vessels and nerves would have entered the brain.

This fragile part of the braincase is usually the first to break off and be lost, so the rarity of this beautiful fossil made it even more special.

While the skull certainly looked like *H. erectus*, two things about it immediately struck us as a bit odd. The first of these was its diminutive size and its slight delicate features. It did not have very pronounced brow ridges, which gave many people the first impression that it must have been a child. Louis's OH 9, which he dismissed as too small-brained a creature to be anything worth bothering about, completely dwarfed Kyalo's skull, which was less than two-thirds the size. Indeed, our new skull, KNM-ER 42700, later proved to be the smallest known *H. erectus*.

The second thing we noticed was its uncanny resemblance to one of the skulls found in Dmanisi. The first of these Georgian discoveries was a mandible retrieved from the ruins of a medieval castle in 1991. That mandible had also drawn Alan Walker's attention for its striking similarities to the Turkana Boy's jaw when he was making his study of the skeleton at around the same time. Three more mandibles and four skulls followed the first jaw out of the Dmanisi excavation as well as a plethora of primitive stone tools and the remains of elephants, giraffes, gazelles, rhinos, sabre-toothed cats, bears, wolves, rodents, and ostriches.

It was the second Dmanisi skull to be discovered, D2700, that had a very similar profile to Kyalo's skull and shared its small size and gracile brow ridges. This skull, unlike ours, still had its teeth preserved as well as its lower jaw. This meant that its age could be determined quite precisely, which showed us that this individual was not yet fully grown because its wisdom teeth had not erupted. This immediately begged the question of Kyalo's Ileret skull: was it small because it was young or was it just small? According to conventional wisdom, *H. erectus* beyond the shores of Africa was not supposed to resemble its African cousin, so why did these two skulls look so incredibly similar? I was intrigued and thrilled. But personal and professional setbacks were destined to interrupt our plans yet again.

New hardships and new resentments had surfaced since Richard and I had worked these sediments some thirty field seasons previously, and our good relations with the community had long since been forgotten. Sibiloi National Park was in desperate trouble. This unique park was gazetted in 1976 to safeguard some of the richest prehistoric sites, and it was declared a UNESCO World Heritage Site in 1994. Because of its locale, it contained animals found only in the northern part of Kenya: Grévy's zebra, the reticulated giraffe, and the beisa oryx as well as a number of the large cats. In the 1970s, few local people used this area because the Dassenetch from the north have a long-standing enmity with the Gabbra who live to the east. When individuals from these two tribes meet, deadly battles follow, and bouts of cattle rustling are frequent. By everyone's avoiding the area except when extreme drought reduced the grazing elsewhere, the people from the two areas rarely met, and a semitruce prevailed. The buffer zone of the park in turn protected the wildlife and the fossils.

This situation inevitably changed with increased population numbers. Depending on their livestock for subsistence, Dassenetch from both Kenya and Ethiopia traditionally move south during the driest months to seek grazing for their livestock. However, by the year 2000, a large number of nomadic herdsmen had moved vast herds of livestock into Sibiloi National Park on a semipermanent basis. Armed to the teeth with semiautomatic rifles and great long belts of bullets, these herdsmen frequently shoot the wildlife, and many species are now seriously depleted or have vanished altogether. The fighting for access to water and grazing between the Dassenetch and the Gabbra is more and more frequent, and fatal conflicts are far too common. The Dassenetch understandably resent any efforts to protect the wildlife and the fossils in Sibiloi because the immediate benefits they derive from grazing their cattle in the park outweigh any value perceived in Nairobi about preserving a rich national heritage. The Dassenetch of this marginalised area, all but forgotten by Kenya's central government, had for many years been almost entirely excluded from reaping the rewards from having a national park nearby. The area is so remote, serviced by such poor roads, and often rife with insecurity that few tourists reach the park. The collections of fossils and artefacts resulting from field expeditions were always packed up and transported back to Nairobi to the National Museum of Kenya following Kenyan government policy and requirements. The local communities had little involvement in the field research apart from the few individuals employed by us as temporary help and no knowledge of the significance and importance of their area revealed by this research.

It was increasingly apparent that if there was to be any long-term future for the fossil sites and the wildlife, we had to have the cooperation of the local people. In order to achieve this, they would have to be made aware of the potential benefits these assets could generate. If we were unable to accomplish this, it would become all but impossible to continue field research in the area. There was a real risk that both the animals and the fossils might be irrevocably lost. And in 2003, we found ourselves completely thwarted by the resentments of a people who took out their grievances on the only outsiders there to hear them —us.

As soon as we resumed our work on the east side of the lake in 2000, Louise immediately set about raising funds for the school and the local clinic, which at the time was a mere shell with neither clinical officer nor medicines. But the young people, who desperately wanted employment and cash, believed that we should be doing far more to help them. Our small expedition simply did not have the resources to satisfy the huge demand, and as a consequence of the pressure that the Ileret people exerted, it was all but impossible for us to work there. The animosity of some community members forced us to postpone our fieldwork in 2003, which gave us time to initiate additional community help—funds for a school extension and a reactivated clinic with regular medical supplies provided by AMREF and a full-time clinician with some training in basic clinical techniques. By 2004, with these measures in place, we felt that we would be able to safely work at Ileret again. But although we had the support of the local leaders, it quickly became clear that we were still unwelcome among the youth. We reluctantly moved back to Koobi Fora to work the sediments farther south. All the solutions involved the need to generate very large sums of money, which seemed to us impossible and were far outside the purview of palaeontological exploration and the shoestring research grants that we operated on. It was also increasingly difficult for Louise and me to be based at the Koobi Fora research camp because it was being used for tourists and field-school students organised through the National Museums of Kenya, and we were eventually forced to leave the camp Richard had built all those years before. This made the difficult logistics for vehicles, supplies, and accommodation for our field crew even more complicated.

Added to these already formidable challenges, Richard's health was again failing rapidly. The kidney that his brother Philip had donated a quarter of a century earlier, although having far surpassed all expectations of how long it might last, was finally packing up. Then, all of a sudden, it failed altogether in 2005 after a reaction to some drugs Richard had been given when he was in Stony Brook, New York. After receiving an honorary degree in 1998, he had struck up a warm friendship with the president of the university, Shirley Strum Kenny, who shared his vision of furthering research in the earth sciences in Turkana and had offered him a professorship to develop ideas about how this could be achieved. Richard chaired a series of human evolution workshops and was in active discussions as to how Stony Brook University could partner with Kenyan institutions. But all of this was put on hold as we grappled with his new illness.

Richard phoned me from Stony Brook as I was preparing for our fieldwork that year.

"My kidney has failed," he said without preamble. "The only way to stay alive will be a second transplant, and fast."

Shaken, but without hesitation, I offered to give him one of my kidneys and left immediately for New York. Upon my arrival in Stony

Brook, I was poked and prodded and subjected to a battery of medical tests that measured every imaginable indicator of my health. In spite of being in my early sixties, I passed these, and the rather poor match between our tissue types was no longer such a problem because the antirejection medicines were greatly improved since 1980. The doctors established that I was a viable donor, and we were soon both on the operating table in early 2006. This time, the transplant was performed using a tiny incision and keyhole surgery. Nevertheless, the operation was excruciatingly painful, and I spared many a thought for Philip's far more invasive procedure that resulted in a lengthy scar running from his spine all the way round to his belly button.

Samira came to stay with us and played nursemaid to both of us. Although the massive advances in medical technology and transplant surgery over the twenty-five years since his first procedure certainly smoothed the way, Richard was again desperately ill, and his stubborn perseverance that makes him such a dreadful patient is also what helped him to survive the ordeal. Unbeknownst to us, amputees need special provisions during an operation like this because of fluid accumulation in the legs. Poor Richard swelled up like a balloon and was incapacitated for several weeks of avoidable suffering.

Louise, newly pregnant with her second child, ran our fieldwork while enduring severe morning sickness. In contrast to when she ran the expedition for me in the aftermath of Richard's plane crash when the only communication possible was by sporadic radio call, we were now equipped with e-mail and satellite phones. Samira was able to keep her updated on our medical progress, and I was able to follow the fieldwork in much more detail.

Within the space of just a few months, both of us were back on our feet again, and we returned to Kenya to resume our lives, once again thankful for the extraordinary feats that modern medicine can perform. Although Richard was officially "retired," he still had many interests and projects. One of these was a burgeoning idea about how to help us to address the challenges thwarting our fieldwork at Ileret, which had been put on hold due to the very real question mark as to whether he would survive to execute it. When Louise and I asked if he could help us raise funds to build a new and independent research centre, his response was classic Richard: "Yes, but only if I can build something big!"

And thus the Turkana Basin Institute, an ambitious research and teaching facility, was born. The idea was to create an institute with permanent infrastructure that would enable year-round research rather than the piecemeal short seasons we had always relied on. Richard outlined his vision to Stony Brook University, and they enthusiastically endorsed the idea of TBI, committing funds and administrative support and becoming TBI's main partner. We planned two centres on either side of the lake as well as an administrative office in Nairobi and a New York office hosted by Stony Brook.

Before long, we had constructed temporary facilities near Ileret that were conveniently close to the airstrip, and this was fully operational by the end of 2007. Construction of the first full field centre on the west side of the lake was completed in 2012, and thereafter the construction of permanent facilities began. At last, Richard and I could work together again on prehistory.

We had the capacity to conduct our own research year-round and host research projects by other scientists in fields that complemented our own interests. We also established a field school that welcomed international students who wanted to learn about human evolution, geology, and ecology in the field while earning credits towards their university degrees. At the same time, the field schools allowed us to invite Kenyan students on scholarship to foster local knowledge and talent in the earth sciences. With greater financial muscle and a permanent presence in the field, we have also built successful community projects that support health and education and employ some of the local youth whose resentment and hostility had been severely thwarting our work.

• • •

WITH OUR FIELDWORK up and running again under the auspices of TBI, we were finally able to turn our attention back to Kyalo's skull. When Richard had first described the beautiful 992 mandible that Ngeneo had discovered, he noted its gracile slender form and its affinity "morphologically and in dental proportions" to at least some of the *Australopithecus africanus* specimens in South Africa. This would turn up a red herring: rather than comparing the mandible and the rest of the *erectus* collection to the Asian fossils, they were compared in detail to *A. africanus* by Colin Groves and Vratislav Mazák in 1975. Because the mandible did not resemble *A. africanus* that closely after all, the pair rather bizarrely and unthinkingly reclassified the entire East African collection of *Homo erectus* as a new species, which they named *Homo ergaster* ("working man"). Their argument for doing so never included a detailed study of the whole *erectus* sample, but because the mandible came to be viewed as a good example of East African *H. erectus*, the name stuck and came into common usage. I was aghast at the new name and longed to find the evidence to dispel the notion that the African *erectus* hominins were a different species than the Asian ones. This was my first order of business.

Fred Spoor, Louise, and I undertook a meaningful comparison of not only the African *H. erectus/ergaster* skulls but also the vast Asian collection of *H. erectus* that none of us were at all familiar with. We immediately realised that we were in over our heads, so we turned to Susan Antón, one of the very best in her field. Fred had known Susan for some time and had visited her at her farm in New Jersey that she shares with geophysicist Carl Swisher and a large family of canine friends. Not only is Susan a tremendously likeable person with a great sense of humour, she had also worked for many years in Indonesia and her forte is the Asian *H. erectus*. Susan readily agreed to join us in our study of Kyalo's skull, ER 42700.

Our first task was to figure out if 42700 was an *erectus/ergaster* or some other early *Homo*. It was clear to us all that we needed to know just how old 42700 was when he or she died before any meaningful comparisons across the collections could be made. This would affect our interpretation of both the brain capacity and the morphology of the skull because both the brain and the brow ridges would be less developed in a subadult or juvenile. Without teeth, this task would have been all but impossible but for the beautifully preserved base of the skull. Unlike the bones in the skeleton, which fuse at very specific times in childhood development and give a good indicator of age, the degree of fusion of the skull bones is a more difficult indicator of age. Luckily for us, however, there is one area on the base of the skull that is a good yardstick, and this was immaculately preserved.

This small telltale area is on the part of the skull called the basioc- cipital, and it's where two bones fuse together. One of these is called the sphenoid—a very important little bone right on the bottom of the skull that resembles a butterfly with outstretched wings and articulates with all the other bones in the skull to bind them together. The second is called the occipital, and it's the bone that encompasses the bun at the back of the skull and curves round underneath it to include the foramen magnum. Spheno-occipital synchondrosis is the tongue-tan- gling name of the area of fusion between the two, and we can tell quite accurately how old a subadult is based on how fused they are.

With the naked eye, we could see quite clearly that the bones were not quite fully fused. The big question was: how much longer before they would have been? Our friends at Diagnostic Centre Kenya just up the road from the National Museum in Nairobi once again came to our rescue. In 2006, Fred and I worked on the skull when human patients had no need of the CT machine. Paying special attention to the sphe- no-occipital synchondrosis, we took a series of one-millimetre cross sections of the skull. Fred converted these data so we could view them on a computer screen. Then we stacked them together digitally, build- ing up a 3-D picture of the whole cranial vault on the computer. When we enlarged these images and zoomed in on the spheno-occipital syn- chondrosis, we could see exactly at which point the fusion stopped. It was two-thirds complete when the individual died.

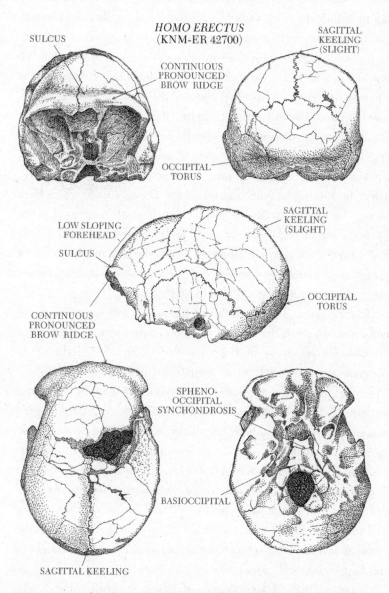

HOMO ERECTUS
(KNM-ER 42700)

SULCUS

SAGITTAL KEELING (SLIGHT)

CONTINUOUS PRONOUNCED BROW RIDGE

OCCIPITAL TORUS

LOW SLOPING FOREHEAD

SAGITTAL KEELING (SLIGHT)

SULCUS

OCCIPITAL TORUS

CONTINUOUS PRONOUNCED BROW RIDGE

SPHENO-OCCIPITAL SYNCHONDROSIS

BASIOCCIPITAL

SAGITTAL KEELING

Homo erectus is found across most of the Old World, and yet, there are far more differences through time than physical space. Nevertheless, our study did identify key features that differentiate this species from *Homo habilis*.

The spheno-occipital synchondrosis is already 50 percent closed by the time humans reach puberty and the second molars are through and the wisdom teeth are not yet erupted. In chimpanzees, fusion is 50 percent complete after the wisdom teeth are already through. We don't know whether *H. erectus* would have aged like a chimpanzee or a human (although we have hypothesised that it might have been half-way between), but this was enough to tell us that the individual was already a young adult or at the very least a late subadult and that its cranium would have already reached its adult size. Thus the diminutive skull held a fully grown brain that weighed in at a modest 690 cc. The spheno-occipital synchondrosis of the Dmanisi skull, which blew us away for its resemblance to 42700, is less fused than 42700's but not by much, and its brain capacity would probably have increased slightly had it lived to be an adult. Knowing that 42700 was an adult, albeit a young one, we could be confident that our comparisons across the *erectus/ergaster* collection would be meaningful.

In the years since Dubois found the first *H. erectus* in 1891, the collection had grown into one of the most comprehensive samples for any hominin species. From Africa, there were four other good skulls to compare 42700 to as well as four partial ones. In addition to the Dmanisi skulls, there were more than thirty specimens from Asia. The skulls range in age from the oldest at around 1.8 million years to the youngest, which could be a few hundred thousand years old. The whole sample spans nearly 1.5 million years — and within this span, most of the African skulls are clustered around the old end of the spectrum while the majority of the Asian ones are considerably younger. How ever were we to sort out which were valid characters that define the *erectus* species and which characters were inevitable differences that arose because of the enormous range — both geographically and chronologically — of this hugely successful hominin?

Looking closely at the collection of African skulls assembled under the appellation *H. ergaster,* we found many variations in size and features. When *H. ergaster* was reincarnated as a separate species based on comparison with the South African australopithecines and on a mandible alone, there was never a gold standard of defining characters put forward for the skull that could be used to determine whether a specimen belonged in this assorted group. People recognised *H. ergaster* skulls from informally agreed-upon characteristics. These features included thinner cranial bones and less pronounced brow ridges than those present on the Central and East Asian *erectus*. A third feature considered key is known as keeling: the prominence of a ridge running along the top of the skull to its base. This feature was believed to be unique to the non-African hominins and was a strong basis for separating *ergaster* from *erectus*. Common to skulls in both the African and Asian collections is the distinctive ridgelike torus at the back of the skull that differentiates *erectus* (or *ergaster*) from other early *Homo*. Lastly, the cranial capacity of the *erectus/ergaster* group is larger than that of other early *Homo*.

Other parallels made the Dmanisi skull very relevant to our study. When this Georgian skull was first found, some scientists latched on to its small size and gracile features to contend that it was an early intermediate hominin that linked an ancestral *Homo habilis* in a linear line of evolution to *H. erectus/ergaster*. This view was pervasive—even Mary Leakey, whom I admired as one of the most thorough and careful scientists I have ever had the privilege to work with, bought into it because *H. habilis* and *H. erectus* are in completely different beds at Olduvai. Although still cautious, Mary wrote that "it is possible that they represent two stages of human evolution." However, this line of thinking ignores evidence that has been with us since the 1970s when we found fossils of both *H. erectus* and *H. habilis* in sediments of the same age of 1.78 million years. These fossils from Turkana suggest that *H. erectus* and *H. habilis* rubbed shoulders in the same habitat for several hundred thousand years.

There is another non-African skull that lacks the "classic" *erectus* characters of a pronounced occipital torus and a heavy continuous brow ridge. This skull recently made headlines after it went on a little unauthorized adventure to New York City in 1999, where it surfaced in a shop called Maxilla & Mandible, a natural-history establishment. The proprietor, Henry Galiano, was uniquely qualified to recognise the import of his acquisition from his days at the American Museum of Natural History. After cleaning the skull, he took it with him for a visit to his colleagues at the museum to see if his suspicions of its importance were well-founded. A massive hunt began for clues as to where this beautifully preserved skull could have come from. Eventually, they got to the bottom of the mystery. The skull was discovered by sand miners in 1977 on the banks of the Solo River in central Java not far from Sangiran. It was then purchased from the miner who found it, and after passing through several hands, it eventually ended up in an antiquities shop in Jakarta, where it gathered dust for some twenty years. In 1998, an Indonesian palaeoanthropologist examined the skull and published a brief description. It was this description, together with a tip-off that an Indonesian antiquities dealer had tried to sell just such a fossil, that allowed the team at the Museum of Natural History to finally put two and two together. The skull was returned to the Indonesian authorities, who gave it its rightful place in the Indonesian collection and its own accession number, Sambungmacan 3 (Sm 3).

Susan Antón works in New York, and she had followed the twists and turns of this story with great interest. When we showed her 42700, she was immediately struck by some similarities of Sm 3 to both the Dmanisi skull D2700 and ER 42700. These three skulls belonged to individuals who lived thousands of miles apart, and in the case of ER 42700 and D2700, several hundred thousand years apart. If we believe the literature, these three were most emphatically not supposed to look alike, yet they could very easily have been triplets. It was clear to us all that we would need to go over the hodgepodge collection of accepted *H. erectus* characters and figure out which ones held true over the huge and diverse range of specimens. This would enable us to answer two confounding questions: is *H. erectus* the same thing as *H. ergaster*, and what makes *H. erectus* different from *H. habilis*?

As he had with Flat Face, Fred set to work building a spreadsheet with all the standard measurements that defined supposed *H. erectus* characters. He and Susan plugged in the measurements of every skull and partial skull in the entire collection from Africa to Asia, giving us a total sample of forty-five specimens to work with. Statistically speaking, this might not amount to much, but compared to many hominin species, this is, in fact, a relatively robust sample.

Next, we divvied up the sample in different ways. To see if there was any pattern to the variation in specimens, we compared the African specimens to the non-African specimens and the younger specimens to the older ones. Then we looked at the whole sample to see if any characteristics held true for young and old and African and non-African. Fred crunched the numbers for us using bivariate analysis — comparing the relationship between two related measurements such as the height of the skull and the width between the ears — to see what simple patterns emerged. Then he conducted more complicated multivariate analyses by throwing all the measurements at the computer program at once.

Much to our delight, some very interesting patterns were evident, and they upturned conventional wisdom about *H. erectus*. We looked at the features that were informally considered to separate *erectus* from *ergaster*. Having a thick skull (vault thickness), prominent brow ridges, a distinctive ridge running along the top of the skull (keeling), and a rather pointed occipital torus at the back of the skull were thought to be Asian traits, but we didn't find a single feature that was unique to African specimens, nor could we discern any that existed solely outside of Africa. Different combinations of features cropped up across the board. The conclusions were unequivocal: there didn't seem to be any justification for maintaining *H. ergaster* as a separate species.

In a nutshell, there are far more differences through time than across geographical distances. In addition to changes seen over time, there are further features that have to do with the developmental age and size of the individual (which scientists call allometric differences).

So a big old African male *erectus* who lived some 1.5 million years ago, such as OH9, will look far more like an old fellow from Indonesia or China who lived around the same time than he will a youth or maiden such as ER 42700 who lived in roughly the same neighbourhood at around the same time. Small skulls such as ER 42700 and D2700 do have features normally associated with *H. erectus,* but these are far less pronounced than in the bigger skulls. Without postcranial material associated with the skulls, we can't tell precisely if there are big differences based on the sex of the individual, but sexual dimorphism likely accounts for a lot of the variation in *H. erectus.*

The "classic" features long held to differentiate *erectus* from *ergaster* hold true across the entire sample although they do not all appear in every individual. A thick skull and pronounced brow ridges don't necessarily make an *erectus,* as these features also show up to some degree in other early *Homo.* There are two features on the base of the skull that do seem to crop up every time, however. The first of these is the area on the skull that makes up the skeletal structure of the ear — the tympanic, which is basically the ear hole, and the petrous, which holds both the bony labyrinth that we use for balance and the three ear ossicles (the bones that vibrate in response to sound frequencies and allow us to hear). The tympanic and petrous are oriented in a very distinctive angle in *H. erectus* that is different from the angle in other early *Homo.* The second distinguishing feature is the surface where the lower jaw articulates with the skull. In *erectus,* this is rather small and narrow from side to side. ER 42700 had both of these latter unique features along with the keeling that used to be thought of as an Asian character and a small gentle occipital torus. Susan, Fred, and I unanimously agreed that there could be no doubt that this was a *H. erectus.* The only potentially controversial thing was that we had demolished *H. ergaster* in the process.

The second important fossil found that year, John Kaatho's maxilla, also gave us a very interesting new insight. The maxilla, which was given the accession number ER 24703, had a row of rather worn teeth from the canine to the third molar on the right side, and the angle and height at which the cheekbone (zygomatic) arched gently away from the upper jaw looked unusual. Because of its very young stratigraphic age of 1.44 million years, I had simply assumed that it must be *H. erectus* in spite of its rather sizable wisdom tooth (third molar). But Fred took one look at that tooth and the angle of the zygomatic arch, and vehemently disagreed with me. Sure enough, when we compared the *erectus* sample to all the other early *Homo,* we found that both the third molar and the shape of the zygomatic were good markers to tell *habilis* apart from *erectus.* What was really astounding about this was that no *H. habilis* had been found at such a young age before. It pushed the range of time during which *H. habilis* and *H. erectus* shared the African savannah habitats to close to a half a million years and made it even more unlikely or even impossible that *H. habilis* was ancestral to *H. erectus.*

The fact that the last *H. habilis* lived for so long alongside *H. erectus* within Africa but had died out whereas *H. erectus* conquered all of the world it could reach by land makes one wonder what *erectus* had that *habilis* didn't. While we've speculated that persistence hunting was the key, we won't know for sure until somebody is lucky enough to stumble across an *H. habilis* skeleton. We still know frustratingly little about this elusive and mysterious hominin, and for now, our story continues with the evolution of *H. erectus* into *Homo sapiens.*

18

||||||||||||||||||||||||||||||||||||||||||||||||||||||||||||||||||||||||||||||||||||||||||||||||||||||||||||||||

THROUGH THICK AND THIN

ONE DAY IN 1980, TWO VISITORS ARRIVED OUT OF THE BLUE AT the National Museum carrying a loaded old *kikapu* basket of woven palm leaves and insisted on seeing Richard urgently. Upon admitting them to his office, Richard had a few choice words for the meteorologist and his wife, for their *kikapu* was chockablock with sturdy black fossils of every description. "What on earth are you doing with all those fossils!" he snapped. "Don't you know that's illegal and wrong?"

The couple had been exploring the western shore of Lake Turkana near a lush, palm-laden oasis called Eliye Springs. They obviously didn't know about Kenya's Antiquities and Monuments Act, which prohibits the collecting of any fossils without a permit. Right at the bottom, unwrapped and unprotected, sat a rather impressive skull. It bore all the hallmarks of a collection of Middle Pleistocene hominins from both Europe and Africa that are early versions of *Homo sapiens*.

These skulls have even more pronounced handlebar brows than any species known from earlier times. They also have a large nasal aperture and a low thick boned cranial vault. They lack the hollow seen in our cheeks (in scientific parlance called the canine fossa) so their cheeks probably would have looked stuffed as though they got caught in the kitchen before dinner. Some of them also have a vestige of the *erectus* occipital torus. Moreover, their brains have finally begun to enter the "human" zone: the largest of these skulls from Africa, the Bodo cranium from Ethiopia, has an estimated brain size of 1,300 cc.

The average brain capacity for *H. sapiens* is 1,350 cc. The skull in the *kikapu* was complete but for its handlebar brows and its upper jaw. Richard could see right away what an unusual hominin this was. He didn't know what to do first—castigate his excited visitors even more severely or thank them grudgingly while admitting to the huge excitement welling up inside him. Here, perhaps, was Kenya's first example of a very early *Homo sapiens.* Because they were of a scientific bent, the couple had not only brought the bones to the museum but were also able to give us a remarkably accurate description of where they had found them, and they had left a conspicuous marker behind on the beach.

Richard and I were soon flying to Turkana, Alan Walker in tow. We had asked Kamoya to meet us there with a vehicle and mark out a makeshift airstrip for us to land on. But as we flew north, huge banks of storm clouds were gathering in all directions. We dodged them with increasing trepidation as large sheets of water accumulated below us and we passed river upon river in full spate. Kamoya would be lucky to get across all of these normally dry rivers—let alone the mighty Turkwell, which lay between the main road and Eliye Springs. Sure enough, as we flew over the Turkwell at Lodwar, we made out a familiar Land Rover with Kamoya beside it waving energetically at us in dismay from the near bank of the swollen river. Richard, rather typically, had no intention of turning around after having got this far. "Well, we'll just find our own place to land and walk from there," he said. I was rather alarmed about this, since every potential exposed flat area that I could see was heavily waterlogged, and if a plane sinks into soft ground on landing, it digs its nose in the sand and cartwheels over the propeller. But being able to arrive within metres of the place he wishes to go with nothing but his highly attuned sense of direction to guide him is one of Richard's most useful—and sometimes most annoying—talents, and we soon found ourselves safely on the ground a short walk away from a well-marked hole in the sand exactly as the couple had described. The hole from which the skull had been extracted was very close to the shoreline, and the only reason that these bones had been exposed was because the lake was unusually low that year. Poor Kamoya had driven in great anticipation all the way to the Turkwell in vain, but at least we had not been likewise thwarted. And I suppose there is some consolation in that he was saved the backbreaking trouble of preparing an airstrip!

Since the brow ridges appeared to have come off very recently in a clean fresh break, we hoped to be able to find them at the site. We also wanted to determine the age by looking at the geology. What we found was rather disappointing. There were bones everywhere — but the brow ridges were not among them. All the bones we saw were very rolled and weathered, and had clearly eroded out of the ground elsewhere and been redeposited in the current location much later. We still don't have a date for this specimen, but its features suggest that it is one of these early modern humans and that it lived anywhere between 600,000 and 200,000 years ago.

Only two of these curious chimeras, which are transitional between *H. erectus* and modern *H. sapiens,* have been found in Kenya. The second was discovered by Wambua in 1976 on the other side of the lake at Ileret. This skull has been dated at 300,000 years and looks really modern in spite of being really rather old. A 270,000-year-old femur that Kamoya found close to the skull is similarly modern-looking although it is notable for its thick-set and robust proportions.

The fossil trail of our ancestors is decidedly feeble between *H. erectus* and *H. sapiens,* and its interpretation is made more difficult by the absence of associated volcanic material suitable for dating. Although at least six alternative dating methods have now been developed for this time interval, they are prone to error in various conditions. The oldest evidence is a handful of skulls of early *Homo sapiens* from around Africa that range in age from about 700,000 years to 200,000 years.

Like we saw in the herring gulls of northern Europe and the *anamensis-afarensis* lineage, we have a series of gradually evolving fossils that grade from typical *H. erectus* to anatomically modern *H. sapiens* with no obvious point when one clearly transitions into the other. As a result, many names have been given to these specimens, and their true affinities are unclear.

Nevertheless, these skulls have been named *Homo heidelbergensis* in Europe after the first of these, a mandible, was found at the village of Mauer near Heidelberg, Germany, in 1907. In Africa, these archaic hominins are often grouped under the name *Homo rhodesiensis* after the most complete and best-known example was found in 1921 in Kabwe in what was then Rhodesia (now Zambia). It is by no means clear, however, that *H. heidelbergensis* and *H. rhodesiensis* are two different species. What is clear is that by 200,000 years ago *Homo sapiens* had evolved to become anatomically identical to our own modern form.

Another feather in Kamoya's cap is that he was the one who found the earliest currently known anatomically modern *Homo sapiens* in 1967. At the time, however, nobody apart from Richard recognised its age or its import. Richard had invited Kamoya to join his first expedition to the Omo River, which he was leading on Louis's behalf. The expedition was a joint collaboration with several teams from Kenya, France, and the United States, and Richard found himself allocated with the most logistically challenging of all the areas they planned to prospect. Most of his problems revolved around the Omo River, which was at that time seething with enormous and ferocious crocodiles. On one occasion, Louis visited both the American and Kenyan sectors to see some fossils and was obliged to go up the river to reach the Kenyan camp. And he reported counting no less than "598 crocs, none of them less than seven or eight feet long, and some nearly twenty feet." Richard and his field crew had to cross the Omo to reach large parts of their survey site, and in all probability, they were the team with the greatest chance of overcoming the formidable obstacles involved, though I believe that Richard felt at the time that he hadn't been dealt a particularly good hand. The raft they had built to ferry the cars across the river was a ramshackle affair, but it would have been perfectly adequate if Louis had provided an engine with enough horsepower to contend with the strong currents and the lightning speed of the attacking crocodiles.

Richard was disappointed to find few fossils of any import in the limited area of older sediments allotted to their group and turned reluctantly to the far younger Kibish Formation, which he was less interested in. But here they were soon rewarded with Kamoya's discovery of an almost complete skull and a partial skeleton. Soon afterwards, another member of their team, Paul Abell, found a second skull from sediments of the same age on the opposite side of the river. At the time, these clearly identifiable remains of *Homo sapiens* were dated at 130,000 years based on the uranium-series dating of snails found in the same level. This was much older than 60,000 years, which was the age when people used to think *Homo sapiens* first appeared.

The full significance of this find would be understood only much later when Frank Brown returned to the Kibish with primate expert and palaeontologist John Fleagle and re-examined the sites and geology of these two ancient *sapiens*. More than three decades after his student days on the original Omo expedition, Frank's understanding of the Omo-Turkana basin geology had radically improved. He had questions to settle on several fronts. Sceptics had questioned whether the two finds were really the same age because Abell's skull looked more primitive than Kamoya's. The original age had also been suspect, and some considered it to be too old. Frank and John were able to locate the exact site of both discoveries, and they even retrieved more parts of Kamoya's skeleton that had eroded out of the ground in the intervening years. Frank found a tuffaceous layer just below the level of the hominins that had gone unnoticed in 1967 and another new tuff some way above the hominin layer. From these two tuffs, the age of the hominins was securely sandwiched between an older limit of 198,000 years and a minimum of 104,000 years.

With about fifty metres of sediments between the two tuffs, the big challenge was to narrow this wide age range down and pinpoint a precise date. Frank correlated the known dates of the two tuffs with well-dated sapropels and palaeomagnetic reversals in the Mediterranean cores so he could count the bands marking wet spells when the lake level rose and match them to the Mediterranean sapropel sequence. The result was an astonishing age of 195,000 years. This makes the Kibish fossils the oldest modern human known in the world. With these reliable dates, the Kibish finds supplanted Tim White's remarkable 1997 discovery of one immature and two adult *H. sapiens* in Herto, Ethiopia, that had hitherto been considered the oldest known example —between 154,000 and 160,000 years.

We are now into such a recent part of our prehistory that we can more precisely track the climate as it is minutely recorded in ice that has not yet melted since it first formed. Antarctica has been covered in ice permanently for at least the last five million years. This ice builds up incrementally in successive layers each winter season, and each summer, dust from the exposed land below the ice cap is whipped up by strong winds and settles in a layer over the last season's snowfall, and these layers accumulate in massive ice sheets.

The very oldest layers get so compressed that they melt under the combination of the huge pressure exerted by the miles-long mass of frozen water above and the thermal heat rising from the ground below. Slices of time are therefore removed from the bottom, and the record is preserved only for the relatively recent past. Still, we have a year-by-year record in the ice for hundreds of thousands of years. The ice cores allow us to build a picture of our recent climate in much more precise detail than we can for the time of *H. erectus* when we have only the foraminifera and other deductive clues in deep-sea cores to rely on. The fine-textured ooze of sea cores is formed through such slow deposition that the resolution of detail is quite low—and a single worm can disrupt the work of hundreds of thousands of years of accumulation!

One of the oldest ice cores, which goes back at least 740,000 years and is a mind-boggling two miles long, was recovered from the Antarctic ice sheet from a site called Dome C. Extracting these ice cores is no mean feat. In his book *The Weather Makers*, Tim Flannery rightly calls the extraction of the Dome C core as one of science's greatest triumphs and describes the inordinate challenges involved.

> The drill site was bitterly cold: -58°F at the beginning of the drilling season and -13°F in the middle of the Antarctic summer. The drill itself is just four inches wide, and as it grinds its way downward, a slender column of ice is separated and drawn to the surface. The first mile was especially difficult, for there the ice is packed with air bubbles, and as the core was drawn up, these tended to depressurize, shattering the ice into useless shards. Worse, ice chips can clog the drill head, jamming it fast.

The very air bubbles that make extraction of the cores such delicate work are part of the reason that the ice cores are such a good record of climate change. It all has to do with the way the ice sheets form. The top layer of the ice sheet is fresh winter snowfall. Immediately below this loosely packed snow is a layer that has been newly transformed into exquisite ice crystals. Farther down the ice sheet, these crystals become compacted by the weight of the layers above until, approximately one thousand years after the snow first fell, they have become recrystallized into solid ice (which resembles the ice cubes we make in our freezers for drinks). Tiny bubbles of air become trapped in the new ice when these snow crystals recrystallize and encapsulate a micro sample of the atmospheric air at the time that they formed. Because the chemical composition of atmospheric air depends on local temperature conditions, analysis of the carbon dioxide levels, the oxygen isotope ratios, and the proportion of methane that is trapped in the tiny bubbles gives an accurate barometer of the climate conditions under which they formed. Layer by layer through the ice sheet, the composition of the air in the bubbles varies, tracking the rise and fall of both atmospheric temperature and sea levels.

The dust separating each ice layer also provides valuable information about annual cycles. During colder glacial periods, there is less precipitation around the world and the dust layers are thicker—and the amount of dust is a proxy for windiness. One of the most well-known ice cores, the Vostok core from Antarctica, is 3,600 metres long and covers a time interval of 420,000 years. Four complete glacial and interglacial cycles are recorded in the Vostok core, each with a periodicity of 100,000 years. The core clearly shows that the windiest, dustiest time is at the height of glacial periods when temperatures are at a minimum. The data are consistent with other evidence that shows that the global climate is cold, arid, and windy during glacial maxima and that there are more grasslands and expanded deserts and less forest cover. Since the work at the Vostok ice core site, scientists have recovered longer sequences—800,000 years of climate history preserved at Dome C and a tantalizing 2.7-million-year sequence.

At first, climatologists concentrated on extracting and analysing ice cores from the polar regions. But to get a truly global climate reconstruction, the ice cores from tropical glaciers also needed to be part of the picture. As you can imagine, extracting a glacial core from the top of a mountain is even more challenging than in Antarctica because of the dangers of working at high altitude for prolonged periods and the logistics of getting heavy machinery and a source to power the drill up precipitous mountain slopes—not to mention the difficulties involved in bringing the heavy cores down without melting them. There is an extraordinary account in *Thin Ice* of Lonnie Thompson's lifelong endeavours to get ice cores out of the tropical glaciers of mountains before they are all gone. This book by Mark Bowen vividly details many of Thompson's greatest triumphs, including the successful extraction of an ice core from the Guliya Glacier on the Tibetan plateau that goes back 760,000 years. Over several decades, Thompson's late friend and colleague Bruce Koci developed lighter and better equipment to use at the top of tropical mountain glaciers. Thompson now has cores from Latin America, China, Bolivia, Alaska, and Africa. He is credited as the

man largely responsible for accumulating much of our body of knowledge on low-latitude, high-altitude ice cores. And ice cores permit us
to reconstruct the global climate at the time *H. sapiens* evolved and
moved out of Africa.

When we compare the data from both alpine and polar ice cores, it
is clear that the greenhouse gases in the atmosphere—carbon dioxide
and methane—have varied in lockstep with the cycles of glaciation and
deglaciation and that these are heavily determined by the Milankovitch
cycles. The ice cores also show very sharp transitions from cold to warm
periods, with sudden temperature changes of twelve degrees Celsius,
whereas cooling happened much more gradually. Although far more
detailed than the sea cores, the ice cores are in agreement about global
climate swings. The composite picture that we can build shows that
Homo sapiens, like *Homo erectus,* evolved during a glacial-interglacial
icehouse world—one that was both more pronounced in the extremes
of cold and warm and more erratic.

This information now makes it possible to trace the last chapters of
human evolution with a vastly improved understanding of the climate.
H. sapiens evolved just under 200,000 years ago, and this coincides
with a warmer interglacial period that started some 245,000 years ago
and ended some 185,000 years ago. Then the ice returned for 55,000
years. All this leads to one unavoidable conclusion: for the large part of
the evolution of *H. sapiens,* the whole planet was perishingly cold and
the tropics were parched, arid lands. During the most arid phases, as
evidenced by cores taken from Lake Malawi, vegetation was so sparse
that the usual bushfires did not occur, for there is a marked decrease
in charred particles and very little pollen present. The level of Lake
Malawi, which today is a staggering 706 metres deep, was at times reduced to a depth of 125 metres, and the salinity and alkalinity rose
dramatically.

There were a few terribly brief intermissions from this deep freeze,
and we are enjoying a relatively long one today. The last time there was
a peak of warmth similar to global temperatures today was 130,000

years ago near the beginning of an interglacial period that ended 75,000 years ago. Then there was another glacial maxima between 70,000 and 60,000 years ago. This period of lower sea levels and increased aridity coincides with the main exodus of *H. sapiens* out of Africa all the way to Australia.

What's more, the original Europeans—the Neanderthals—evolved in icy Europe at the same time that *H. sapiens* was evolving in Africa. What is most amazing to me about the Neanderthals is how marvelously cold-adapted they must have been, for they were a very successful species. Compared to the dearth of fossils for other hominin species, they seem to crop up all over the place. The evolution of the Neanderthals began between 300,000 and 250,000 years ago. Like their African cousins, they expanded and contracted multiple times according to the opportunities the shifts in climate offered, and they reached across an area extending from the Iberian Peninsula to southern England and the Mediterranean and across the Caucasus to beyond the Caspian Sea until the Altai Mountains and the Hindu Kush blocked their path farther into Asia. To the north, the arctic blast of ice age winter on the cold steppes presented an equally impenetrable barrier to further dispersal.

The Neanderthals' prevalence and their big brains fed the European bias of early palaeontologists who had balked at the idea of an African ancestor. The Neanderthals were the first hominins ever found —the first specimen was discovered in 1829 in Belgium, and the first to be recognised as a Neanderthal was found in a quarry in the Neander Valley in Germany in 1856. So for a long time, there was no other candidate for our immediate ancestor, and no reason to suppose that we didn't evolve from these human-looking creatures.

But in common English usage today, calling someone a Neanderthal is usually meant as an insult akin to calling someone primitive and uncivilized. Neanderthals and cavemen are inextricably linked to this day in the popular imagination, with an image of a Hobbesian life that was "nasty, brutish and short," where hairy, brawny cavemen with huge clubs clobbered one another to death. Evolving from a European Neanderthal may have been viewed as preferable to evolving from an African ape, but it still did not sit at all comfortably with the Christian notion of man being created in the image of God. This bias is seen in the writing of the French palaeontologist Marcellin Boule who, along with Arthur Keith, heavily influenced early thinking on Neanderthals. Boule gives free rein to his prejudices and preconceptions, writing most unscientifically in 1920 of the "rudimentary psychic nature" of Neanderthals, "superior certainly to that of anthropoid apes but markedly inferior to that of any modern race whatsoever," and the "brutish appearance of this energetic and clumsy body, of the heavy-jawed skull, which itself declares the predominance of functions of a purely vegetative or bestial kind over the functions of the mind." This compares to his highly poetic praise of modern human contemporaries of Neanderthals "who had a more elegant body, a finer head, an upright and spacious brow, and who have left, in the caves which they inhabited, so much evidence of their manual skill, artistic and religious preoccupations, of their abstract faculties, and who were the first to merit the glorious title of *H. sapiens!*"

How Boule arrived at his conclusions of the inferior quality of the Neanderthal mind is only to be supposed at, for the specimen he was comparing to modern humans, a skeleton known informally as the Old Man from La Chapelle-aux-Saints (for his arthritic bones and missing or worn teeth), had an impressive brain capacity of more than 1,600 cc —considerably bigger than the average modern human's 1,350 cc. He claimed he could tell how rudimentary this brain was from the coarseness of the endocast he moulded from the skull's interior surface, but his interpretation is totally discredited now. Biases persisted in how people viewed Neanderthals, though, and they inevitably led to a high degree of subjectivity in reconstructions. Reconstructions that imply a brawny dimwit unable to think, let alone survive the rigours of ice age winter, belie the reality that the Neanderthals prospered where no other hominin had done before and prevailed against all odds for well more than a hundred thousand years in the most inclement weather imaginable. The actual evidence points to a far more benign and gentle species that is very closely related to our own and superbly adapted to the unforgiving ice age climate.

But life was indeed brutish and quite short—very few Neanderthals (less than 10 percent) lived beyond age thirty-five, and most of them died in their late twenties or early thirties. In modern hunter-gatherer societies and tribal agriculturalists without the benefits of modern medicine, this figure is about 50 percent. So few Neanderthals would have reached what in modern terms is considered middle age. That doesn't leave much time for handing down hard-won knowledge and experience about surviving—or for grandmothering, for that matter. Neanderthals lived a tough life, and this is dramatically demonstrated by the frequent occurrence of lesions on their skeletons. Some of these were the result of quite violent injuries, as evidenced by broken bones. Others are simply the result of malnutrition and disease during crucial periods of growth. But where this picture departs from the brutish caveman savage is that many Neanderthals survived these hardships because the trauma to their bodies healed during life. Rather than clubbing one another to death, they were nursing one another back to health. Neanderthals, like *H. erectus* before them, were hugely social creatures with bonds that extended well beyond that of mother and child. Society had evolved considerably from *H. erectus*'s time to include burials, care for the infirm, and very sophisticated toolmaking.

With all this evidence of intelligence, speculating on the differences in brain function between ourselves and Neanderthals is unenlightened. Our bodies were shaped quite differently, however, and the reasons are mostly due to climate. There are two rules, which are named for their inventors, that pertain to how the shape of animal bodies adapts to climate: Bergmann's rule, which postulates that in colder conditions body weight will tend to be larger, and Allen's rule, which states that in colder conditions body extremities will be relatively shorter. For example, the Inuit tend to have thick torsos on squat bodies with short arms, stocky legs, and shorter fingers and toes. Desert-dwelling nomads are on the other end of the spectrum. When Alan Walker was studying the Nariokotome Boy's skeleton, he looked at the body proportions with Chris Ruff, an expert on body shape and its relationship to ambient temperatures. They found that the Nariokotome Boy's long slender frame was ideally suited to living in a tropical climate where temperatures hovered around 30°C (86°F) and that his body shape very closely resembled that of modern nomadic Turkana tribesmen.

Neanderthals had very short, stout bodies and differ most markedly from the tall slender form of desert-dwelling nomads like the Turkana and Masai. Most impressive are the strong, sturdy, heavy bones that allowed for the attachment of massive muscles. In the icehouse climate, persistence hunting was probably much less effective than in the tropics, and running an animal to hypothermic exhaustion was not likely to be a successful hunting strategy. Indeed, we see evidence that Neanderthals fashioned and used fire-hardened and weighted throwing spears. To kill their prey, they would have had to get close to them, which probably accounts for most of the bone injuries they suffered as well as their need for superhuman strength. No modern athlete comes close to the phenomenal power of their bodies, which were built to withstand long periods of heavy physical activity in fiercely cold weather. Chris Ruff and Alan Walker plugged in the Neanderthal body proportions to their temperature model when they were studying the Nariokotome Boy. They found the mean annual temperature for the Neanderthals' Europe to be -1°C (30°F)—on par with temperatures found above the Arctic Circle today.

The skulls of Neanderthals are also unmistakable. Like *H. erectus*, they have a long, low, flattened braincase, but the handlebar brow ridges of solid bone had evolved into two thick prominent arches that are hollow to enclose the frontal sinus. Even more distinctive is the large nasal opening—it looks like somebody got hold of the Neanderthal nose and pulled hard, giving it a pronounced elongated and forward-jutting shape quite different from ours (a *sapiens* nose is both smaller and tucked in below the brows).

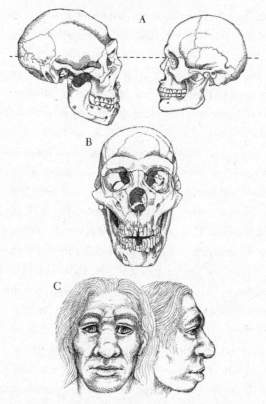

Neanderthal skulls are unmistakable with their long, low, flattened braincase, thick prominent brow ridges, and large nasal opening:
A) Profile comparison of *Homo neanderthalensis* (left) and *Homo sapiens* (right)
B) Frontal view of *Homo neanderthalensis*
C) Reconstruction of *Homo neanderthalensis*

Physical differences notwithstanding, just how different are we from the Neanderthals? Should they be banished from the exulted company of modern humans or admitted to our privileged circle as an immediate family member? This has long been a matter of fierce contention, and the battleground has shifted from bone morphology to include genetic analysis. Both sides agree on at least one thing: Neanderthals and modern humans both evolved in ways that responded to climate: *H. sapiens* to suit the arid, dry tropics of Africa, and the Neanderthals to survive the glacial and wildly fluctuating conditions of Europe. But the two schools of thought reach widely different conclusions.

The first group ("splitters") holds that we evolved in Africa through archaic *H. sapiens* (sometimes called *H. rhodesiensis*) and that the Neanderthals evolved in Europe from a different population of archaic *H. sapiens* (*H. heidelbergensis*) that had already moved out of Africa. Our separate paths continued as an example of parallel evolution until a full-fledged modern *H. sapiens* moved out of Africa into Europe some 60,000 years ago. The Neanderthals were a distinct species, and as we have seen, they had impressive qualities in their own right. But as a separate species, they either could not interbreed with *H. sapiens* — or did so relatively rarely. Thus, when *H. sapiens* eventually invaded their territory, they were rapidly replaced and quickly thereafter became extinct. Quite simply, we outcompeted the Neanderthals, and the most likely explanation is because we had evolved unique cognitive talents so the Neanderthals' cold-adapted bodies no longer had the advantage over our puny desert-toned ones — and we outsmarted them. This view is known as the "out of Africa" or "single origin" hypothesis.

The second school ("lumpers") believes that there was a lot more gene mixing through interbreeding between the populations. A lead proponent of this alternative theory, the "multiregional hypothesis," is a man named Erik Trinkaus. In his view, *H. erectus* evolved into archaic *H. sapiens* simultaneously in Africa, Asia, and Europe. In Africa and Asia, this intermediate creature evolved into modern *H. sapiens* while in Europe it evolved into the Neanderthals. A good deal of breeding ensured that the non-African archaic *sapiens,* and later the Neanderthals, regularly inserted their genes into the evolving African human pool. Eventually, *H. sapiens* also emerged in Europe and busily bred with the Neanderthals and Asian *sapiens* to produce the modern homogeneity of our single species. In this scenario, all the larger-brained hominins of the Late Pleistocene were part of a single evolving species — one widely dispersed, interbreeding population that stretched from Africa through Europe to the Far East with some regional and temporal variation. How they managed to traverse the mountains and miles of frozen terrain at such frequent intervals as to keep up a respectable gene flow is anyone's guess, but the lumpers do not regard the frigid climate as an impediment to their model.

The fossils, rather unhelpfully, can provide cannon fodder for both theories depending on interpretation. For splitters, the features of the Neanderthals are so distinctive as to clearly represent a well-established species that would not have easily been able to interbreed and produce fertile offspring. This is emphasized by the very distinctive characters evident even in very young children. Usually, a baby is a baby is a baby —features that are evident later in life are not yet distinguishable among the infants of closely related species. But baby Neanderthals do show classic Neanderthal features and a different growth trajectory that is visible from birth. These are clearly manifest on a nine-month-old Neanderthal baby from Amud Cave in Israel and a child from El Sidrón in Spain.

Tool technology at this time rather supports the splitters. After little to no change in the widespread Acheulean industry used by *H. erectus* and its descendants in both Africa and Europe for well more than 1.5 million years, a revolutionary breakthrough in production techniques took place in Africa around 300,000 years ago. In contrast to the production of artefacts that rely on visual feedback to guide the maker, the new techniques included a much surer method called "prepared core." Named after Levallois, the suburb in Paris where the technique was first documented, the Levallois technique entails first preparing the flint nodule with one domed surface. Next, a striking platform is prepared so a flake can be struck off to form a ready-made tool. The process is repeated until the core is too small to remove further flakes. This "prepared core" technique therefore relied on a strong mental template, forethought, and an ability for complex abstract thinking.

The subsequent appearance of blades, points, microliths, and bone tools as well as evidence of long-distance trade and the use of pigments and ochre appear progressively in Africa over time and at sites widely separated geographically. In Europe, Neanderthals also made stone tools with the Levallois technique, and the Neanderthal and early *sapiens* toolkits are virtually the same. However, most of the innovations that appear in the African archaeological record of the last 100,000 years never occurred among Neanderthals and do not appear in Eurasia until tens of thousands of years later when *H. sapiens* replaced the Neanderthals. If the lumpers are correct about the repeated intermixing of the two populations, the archaeological record in Africa should match the archaeological record outside Africa much more closely, and we should not see the delay of tens of thousands of years before the sudden appearance of these modern technologies in Europe.

Splitters and lumpers alike must explain the fact that the earlier Neanderthals who first appeared in Europe between 130,000 and about 70,000 years ago are much more variable than the many examples from the Neanderthal's heyday between 50,000 and 40,000 years ago. Splitters do not see this as evidence that interbreeding progressively reduced differences over time as the lumpers do. Instead, the splitters suggest that the earliest Neanderthals in Europe frequently became isolated in remote areas because of the vagaries in climate. These refuge populations each developed slight differences in features through their own small evolutionary experiments, which would explain the large degree of variation much as we have hypothesised for *H. erectus*. Over time, the Neanderthal recipe for a successful adaptation to the European ice age conditions evolved, and it was representatives of these improved genes that survived and moved into previously isolated areas, mingling with the refuge populations and smoothing the entire gene pool.

The lumpers, on the other hand, see mixed features in several Neanderthal specimens that they interpret as the result of repeated interbreeding. One such specimen is a child from Lagar Velho in Portugal that has been dated at 24,000 years—after the last of the Neanderthals disappeared around 40,000 years ago. This skeleton is perceived by the discoverers as a Neanderthal-human hybrid, and the presence of some classic Neanderthal features—notably in its body proportions—is interpreted to suggest that there had been interbreeding between Neanderthals and true modern humans for a long period of time. Additional evidence that is advanced to support the lumper interpretation comes from eastern Romania where a mandible and a skull were recently excavated from Peştera cu Oase ("the cave with bones") and dated between 37,000 and 42,000 years ago, which is also the earliest fossil record of modern *H. sapiens* in Europe. These specimens have been interpreted as displaying mixed *H. sapiens*–Neanderthal characters.

Scientists are beginning to agree that the splitter view is correct by and large, although most now acknowledge that some interbreeding also occurred. There probably is an inner Neanderthal within us — but whether this is just a dim and distant ghost or an intrinsic and big part of who we are depends on the degree to which this intermixing happened. For my part, though I find it fascinating, I defer this argument to the experts as it is so removed from my own field of expertise. But this is now the intense focus of new research and debate. It is genetic studies that are slowly bringing convergence between the two opposing views in a way that fossils cannot hope to. Through genetics, we can trace our ancestors' path out of Africa and across the globe, and we can peer into the chapter of our history that we shared with the Neanderthals to try to figure out what really went on in the sex lives of our ancestors.

19

MIGRATING MUTANTS

EVERY HUMAN ALIVE TODAY — MORE THAN SEVEN AND A HALF billion people — shares ancestry from a group of about ten thousand African people. That's all, and it seems simply impossible. What could have possibly happened to reduce our number to such a rare breed? For every living human today to originate from this ancestral population, one of two things had to have happened: either this was the only remnant population of *Homo sapiens* left alive or this population was isolated — what in genetics is called "structured' — for long enough that we developed separate, private mutations that differentiated this group from all the others. In other words, even if Africa had other populations of early people, they didn't have any sex with the ten thousand we descend from.

Either way, we clearly overcame many hardships to become what we are today — inhabiting every corner of every continent and altering habitats to such an extent that no other species on land or sea remains untouched by our reach.

Only two things could have decimated the population so dramatically: a ferocious outbreak of disease or rapid and extreme climate change. Of the two, climate change is more likely because a disease can no longer spread when it becomes so virulent that it wipes out its host (outbreaks of lethal diseases such as Ebola tend to fire up quickly and then peter out). We've already witnessed the huge role that the earth's usually erratic climate has played in our evolution up to this point, and it is certain that climate change will remain a major force in our future too. In 2019, the United Nations released a report on the state of biodiversity on earth that summed up 15,000 scientific papers. The staggering findings are that the earth is set to lose one million species within decades due to human activity unless we change our habits. Man-made climate change is a huge and inexorable part of this equation.

On a rushed visit to Washington, D.C., in March 2007, I bumped into Spencer Wells during a brief break from an annual symposium of National Geographic's Explorer-in-Residence program. Louise and I had become "explorers" in 2002, and the symposia offer an opportunity to meet and learn more about the varied and extremely interesting people and projects under the Explorer program. Spencer, who is completely committed to his work and always gushing with new and exciting ideas, was even more buoyant than usual.

Spencer is a geneticist who has had an illustrious career, and he is the mastermind behind the National Geographic's Genographic Project, which was launched in 2005. This was an ambitious enterprise that looked at the bloodlines of people to trace the migration of *Homo sapiens* around the world. As Spencer is wont to say, "Every drop of human blood contains a history book." This history book is a genetic record of our past encrypted in an elaborate code that we pass down to later generations through our DNA. These days, so many words are bandied about concerning our human genome and our DNA, that it is all too easy to casually use these terms without fully comprehending them. Permit me a brief digression into the nitty-gritty of cellular biology courtesy of my discussion with Spencer because deciphering this genetic code is now a mainstay of our tracing our recent past.

The makeup of our cells is fascinating. More than a billion years ago, a wayward bacterium found itself enclosed within the walls of another single-celled organism, and so began a long and fruitful association that facilitated the development of complex life-forms made of multiple cells. The descendants of that original bacterium prisoner are mitochondria, and they serve as an army of battery packs that literally give us life as they use the nutriments we ingest and the oxygen we breathe to generate energy. In return, they are housed in a safe, warm environment where these raw materials are conveniently delivered in a constant stream for processing. But this endosymbiosis has changed little over time: mitochondria still operate like quasiseparate organisms complete with their own DNA, RNA, and ribosomes, and their own schedule for reproduction quite independent of the cell they are enclosed in. Thus there are two independent sets of DNA within each of our cells. The cell nucleus houses our own much more complex set of nuclear DNA, which is the sequence or code that controls how our bodies form and function (more on this later). And outside the nucleus of each cell, thousands of mitochondria share a mitochondrial DNA (mtDNA) sequence that is likely to be identical across all the cells in one human body.

Two qualities of mitochondria help us trace historical populations. The first is that mitochondrial DNA mutates far faster than our own nuclear DNA: it contains roughly ten times as many polymorphs (mutation errors that alter the sequence on the DNA chain) as our nuclear DNA. While all the mtDNA in one human is likely to be identical, there is so much variation between humans that there may even be mutations or sequences specific to individual families. This is a good thing for geneticists because the greater the degree of variation, the better our ability to distinguish between individuals. The second quality is that mtDNA is passed down only through the mother because the mitochondria in sperm are concentrated in the tail region of the cell. Once the tail has propelled the sperm on its epic swim up the Fallopian tube to meet the egg, it is summarily dispensed with along with its mitochondria. If you are born male, you are basically an mtDNA evolutionary dead end. These two characteristics of mtDNA make it possible to trace the ancestry of every single woman alive back in time to an original woman. It was she who provided the mitochondria we carry in our cells today.

This woman has been anointed with the loaded name of "Mito-chondrial Eve"—and it would be easy to suppose that she was the only woman alive when she began propagating her mitochondria. This is a mistaken deduction. It is mathematical logic that leads us to the con-clusion that we all share an original "mother." All the women from the same generation descended from a smaller number of women—as some of our mothers' generation, for example, either had no offspring or only produced boys, and this happened in their mothers' generation and the one before that. This logic forces us inexorably up a pyramid of ever-diminishing numbers of ancestral women, a process called coales-cence, which cannot continue indefinitely. Eventually, there will be just two women, and they must have shared a single mother, the so-called Mitochondrial Eve. All the earlier lineages of mtDNA—including Eve's mother's—are now extinct. But Eve obviously did not live alone; she just happened to be the one whose mtDNA made it through.

Specific mutations are like flags on the mtDNA sequence. Although these mutations are random errors in the copying of DNA, they can be averaged over time to give a mutation rate for a gene (such as mtDNA) that can be used as a molecular clock. By comparing the sequence of two individuals, or two species, scientists can establish how different they are, and using a known historical event (such as a fossil), they can calculate how long in years those differences took to accumulate—in other words, scientists can calibrate the clock. Using a calibrated clock, researchers are able to map the history of individuals, populations, and species.

By looking at which populations share particular markers, geneticists can decipher the timing and path of the dispersal of *Homo sapiens* out of Africa, and this work began using mtDNA. Two very early markers on mtDNA are called the L0 and L1. These flags trace migrations of African peoples before the exodus out of the continent. Both groups lived in Africa more than 130,000 years ago, and today L0 is found in highest frequencies among geographically diverse populations such as the Cen-tral African Pigmies and the Khoisan of Southern Africa, but it has low frequencies in West and North Africa. L1 markers, on the other hand, are found in higher frequencies in West Africa but exhibit the greatest diversity (and thus age) in Central and East Africa. These two ancient markers only made their way out of Africa with the forced migration of Africans to the New World in the barbarous Atlantic slave trade.

By tracing the mutations on mtDNA back in time, geneticists now believe that Eve lived in Africa sometime around 170,000 years ago. It is interesting to note that the process of coalescence is much more likely to occur when a population is small and their genes are passed through what is called a genetic bottleneck—as when our population was reduced to a mere ten thousand souls.

Unlike mtDNA, nuclear DNA undergoes a huge shake-up with each generation—rather like a shuffling and splitting of two decks of cards so the child ends up with a random half of each parent's genetic makeup. When we procreate, we make unique reproductive cells— the eggs and sperm. These gametes contain only one of each of the twenty-three pairs of chromosomes, so that they have twenty-three single chromosomes, rather than the usual paired forty-six. Genes are sequences of nucleotides (basic building blocks) on a strand of DNA, and human genes are located along the strands of DNA in twenty-three different pairs of chromosomes. We inherit half of our genes from our mother and half from our father, but the combinations are almost end-less—the twenty-three chromosomes in each human sperm or egg cell contain about three billion nucleotides, which are the basic building blocks of genes.

There are roughly 1,500 active genes on the average human chro-mosome, but the Y chromosome has only twenty-one genes, and all of them are involved in making a boy instead of a female, the mammalian default. The rest of the Y chromosome appears to be biological junk with no discernable function. Spencer Wells calls this junk DNA "gold dust to population geneticists" because mutations in the Y chromosome are handed down through the generations and can serve as markers to trace our gene tree.

Since women have two X chromosomes and men inherit a Y chromosome only from their father, we can trace an ancestral Adam from mutations in the Y chromosome just as we followed mtDNA mutations back in time to a Mitochondrial Eve. The Y chromosome equivalent to the earliest female markers, Lo and L1, carry particular mutations such as the M91 and M60 markers. Men bearing the M91 marker are found today in non-Bantu populations in Ethiopia, the Sudan, and the southern regions of Africa. The M60 marker is widely dispersed across the African continent today and shared by many different African people. Our common Y chromosome ancestor lived in Africa between 160,000 and 300,000 years ago, although it is unclear whether Adam and Eve ever had the pleasure of making each other's acquaintance! The discrepancy in age sounds implausible at first—for where were the men that Eve and her group were procreating with?

Spencer explains this as a product of the hierarchy of traditional patriarchal societies. Having many wives and children is an expensive privilege, and in many instances, polygamy is reserved for an elite few, and some men cannot afford any wives at all because of the costly bride price. The current king of Swaziland, Mswati III, has 15 wives and 35 children at last count, and his father, King Sobhuza II, had 70 wives and 210 children before his death at age 82 in 1982.

We obviously can't speculate on whether Adam was a rich chieftain who paid a generous bride price for his mates, but in most primate societies, it is the dominant males who enjoy most of the sexual action and pass down their genes. There is, therefore, far more equal opportunity in the evolutionary fate of mtDNA than there is for Y chromosomes because many more females will pass down their genes than their male counterparts. Effectively, most of the older lineages of Y chromosomes have become extinct, although a few survive and give us precious insights into earlier periods in our species African history. So far, the molecular evidence is firmly in the single-origin court of the splitters.

At this point, deciphering DNA gets much more complicated. A vast amount of information is contained on our chromosomes, but until only very recently, this library has been inaccessible to us. DNA is a double-stranded molecule with each strand consisting of a linear sequence of nucleotides that can be one of four types: adenine (A), thymine (T), cytosine (C), and guanine (G). The combination of A, T, C and G nucleotides is remarkably uniform in all people—it is 99.9 percent identical throughout the world. The remaining 0.1 percent of variation is responsible for the individual characteristics that make each of us unique. This sequence in our nuclear DNA and our mtDNA makes up the human genome, and when scientists succeeded in mapping it in 2003, they at last found a key to unlock the vast secrets contained in our nuclear DNA.

Copying errors that alter one of the nucleotides in the sequence—changing an A to a G, for instance—are what make evolution possible. These mutations account for the 0.1 percent of variance in the genetic makeup of the more than seven and a half billion people alive today. What happens after a mutation occurs depends partly on how it affects the body. Some of this DNA is responsible for individual characteristics—the most glaringly obvious physical trait in my family is the "Epps chin," which is about as far from a chinless Neanderthal as one can get. The Leakeys don't have very distinctive chins, but they carry a particularly dominant gene for very prominent ears that on young Leakeys seem to be placed at right angles to the side of the head rather than laid flat against it. Other famous Leakey genetic traits include what could politely be called extreme tenacity, a zealous missionary streak, and an obsession with punctuality that makes the military a sound career option. Richard has every single one of these genetic traits.

Our cells have complex ways of fixing mistakes that may occur when DNA is being copied, and the most common mutations that persist are those that have no effect on us. Such mutations carry on from generation to generation as harmless variants. Second-most common are mutations that are harmful, and nature's solution is usually to terminate the pregnancy. At least one-fifth of all conceptions end in miscarriages, many of them because the genetic mutations are fatal to the foetus. Much rarer are mutations that are beneficial to an individual in a given environment. These mutations confer a reproductive advantage and are the basis for Darwin's theory of natural selection as a driving evolutionary force.

Mutations in mtDNA and the Y chromosome are easy to trace because they are passed down intact from generation to generation. If the same mutation can be identified generations later in two individuals, it indicates a common ancestor. By comparing markers in different populations, scientists can trace ancestral affinities along a single line of descent (maternal or paternal) until all living lines can be traced to the mtDNA Eve or Y chromosome Adam (coalescence). But for the vast majority of our genes, the number of ancestors grows and grows (two parents, four grandparents, sixteen great-great-grandparents, and so on), and we receive a random configuration of their genes. This is made all the more variable by the fact that there is some gene mixing—known as crossing over or recombination—before an egg or a sperm is produced. In other words, not only does each child receive a random set of genes from each parent, but each set is a reshuffle of the child's maternal and paternal grandparents' genes. And this reshuffling is repeated at every generation. This makes it all but impossible to trace mutations along ancestral lineages because it is hard to sort out which novel combinations have arisen from gene reshuffling and which are mutations.

But scientists have discovered that there are small chunks of nuclear DNA that don't get broken up by recombination very often and are passed down from generation to generation like mtDNA. Any changes in these sections, called haplotypes, will be the result of mutation, so they too can be used to build gene trees. One such haplotype lies within a gene called PDHA1 (which with a gene called PDHB produces two proteins that combine into an enzyme that is involved in the chemical pathway that converts energy from food into something the cells can use), and in the modern population, there are a number of variations that can be grouped into two basic types of sequences. It is interesting to note that these last shared a common ancestor 1.8 million years ago—when *Homo erectus* first appears in Africa. An even more startling revelation followed: one of these two lineages then split a second time at 200,000 years, which coincides with the appearance of *Homo sapiens* in Africa.

This is just one of a growing number of genetic studies that have found genes that mutated long ago. An intriguing finding relates to a gene called microcephalin—mutations in this gene lead to microcephaly, which is the explanation some attributed to the very small brain of the Flores Hobbit. There are two variants of this gene, and they are so different that Bruce Lahn, the scientist who studied them, concluded that they must have diverged about 1.1 million years ago—either in *H. erectus* or in the ancestor of *H. sapiens* and Neanderthals. The second variant found in modern humans can be traced back only about 37,000 years. As most of our genes are there to make us function—genes that make us a multicellular organism, a vertebrate, a mammal, a primate, an ape—we share them with other creatures at each ring in our particular evolutionary chain. Now we are beginning to discover those genes that mutated at the time of *H. erectus* and, through them, some of the genetic code for being *Homo*.

It is remarkable that scientists have managed to recover DNA from Neanderthal remains. Extracting DNA from fossils is extremely difficult because so little remains and contamination from modern DNA is difficult to contain. Scientists work in hermetically sealed labs and endure lengthy procedures so every wayward cell from their bodies is tightly contained in protective clothing and masks. But this work on ancient DNA has transformed the study of recent human evolution.

Since those early discoveries using mtDNA and Y chromosome markers, geneticists have moved in giant leaps. The entire human genome was sequenced in 2003; the chimpanzee genome in 2005; and, most astonishing of all, the Neanderthal genome was pieced together from tiny fragments of ancient DNA in 2010. Since the mapping of the first complete human genome, geneticists have sequenced thousands of others from across the world as well as those of hundreds of ancient humans, twelve Neanderthals, and six Denisovans, an elusive new hominin. This extraordinary library of genetic variation has given us unparalleled insights into the history, adaptation, and behaviour of recent hominins.

Geneticists have been able to compare our DNA to that of Neanderthals and Denisovans in detail. These comparisons show that the "out of Africa" model was mostly correct—all *H. sapiens* have their origin in Africa between 300,000 and 200,000 years ago, and the Neanderthals (and Denisovans), with whom we share an African ancestor around 500,000 years ago, are our closest relatives. This means that African populations expanded and colonised Eurasia at least twice in the last million years—once 500,000 years ago, which gave rise to the Neanderthals in the west and the Denisovans in the east, and once 70,000 to 60,000 years ago when humans progressively settled the entire world. And there are genomic signatures for possible further "out of Africa" events in between. It is easy to see these expansions as something uniquely special to hominins, but they are not. They were fuelled by climate change. The glacial cycles of the last 800,000 years had hotter and colder extremes than earlier climatic shifts. One consequence of those extremes was a very wet beginning to interglacials, the "green Sahara" events. It was during these that the African ancestors of Neanderthals and modern humans increased in numbers, crossed the Sahara, and dispersed into Eurasia.

When our ancestors reached Eurasia 70,000 to 60,000 years ago, the Neanderthals and Denisovans were there. We did not know the Denisovans existed until their genome was extracted from unidentifiable fragments of bone, but there are dozens of Neanderthal skeletons, some of them found in the same caves that were later occupied by *H. sapiens*. For 150 years, scientists have pondered and argued about what their relationship to humans would have been, and geneticists have now given us some of the answers. When a human population first invaded Eurasia, they met a group of Neanderthals, and from that encounter, some of the children in the human population had a Neanderthal parent. From those children, every non-African person alive today —some 6.5 billion people—carries small fragments of Neanderthal DNA that make up 1 to 2 percent of their genome. And in the process of expanding further into Asia, some people met Denisovan populations, and a few modern groups—Aboriginal Australians, Papuans, and groups of hunter-gatherers such as the Aeta in the Philippines—can trace a percentage of their DNA to the Denisovans. Given that the different groups obviously could have children together, the real mystery is not that they interbred, but that the admixture was so rare. But we had been evolving separately from Neanderthals for half a million years during which each lineage acquired its own unique features, some of which made us genetically incompatible. Geneticists have been able to show that there was strong selection against some Neanderthal genes in those hybrid children, most of whom probably did not survive. Indeed, the ancient genome of a 38,000-year-old fossil from Romania shows that one of his great-great-great-great-grandparents was a Neanderthal, but those Neanderthal genes are not found in anyone alive today.

Pulling together the information from ancient and modern genomes, scientists have been able to map the dispersal of humans out of Africa. Remember, this was a time of extreme cold. Sea levels were low, and northern Eurasia would have been blocked by arctic deserts while Europe had the resident Neanderthals. Early Eurasians (after mixing with Neanderthals, presumably in the Near East) split into several small groups of hunters. One of these ventured east right away, probably following the coastline around India to the land bridges that emerged to connect many of the Indonesian Islands to the continent of Australia. It is thought that they arrived in Australia about 60,000 years ago, and this is corroborated by the earliest evidence of human occupation in Australia at this time found at Madjedbebe in the Northern Territory. This small founding Australian population included several individuals whose ancestry was part Denisovan as the genome of today's Aboriginal Australians and Papuans has 6 to 8 percent Denisovan genes (as well as the 1 to 2 percent Neanderthal genes that all non-Africans have). Additional genetic mutations and differences arose through time and allow many details to be added to the times and routes that explain how people spread around the globe.

It's been an honor to participate in the National Geographic's Genographic Project — an exciting endeavour that began in 2005 and combines the number-crunching power of IBM with the financial muscle of the Waitt Family Foundation under the direction of Spencer Wells. I have already learned much from my association — including my own genetic trail. My ancestors belong to mtDNA haplogroup H. Everyone in my haplogroup has an H marker — we are a single branch of the human family with a common ancestor who first exhibited a new mutation labelled H. Tracing the H marker backwards, I can reconstruct much about my ancestral migratory route. It begins with the group of people who first moved out of Africa some 60,000 years ago, probably following an expanding food source much as we have speculated that the first *Homo erectus* migrants did over a million years previously. We don't know the exact route: once again, the Nile must have played an important part.

These early African ancestors moved into the Middle East, where they had children with Neanderthals and hung out for several thousand years. At this time, the ancestral African mtDNA mutated, creating two new lineages, M and N, which in turn gave rise to new mtDNA lineages that spread through most of Asia. My own early ancestors were a sturdy bunch, for among them are humans who traversed the Caucasus Mountains of Georgia and southern Russia and colonised Europe. These were the first Cro-Magnons, and their arrival heralded the end of the Neanderthals. Ancient DNA from Cro-Magnon fossils shows a history of successes and failures shaped by the extreme climate of glacial Europe. Early Europeans carried mtDNA haplogroups M and N as well as already differentiated N lineages such as U and R, which would eventually spawn my own haplogroup H. However, a new and brutal cold spell pushed these ancient hunters into small refugia, and only some of them survived. My ancestors waited out the cold spell in southern Europe, most likely on the Iberian Peninsula. As the current warm, balmy weather set in some ten thousand years ago, they recolonised land that had previously been swathed in thick sheets of ice. But they soon had to compete with newcomers—farmers and Bronze Age peoples from Western Asia. I share the H marker with some 40 to 60 percent of the gene pool of most European populations.

The New World was the last to be populated. Southern Siberia was invaded about 40,000 years ago, but these early migrants were blocked from reaching the New World until the last glacial maximum lowered sea levels sufficiently to provide a bridge across the Bering Strait. Humans probably first arrived in North America some 15,000 years ago before they migrated south. There are few fossils to corroborate the genetic findings because these migrations took place when the world was swathed in ice and the continental shelves were exposed. Most of the physical evidence of their epic trek has long been submerged under the oceans.

The genetic makeup of people today gives us critical clues about our ancestral histories. However, the details of those histories—when and where particular populations lived or changed—can only be inferred from where people live today. We know that our very recent past is characterized by massive population movements as farmers and pastoralists expanded throughout vast parts of the world in the last 10,000 years. Ancient genomes give the time-and-place stamp for these movements as well as giving us extraordinary insights into the history of humans in Eurasia and the Americas. Unfortunately, there are extremely few human fossils in Africa from 200,000 to 10,000 years ago—only about ten! On top of that, DNA preservation is affected by heat and humidity, and attempts to retrieve genetic information from those few precious fossils have failed. And yet the most revealing information has to come from Africa.

Africa is a vast continent—30.37 million square kilometres, large enough to hold three Europes. But it is not only large, it is also where humans have lived some 150,000 years longer than anywhere else. It is not surprising that the people of Africa today have the greatest genetic diversity since they have had the longest time to build up mutations. Most African populations descend from the same group that expanded out of Africa 70,000 years ago and also from small groups of their descendants who expanded as farmers and herders in the last few thousand years. But a few groups of hunter-gatherers, such as the San of southern Africa, the Bayaka pygmies of Central Africa, and the Hadza and Sandawe of Eastern Africa, trace their genetic ancestry deep into the human past. These hunter-gatherer peoples have languages that include unique clicks that are believed to have been initially used when hunting to avoid alerting animals to their presence. They are also considered to be the most closely related to the ancient people who left their beautiful paintings on the rock faces of Southern and Eastern Africa, and through their genetic uniquenesses, they provide us with insights about our remote past. Yet, we must await the joint effort of fossil and ancient gene hunters to tell us where in Africa our ancestors lived, when they spread throughout the continent, whom they met while doing so, and what selective pressures shaped our shared humanity.

Tracing the myriad paths through the genetic footprints of our ancestors has taken on a new urgency because these DNA trails are rapidly becoming tangled as people move from their ancestral homes in search of economic opportunity. Teams of geneticists from all over the world are racing to analyse the DNA of indigenous populations to trace the genetic markers that characterise them. Collecting samples from as many indigenous people as possible is literally a race against time. More and more people are abandoning their traditional lifestyles for the melting pot of the big city, and the genetic trail that has persisted in their bloodlines for generations will soon be removed from its geographical context and forever lost in subsequent generations.

While many of these markers are hidden clues to our past buried in intricate codes in our cells, other mutations—those rare beneficial ones—have led to tangible differences between the peoples that have populated different parts of our planet. One of the most visible of these is skin colour. This highly visible identifying feature has historically been loaded with tragic implications, falsely signifying differences in superiority that have had—and continue to have—devastating consequences for those seen to be lower down the racial power ladder. The kaleidoscope of skin colours we exhibit ought to be seen as a celebration of our collective versatility and adaptability in an unforgiving world.

We have already suggested that the first move towards our modern cooling system—proficient sweating through bare skin—began when *Homo erectus* became a successful predator through persistence hunting. Being upright in hot sun and coping with higher body temperatures during marathon runs necessitated this. But it is unlikely that our ancestors permanently lost all their fur very early because we know they survived frigid weather when they migrated northwards. It is impossible to tell when exactly humans lost their protective fur, but I believe it was relatively recently and had to do with the acquisition of clothing.

Like the tapeworm that hitched a ride out of Africa in the gut of *Homo erectus*, another parasite, the lowly louse, latched on to *Homo sapiens* and gives us some clues as to when we might have abandoned the naked state. African lice are the most genetically diverse in the world, so we know that the louse originated in Africa. Of the two most common kinds of human lice, the head louse (*Pediculus humanus capitis*) exhibits the most diversity in its genes, suggesting that the body louse (*Pediculus humanus humanus*) evolved from the head louse. Mark Stoneking and his colleagues at the Max Planck Institute for Evolutionary Anthropology have hypothesised that this new kind of body louse must have evolved when a new habitat became available —clothes (or probably furs at first). They looked at the sequences of mtDNA and segments of nuclear DNA from the lice, and their molecular-clock analysis suggests that body lice originated 72,000 years ago

—right before *Homo sapiens* started to move around. Their date has a huge margin of error of 42,000 years. But this is certainly around the time we would expect the advent of clothing if *Homo sapiens* was to survive the northern cold. The first clothes were almost certainly un-shaped skins and furs, but it is unclear whether the first body lice could have lived in furs as they infest woven fabrics today.

If you aren't a naked mole rat—living a protected subterranean existence far from the reach of the sun's harmful rays—being a terres-trial animal with a lot of exposed skin is a risky business. Skin is easily damaged by the ultraviolet radiation in sunlight, and this damage can eventually lead to skin cancer that can sometimes be fatal. Chimpan-zees and other apes have pale skin beneath their protective fur, and the areas of exposed skin—the face, hands, and feet—darken over time with progressive exposure to sunlight (those kept in cages out of the sun never darken). It is generally believed that the common ancestor of humans and other apes had light skin covered with dark hair. It was long assumed that darker skin evolved in humans to protect against skin cancer or conversely, later in our evolutionary history, lighter skin evolved as those in less sunny climes had a reduced need for this UV protection.

Darker skin does offer better sun protection because it has more of the pigment melanin, nature's sunscreen. Melanin works both phys-ically and chemically to filter the harmful UV radiation in sunlight. It absorbs UV rays so they lose their powerful energy, and it neutralizes the cancer-forming free radicals that develop in sun-damaged skin. Nina Jablonski, whom I first met many years ago through a common passion for monkeys, is an anthropologist with an abiding interest in all things skin. Nina is an intelligent, focused, and grounded person. I knew that she had long puzzled about how changes in skin colour came to be selected for, because skin cancers only develop later in life —after women would have had the chance to have babies. Different amounts of melanin could not have been selected for their protection against cancer. There was something else going on.

Nina stumbled on part of the elusive connection quite by accident in a paper written in 1978. This paper reported that when very fair-skinned people were exposed to large amounts of strong artificial sunlight, they had abnormally low levels of folic acid in their blood afterwards. Sticking a container of human blood serum in the same artificial sunlight reproduced the same result within an hour: the folic acid in the blood serum was reduced by an astounding 50 percent. A second clue turned up in the late 1980s when some of Nina's colleagues at the University of Western Australia discovered that folic acid deficiency in pregnant women can lead to abnormalities such as spina bifida, a neural tube defect that prevents the spinal vertebrae of the foetus from closing around the spinal cord. They then discovered the huge role that folic acid plays in any process of cell proliferation because it is essential for the synthesis of DNA. It has since been found that folic acid treatment can boost the sperm count of men with fertility problems. Dark skin likely evolved to protect vital B-complex vitamins.

If dark skin is so much better than light, why did those migrants to higher latitudes lose it? This was the next conundrum for Nina and her colleagues. Nina knew about another relationship between the pigment melanin and UV light. It has long been known that the body needs vitamin D for various essential processes ranging from calcium absorption for bone development to maintaining a healthy immune system. It had also been established that the shorter wavelength UV radiation, UVB, helps the skin to manufacture vitamin D. Nina's brilliance was in putting all these parts of the story together.

Nina is married to George Chaplin, a British expert on global information systems. George and Nina combined their considerable intellectual talents and produced some illuminating maps. First, they constructed a map of the intensity of UV radiation levels on the earth's surface based on data from NASA satellites equipped with specialized spectrometers that were measuring ozone values. From this map, they produced a second map predicting the skin pigmentation of indigenous people from all over the world based on UV radiation levels. Then they looked at how close actual pigmentation was to their predictions. They found that the skin colour of indigenous people in the Old World closely matched their predictions. But the skin colour of long-term residents in the New World tended to be lighter than expected, which they attribute to more recent migration and factors such as diet.

Nina and George found that the world could be divided into three vitamin D zones. For dark-skinned people living in tropical latitudes, enough UVB gets through their protective melanin coating year-round to initiate the vitamin D synthesis their bodies depend on. For light-skinned people in the second zone, the subtropics and temperate regions, there is insufficient UVB radiation for them to synthesize vitamin D for at least one month of the year (for example, vitamin D production by the skin cells in light-skinned people begins in mid-March in Boston while dark-skinned people who move to these latitudes are unable to synthesize vitamin D for many months of the year). For the third zone, the polar regions above 45 degrees north and south, there is insufficient UVB year-round to prompt vitamin D synthesis. This led Nina and George to conclude that there was a strong natural selection for adequate vitamin D synthesis, which was the primary driver for the development of lighter skin colour in higher latitudes. You might be thinking, as I did when Nina first told me about her work, that the Inuit are not particularly light-skinned considering their polar environment of Alaska and northern Canada. Nina explains this as a likely result of two factors. First, they are relatively recent migrants, having arrived only about five thousand years ago. Second, they consume lots of vitamin D–rich fish and marine mammals, and this dietary supplement offsets the selective pressure for lighter pigmentation.

Women all over the world tend to have lighter skin than men by a margin of between 3 and 4 percent. Nina again struck down a conventional explanation for this that has strong racist overtones: that men prefer having sex with lighter-skinned women. Remember the role that vitamin D plays in calcium absorption. Women have a higher calcium requirement over their lifetime than men because of the heavy toll that pregnancy and lactation take on the mother's calcium levels. The reason for women's skin to be paler has to do with the need to allow more UVB rays in and thereby increase their ability to produce the life-giving vitamin D. "In areas of the world that receive a large amount of UV radiation, women are indeed at the knife's edge of natural selection, needing to maximise the photo-protective function of their skin on the one hand, and the ability to synthesize vitamin D on the other," Nina and George conclude. Their eye-opening work has been widely cited and translated into a multitude of languages, and it strikes down some of the worst prejudices that divide us to this day.

One element fuelling our epic migration out of Africa cannot be traced through mutations in our genes or in the physical remains of our fossil bones. This is the mysterious event that led to the evolution of the single feature that most separates humans from every other creature alive on the planet: our tremendous brain power. Our brains reached their present size with the birth of our species some 200,000 years ago. These early humans crafted tools and used fire—but these modest accomplishments are only a small step away from our nearest cousins, the chimpanzees—who have also been known to fashion basic tools from sticks and stones, and who can develop an amazing repertoire of signals for communicating. The leaps and bounds ancestral hominins took—from their first tentative steps as bipeds to the development of an opposable thumb and manual dexterity to the breakthrough of persistence hunting—certainly improved the odds for humans. But there was a quantum leap forward somewhere in the last 40,000 years, long after all these other evolutionary advantages were first united in our species, that led us to become the technologically savvy and supremely dominant species on earth that we are today.

THE LAST DECADE has involved many changes for me, both professional and personal. At home, our life has been enriched by three granddaughters who, like earlier generations of Leakeys, feel at home roaming for fossils in Turkana. Our old field team has retired, and a new guard of sharp-eyed fossil hunters has taken over—I no longer spend months out in the field with them but enjoy my all-too-short field days and the excitement when they find hominins. We have recently found some thrilling 3.5-million-year teeth in South Turkwell that again challenge our views about the diversity in early hominins. But not everything has been smooth. While the kidney I gave Richard all those years ago continues to work, his liver failed a few years ago. This involved a massive operation in the United States that gave him a new liver from a dear friend, and there was another arduous road to recovery afterwards. But recover he did, and through the building and running of the Turkana Basin Institute, Richard and I have once again shared many enjoyable discussions about human evolution with the new groups of researchers TBI has attracted.

The questions have not changed that much. How many different evolutionary paths did early hominins take? How old is our own genus? When did hominins first begin to make stone tools? Can we map the evolution of our own species onto the descendants of the intrepid *H. erectus* in Africa? While these questions were already the subject of dinner conversations between Louis and Mary when Richard was a child, the thousands of fossils discovered in the last fifty-odd years make the discussions at our house all the more fun. But the last decade has also brought entirely new ways of exploring the information held in the fossils we find—micro-CT scanning, geometric analyses, ancient DNA, and even palaeoproteins.

One of my personal highlights of 2018 occurred when, together with Fred Spoor and another long-term colleague, Isaiah Nengo, I was granted the privilege of scanning three fossils at the European Synchrotron Radiation Facility in Grenoble, France. The ESRF is the most intense source of synchrotron-generated light and produces X-rays a hundred billion times brighter than the standard ones used in hospitals. High-energy electrons race around a circular tunnel 844 metres in circumference. The brilliance and quality of the X-rays it produces are unparalleled, and they reveal the internal structure of matter in all its beauty and complexity.

The synchrotron is in high demand, with time allocated on the basis of merit to only nine thousand scientists per year, and it was due to close that same year for upgrading. We were able to use this amazing technology on three of our fossil hominins, including Kyalo's "very good hominin," the *erectus* skull we discussed in chapter 17. These three specimens were chosen because they all have essential information in their inner ear that we had not been able to visualize using either medical-CT or "normal" micro-CT scanning. Thanks to this advanced technology, the analysis of these scans will likely tell us much that was previously unknowable from fossil evidence.

All this new technology has significantly broadened the fields of research, amplified what we are able to glean from fragmentary evidence, and completely changed how we look for patterns and relations in the fossils. The field of genetics has already yielded some giant surprises, such as the existence of creatures whose ancient genomes were extracted from one small finger bone, but whose fossil faces remain undiscovered. In the fifty years that have gone by since I first went to Koobi Fora with Richard, we have learnt so much about our past. And yet this body of evidence remains but a small fraction of the past seven million years of our evolution. It is hard to conceive how much more will be discovered in the next half century or so.

EPILOGUE

As I approach the last few years of my seventh decade, I reflect on the immensely rich and action-packed life that I have had. I have been extremely fortunate to have witnessed firsthand many exciting discoveries and developments beyond my own specialty of human evolution. But I am constantly reminded that my children and grandchildren will live to see a very different world. The dizzying pace of new human inventions and explorations illustrates how humans have penetrated every corner of the planet with increasingly sophisticated technology. The speed with which this is happening is in stark contrast to the millions of years of slow human development that have been the focus of my life's research.

Evolution is constantly at work as features change or are co-opted for a different task at different times and rates. The configuration of our hands and feet, evolved for an arboreal life with grasping fingers and toes, adapted with relative ease to bipedality. Once freed from arboreal constraints, we became still more adept with our hands and acquired a wide range of motions and dexterity. These changes, when coupled with the selective pressures of a highly variable global climate, then paved the way for a strategy for procuring meat with relative ease, which fuelled our growing brains. This in turn pushed cooperation and communication further than the heights achieved in highly social, complex, and hierarchical primate societies. To these accomplishments, we recently added the great improvements to our health and comfort brought on by all our modern technology.

It is to our primate ancestry that we owe the morphological and behavioural patterns that have thus far been to our immense benefit. But today, this heritage is a double-edged sword that could be our undoing. The unfortunate fact is that we are a greedy, acquisitive, and destructive species by nature—like many monkeys. Do not mistake me, for I have loved monkeys ever since my early days caring for them at the Tigoni Primate Research Centre. But when the baboons breach the barriers we have erected and get into our vegetable garden, the destruction is a sight to behold and lament. They invariably leave a trail of devastation behind them. Half-eaten carrots, tomatoes flung about, and maize and potato plants ripped ruthlessly from the ground testify more to a destructive intent and willful gratification than to a pattern of sustainable foraging. The monkeys are doing no more than what we humans are doing on a far grander scale all over the planet—with our depletion of the oceans through overfishing and a wanton disregard for the bycatch, our unchecked logging in forests, our clearing of huge tracts of land for agriculture and urban settlement, our ever-increasing pollution of the atmosphere with chemicals and carbon dioxide, and our profligate, wanton, and senseless overconsumption and dumping of single-use plastic that has resulted in a garbage patch in the Pacific Ocean twice the size of Texas. Microplastics now pollute our water system so completely that they have been found in the ice and snow in remote parts of the Arctic and at the bottom of the Mariana Trench, the deepest point of the ocean anywhere on earth.

Our intelligence and our technology make us so much more efficient at pillaging resources. Where before there was a hand axe, now there is a chain saw or a bulldozer or even a mechanical digger. Where there was once a wooden canoe and a harpoon, now there is a trawler equipped with a sonar detector, refrigeration, and mechanised cranes. Our footprint on the planet is so large that geologists have now designated a new era, Anthropocene ("of man"), for our labours are now indelibly recorded in new layers of sediments and in the scars we have rendered in the earth's surface. If we don't survive, the rocks will bear witness to the havoc that we wrought long after we are gone.

We have learned to cure many diseases that were previously often fatal, and we can overcome almost all of nature's checks and balances to overpopulation. Many more children survive today than in the past, and our life expectancy has been significantly extended. As a consequence, our population numbers have soared and continue to do so. When *Homo sapiens* first left Africa, our numbers were extremely small. At the end of the Second World War, just after I was born and before antibiotics were generally available, the world's population was 2.3 billion. In 1963, I was already concerned by the urban development that was gradually creeping over many wild areas I had known as a child in England. Today, the world population numbers more than 7.7 billion, and by 2050, it is predicted to rise to more than 9.8 billion. With this many humans on one planet, dangers not previously perceived as threats can become lethal unless they are properly managed.

On less sanguine days, I sometimes wonder if we merit the glorious and self-congratulatory name of *H. sapiens* ("wise man"). The remains of our extinct species might one day be discovered many million years hence and be given the alternative appellation of *Homo stupidensis*. For, as Richard remarked to me one evening early in 2008 as we discussed the tragic and worrying postelection violence and killings that were gripping Kenya at the time, "We are most certainly the only animal that makes conscious choices that are bad for our survival as a species!"

Our brains are unquestionably the most exciting evolutionary adaptation of all time. But our cognitive superpower may be the making or the undoing of our species. Will our intelligence, this extraordinary and unique adaptation, ensure our survival? Or, like other extreme specializations (such as the exclusive grass-eating adaptations of the giant gelada baboons or the enormous antlers of the Irish elk), will our great intelligence ultimately drive us to extinction?

As I ponder these unknowable questions about our future, I am sitting on the verandah of our comfortable cottage at TBI Ileret, which overlooks the lake and some of the richly fossiliferous sediments I know so well. The colours expose millions of years of history—from the deep-red beds full of fish to the paler fluvial deposits to the layers of volcanic ash spread over the landscape and the dark sandstones full of fossils laid down on the lakeshore. A chocolate streak darkens the middle of the lake beyond; the Omo River is flooding and bringing yet more sediments into this huge lake basin and continuing the cycle of sedimentation and fossilisation. I marvel once again at this incredible research locality with its unrivalled seven-million-year record documenting our ancestors' evolution from the first bipeds to the technologic masters of the world. Beyond the lake are the Lapurr Hills, where even older sediments have preserved dinosaurs some seventy to ninety million years old. These hills are a constant reminder of our own fleeting presence on this planet.

In the human evolution segment of our field schools, we demonstrate how extraordinarily brief this time is using an unfurled loo roll to represent the full span of our evolution. The earth's formation resides on the first piece of loo paper five billion years ago. A full sixty sheets later, the first life appears in the oceans. Down the path of two hundred segments of loo paper, the students can then count out the history of early organisms, plants, reptiles, fish, mammals, and birds, with many blank sheets representing the sheer amount of time it all took for the families, genres, and species that are alive today to evolve from that first miraculous single-celled organism. In stark contrast, our entire human evolutionary history is crammed onto the two and a half centimetres of a single sheet of two-ply. Our own species appears one millimetre from the end of that final segment. I find this fact at once terribly humbling and immensely reassuring.

Thanks to a career unearthing the fossils of many extinct creatures, I am aware that extinction is an inevitable part of life on the earth and that every species has a limited sojourn on this planet. By some estimates, 99.9 percent of all species that ever lived are now extinct. And there is every evidence to suggest that we humans are currently driving a new wave of extinction. But we are unique: no other species has the capacity to change the course of its longevity. Because we have the intelligence to understand the consequences of our actions and devise solutions to them, we alone can limit the damage that we cause and stop these negative trends. The very intelligence that has brought us so rapidly to this point could most certainly ensure a long and bright future if we choose.

One of the hallmarks of humanity is our ability to combine our knowledge and expertise. It is our collective intelligence that has enabled us to land on the moon and send a rover to Mars. No single person has the expertise to do these things, but large teams of people specializing in different aspects of the project can. Our collective intelligence provides us with the knowledge to now plan for a secure future with fewer wars, a smaller carbon footprint, less exploitation of finite resources, and, above all, a more sustainable population. And where there is a will there is a way. Although we have a lamentable predilection to procrastinate and leave things to the eleventh hour, never before has there been such urgent need to take action to preserve our planet for the future.

Acknowledgements

I am immensely fortunate to have enjoyed the support of family, friends, and colleagues through the years, and those reading this memoir know who you are. My gratitude for all you have contributed to make my life and my work more fulfilling is profound.

My greatest thanks must be to my family. Ours is a family business, and Richard, Louise, and Samira have been part of this journey at key steps of the way. Before them, Louis and Mary blazed a path of believing that humans originated in Africa. I owe a debt of gratitude to them too, and to my parents, my sister Judy, and my brother Roger. My aunt Margs, who looked after me during the war, was also a source of inspiration and long-forgotten facts about the early days.

In writing about a time period of this span, there would be too many people to thank individually than could fit within the pages of a book. Our field is a collaborative one, and our knowledge is gleaned from small pieces of a large and connected puzzle.

It goes without saying that the fossils are all important, and many people have worked as part of the Koobi Fora Research Project since its inception in 1968. Our original fossil searching team, all retired now, were with us as we broke ground, found new sites, built roads, camps, and airstrips, searched, sieved, excavated, and much, much more. Of this group, Kamoya Kimeu, Nzube Mutiwa, and Wambua Mangao deserve a special mention. During the writing process, these three also contributed many mirthful stories of the old days, some of which are recounted in these pages. They have been succeeded by a new generation of talented, dedicated individuals who have contributed much to field and labwork and must likewise be recognized, with special mention to Lawrence Nzuve, Cyprian Nyete, Benson Maina, and Martin Kirinya.

A huge thank you to all the friends and colleagues who, over the years, have worked with me, both directly and indirectly, in pursuit of answers. These include more colleagues than I can name but who have nevertheless played significant roles at the National Museums of Kenya, as well as colleagues from institutions in Ethiopia, Tanzania, and South Africa. Among those who must be named in person are Zeray Alemseged, Chris Dean, Patrick Gathogo, Fred Grine, John Harris, Jean-Jacques Hublin, Christopher Kiarie, William Kimbel, Benson Kyongo, Marta Lahr, Fredrick Kyalo Manthi, Lawrence Martin, Emma Mbua, Ian McDougall, Yohannes Haile-Selassie, Fred Spoor, Alan Walker, and Carol Ward. Special thanks go to the great geologist Frank Brown, who sadly is no longer with us. Frank dedicated his life work to deciphering the geology of the Turkana Basin, and without his long-term collaboration all our finds would be meaningless. He was followed by other geologists and geophysicists, and among them Craig Feibel and Thure Cerling must be singled out for their signature contributions.

We have been privileged to receive research funding from a variety of sources, and of these the National Geographic Society has played a key and enduring role throughout. With the inception of the Turkana Basin Institute, our research is at last on a more secure footing, and this is largely due to the support of Stony Brook University.

For her part, Samira would like to thank many of the same circle of family, friends, and colleagues. In addition, thank you to my special friends for your encouragement, love, and laughter at all the necessary times, along with practical help in school runs, cooked meals, DIY, and other essential and much-appreciated support to keep my household ticking over. I am so lucky to have you in my life. My ex-husband supported my collaboration in this book in its earliest stages. Kika has been inspiration, support, and distraction to us both, and enriched the process with her boundless curiosity and questions.

As to the actual writing process, an enormous debt of gratitude from us both goes to Marta Lahr, who encouraged and corrected us where we most needed it, and helped us in so many ways. Without Gillian MacKenzie, our fabulous agent, the book would likely have forever remained a manuscript languishing on a hard drive. Thank you to the publishing house, Houghton Mifflin Harcourt, and to Bruce Nichols who understood exactly what we were trying to do and believed in our book. Special mention must go to our wonderful and tireless editor, Pilar Garcia-Brown, and to our sharp-eyed copyeditor, David Hough, who was able to add order to our unruly punctuation. We have been fortunate to work with Patricia Wynne, who drew the marvelous illustrations. Huge gratitude is also due to our cover designer, Martha Kennedy, along with our production team: Kimberly Kiefer, Heather Tamarkin, Chris Granniss, Chloe Foster, Katie Kimmerer, Laura Brady. Last but not least, we acknowledge the tireless efforts of our publicist, Michelle Triant, and our marketing expert, Liz Anderson. It goes without saying that any remaining errors are all our own.

Sources

CHAPTER 3: RACING AGAINST THE CLOCK

Leakey, M. D. *Africa's Vanishing Art: The Rock Paintings of Tanzania.* London: Hamish Hamilton Rainbird, 1983.

Leakey, M. G.; Leakey, R. E.; Richtmeier, J. T.; Simons, E. L.; and Walker, A. C. "Similarities in *Aegyptopithecus* and *Afropithecus* facial morphology." *Folia Primatologica* 56 (1991): 65–85.

Leakey, M. G.; Ungar, P. S.; and Walker, A. C. "A new genus of large primate from the Late Oligocene of Lothidok, Turkana District, Kenya." *Journal of Human Evolution* 28 (1995): 519–32.

Leakey, M. G., and Walker, A. C. "*Afropithecus* Function and Phylogeny." In *Function, Phylogeny, and Fossils: Miocene Hominoid Evolution and Adaptations,* edited by Begun, D. R.; Ward C. V.; and Rose, M. D. New York: Plenum Press, 1997, 225–39.

Leakey, R. E., and Leakey, M. G. "A new Miocene hominoid from Kenya." *Nature* 324 (1986): 143–46.

Leakey, R. E., and Leakey M. G. "A second new Miocene hominoid from Kenya." *Nature* 324 (1986): 146–48.

Leakey, R. E.; Leakey, M. G.; and Walker A. C. "Morphology of *Afropithecus turkanesis* from Kenya." *American Journal of Physical Anthropology* 76 (1988): 289–307.

Leakey, R. E.; Leakey, M. G.; and Walker, A. C. "Morphology of *Turkanapithecus kalakolensis* from Kenya." *Amercan Journal of Physical Anthropology* 76 (1988): 277–88.

Leakey, R. E., and Walker, A. C. "New higher primates from the early Miocene of Buluk, Kenya." *Nature* 318 (1985): 173–75.

Morrell, V. *Ancestral Passions.* New York: Simon and Schuster, 1995.

Walker, A., and Shipman, P. *Wisdom of the Bones: In Search of Human Origins.* New York: Alfred A. Knopf, 1996.

CHAPTER 4: CHANGING OF THE GUARD

Brown, F. H., and Feibel, C. S. "Stratigraphy, Depositional Environments, and Paleogeography of the Koobi Fora Formation." In *Koobi Fora Research Project Vol. 3, The Fossil Ungulates: Geology, Fossil Artiodactyls, and Palaeoenvironments*, edited by Harris, J. M., 1–30. Oxford: Clarendon Press, 1991.

Cerling, T. E., and Brown, F. H. "Tuffaceous marker horizons in the Koobi Fora region and the lower Omo Valley." *Nature* 299 (1982): 216–21.

Feibel, C. S.; Harris, J. M.; and Brown, F. H. "Palaeoenvironmental Context for the Late Neogene of the Turkana Basin." In *Koobi Fora Research Project Vol. 3, The Fossil Ungulates: Geology, Fossil Artiodactyls, and Palaeoenvironments*, edited by Harris, J. M., 321–46. Oxford: Clarendon Press, 1991.

Fitch, F. J., and Miller, J. A. "Radioisotopic age determinations of Lake Rudolf artefact site." *Nature* 226 (1970): 226–28.

Leakey, M. D. *Disclosing the Past: An Autobiography*. London: Weidenfeld & Nicolson, 1984.

Leakey, R. E., and Morell, V. *Wildlife Wars: My Fight to Save Africa's Natural Treasures*. New York: St. Martin's Press, 2002.

Lewin, R. *Bones of Contention: Controversies in the Search for Human Origins*. New York: Simon and Schuster, 1987.

Morrell, V. *Ancestral Passions*. New York: Simon and Schuster, 1995.

Ward, C. V.; Leakey, M. G.; and Walker, A. C. "South Turkwell: a new Pliocene hominid site in Kenya." *Journal of Human Evolution* 36 (1999): 69–95.

Wilson, A., and Sarich, V. "Immunological time scale for hominid evolution." *Science* 158 (1967): 1200–03.

CHAPTER 5: WATER, WATER EVERYWHERE

Boisserie, J. "The phylogeny and taxonomy of Hippopotamidae (Mammalia: Artiodactyla): a review based on morphology and cladistic analysis." *Zoological Journal of the Linnean Society* 143 (2005): 1–26.

Brochu, C. A., and Storrs, G. W. "A giant crocodile from the Plio-Pleistocene of Kenya." *Journal of Vertebrate Paleontology* 32 (2012): 587–602.

Feibel, C. S. "Stratigraphic and Depositional History of the Lothagam Sequence." In *Lothagam: The Dawn of Humanity in Eastern Africa*, edited by Leakey, M. G., and Harris, J. M. New York: Columbia University Press, 2003, 17–29.

Gibbons, A. *The First Human: The Race to Discover Our Earliest Ancestors*. New York: Doubleday, 2006.

Harvard University Lothagam Expeditions field records 1966–67, 1970–71, National Museum of Kenya archives.

Lothagam field diaries 1989–1994, Koobi Fora Research Project.

Leakey, M. G., and Harris, J. M., eds. *Lothagam: The Dawn of Humanity in Eastern Africa*. New York: Columbia University Press, 2003.

Leakey, R. E., and Morell, V. *Wildlife Wars: My Fight to Save Africa's Natural Treasures.* New York: St. Martin's Press, 2002.

Maglio, V. J. "Origin and evolution of the Elephantidae." *Transactions of the American Philosophical Society* 63 (1973): 1–149.

Patterson, B.; Behrensmeyer, A. K.; and Sill, W. D. "Geology of a new Pliocene locality in northwestern Kenya." *Nature* 256 (1970): 279–84.

Patterson, B., and Howells, W. W. "Hominid humeral fragment from early Pleistocene of northwestern Kenya." *Science* 156 (1967): 64–66.

Powers, D. W. "Geology of Mio-Pliocene sediments of the lower Kerio River Valley." PhD diss., Princeton University, 1980.

Ward, C. V.; Leakey, M. G.; Brown, B.; Brown, F.; Harris, J. M.; and Walker, A. C. "South Turkwel: a new Pliocene hominid site in Kenya." *Journal of Human Evolution* 36 (1999): 69–95.

CHAPTER 6: A BRAVE NEW WORLD

Aiello, L. C., and Wheeler, P. "The expensive tissue hypothesis." *Current Anthropology* 36 (1995): 199–221.

Bernor, R. L., and Harris, J. M. "Systematics and Evolutionary Biology of the Late Miocene and Early Pliocene Hipparionine Equids from Lothagam, Kenya." In *Lothagam: The Dawn of Humanity in Eastern Africa,* edited by Leakey, M. G., and Harris, J. M., 387–438. New York: Columbia University Press, 2003.

Cerling, T. E.; Harris, J. M.; and Leakey, M. G. "Isotope Paleoecology of the Nawata and Nachukui Formations at Lothagam, Turkana Basin, Kenya." In *Lothagam: The Dawn of Humanity in Eastern Africa,* edited by Leakey, M. G., and Harris, J. M., 605–25. New York: Columbia University Press, 2003.

Cerling, T. E.; Harris, J. M.; Leakey, M. G.; Passey, B.; and Levin, N. "Stable Carbon and Oxygen Isotopes in East African Mammals: Modern and Fossil." In *Cenozoic Mammals of Africa,* edited by Werdelin, L., and Sanders, W. Berkeley: University of California Press, 2010, 941–52.

Cerling, T. E.; Harris, J. M.; MacFadden, B. J.; Leakey, M. G.; Quade, J.; Eisenmann, V.; and Ehleringer, J. R. "Global vegetation change through the Miocene-Pliocene boundary." *Nature* 389 (1997): 153–58.

Cerling, T. E.; Harris, J. M.; and Passey, B. H. "Diets of East African Bovidae based on stable isotope analysis." *Journal of Mammalogy* 84 (2003): 456–70.

Cerling, T. E.; Wang, Y.; and Quade, J. "Expansion of C_4 ecosystems as an indicator of global ecological change in the Late Miocene." *Nature* 361 (1993): 344–45.

Darwin, C. *On the Origin of Species.* Edited by Mayr, E. Cambridge, MA: Harvard University Press, 1964.

Ehleringer, J. R.; Cerling, T. E.; and Dearing, M. D., eds. *The History of Atmospheric CO2 and Its Effects on Plants, Animals, and Ecosystems.* New York: Springer-Verlag, 2005.

Harris, J. M. "Bovidae from the Lothagam Succession." In *Lothagam: The Dawn of*

Humanity in Eastern Africa, edited by Leakey, M. G., and Harris, J. M., 531–79. New York: Columbia University Press, 2003.

Harris, J. M. "Deinotheres from the Lothagam Succession." In *Lothagam: The Dawn of Humanity in Eastern Africa,* edited by Leakey, M. G., and Harris, J. M., 359–61. New York: Columbia University Press, 2003.

Harris, J. M., and Leakey, M. G. "Lothagam Rhinocerotidae." In *Lothagam: The Dawn of Humanity in Eastern Africa,* edited by Leakey, M. G., and Harris, J. M., 371–85. New York: Columbia University Press, 2003.

Harris, J. M., and Leakey, M. G. "Lothagam Suidae." In *Lothagam: The Dawn of Humanity in Eastern Africa,* edited by Leakey, M. G., and Harris, J. M., 485–519. New York: Columbia University Press, 2003.

Hunt, K. D. "The postural feeding hypothesis: an ecological model for the origin of bipedalism." *South African Journal of Science* 92 (1996): 77–90.

Jablonski, N. G., and Chaplin, G. "Origin of habitual terrestrial bipedalism in the ancestor of the Hominidae." *Journal of Human Evolution* 24 (1993): 259–80.

Jolly, C. J. "The seed-eaters: a new model of hominid differentiation based on a baboon analogy." *Man* 5 (1970): 1–26.

Kramer, A. "Hominid-pongid distinctiveness in the Miocene-Pliocene record: the Lothagam mandible." *American Journal of Physical Anthropology* 70 (1986): 457–73.

Leakey, M. D., and Harris, J. M. *Laetoli: A Pliocene Site in Northern Tanzania.* Oxford: Oxford University Press, 1987.

Leakey, M. G., and Harris, J. M. "Lothagam: Its Significance and Contributions." In *Lothagam: The Dawn of Humanity in Eastern Africa,* edited by Leakey, M. G., and Harris, J. M., 626–60. New York: Columbia University Press, 2003.

Leakey, M. G.; Teaford, M. F.; and Ward C. V. "Cercopithecidae from Lothagam." In *Lothagam: The Dawn of Humanity in Eastern Africa,* edited by Leakey, M. G., and Harris, J. M., 202–48. New York: Columbia University Press, 2003.

Leakey, M. G., and Walker, A. C. "The Lothagam Hominids." In *Lothagam: The Dawn of Humanity in Eastern Africa,* edited by Leakey, M. G., and Harris, J. M., 349–56. New York: Columbia University Press, 2003.

Leonard, W. R., and Robertson, M. L. "Rethinking the energetics of bipedality." *Current Anthropology* 38 (1997): 304–09.

Palkopoulou, E., et al. "A comprehensive genomic history of extinct and living elephants." *PNAS* 115 (2018): E2566–74.

Rodman, P. S., and McHenry, H. M. "Bioenergetics and the origin of hominid bipedalism." *American Journal of Physical Anthropology* 52 (1980): 103–06.

Rose, M. D. "Food Acquisition and the Evolution of Positional Behaviour: The Case of Bipedalism." In *Food Acquisition and Processing in Primates,* edited by Chivers, D. J.; Wood, B. A.; and Bilsborough, A., 509–24. New York: Springer-Verlag, 1984.

Shoshani J., ed. *Elephants: Majestic Creatures of the Wild.* London: Weldon Owen, 1992.

Stanford, C. B. *Upright: The Evolutionary Key to Becoming Human.* Boston: Houghton Mifflin, 2003.

Stanford, C. B.; Allen, J. S.; and Antón, S. C. *Biological Anthropology*. Boston: Pearson, 2006.

Stewart, K. M. "Fossil Fish Remains from Mio-Pliocene Deposits at Lothagam, Kenya." In *Lothagam: The Dawn of Humanity in Eastern Africa*, edited by Leakey, M. G., and Harris, J. M., 75–111. New York: Columbia University Press, 2003.

Tassy, P. "Elephantoidea from Lothagam." In *Lothagam: The Dawn of Humanity in Eastern Africa*, edited by Leakey, M. G., and Harris, J. M., 331–58. New York: Columbia University Press, 2003.

Turner, A., and Antón, M. *The Big Cats and Their Fossil Relatives*. New York: Columbia University Press, 1997.

Weiner, J. *The Beak of the Finch: A Story of Evolution in Our Time*. New York: Vintage Books, 1995.

Werdelin, L. "Mio-Pliocene Carnivora from Lothagam, Kenya." In *Lothagam: The Dawn of Humanity in Eastern Africa*, edited by Leakey, M. G., and Harris, J. M., 261–330. New York: Columbia University Press, 2003.

Weston, E. M. "Fossil Hippopotamidae from Lothagam." In *Lothagam: The Dawn of Humanity in Eastern Africa*, edited by Leakey, M. G., and Harris, J. M., 441–48. New York: Columbia University Press, 2003.

Wheeler, P. E. "The thermoregulatory advantages of hominid bipedalism in open equatorial environments: the contribution of increased convection heat loss and cutaneous evaporative cooling." *Journal of Human Evolution* 21 (1991): 107–15.

White, T. D. "*Australopithecus afarensis* and the Lothagam mandible." *Anthropos* 23 (1986): 79–90.

CHAPTER 7: NINE LIVES

Koobi Fora Research Project field diaries, 1993.

Leakey, R. E., and Morell, V. *Wildlife Wars: My Battle to Save Kenya's Elephants*. New York: St. Martin's Press, 2002.

Morell, V. *Ancestral Passions: The Leakey Family and the Quest for Humankind's Beginnings*. New York: Simon and Schuster, 1995.

CHAPTER 8: A NEW EARLY BIPED

Alemseged, Z.; Spoor, F.; Kimbel, W. H.; Bobe, R.; Geraads, D.; and Reed, D. "A juvenile early hominin skeleton from Dikika, Ethiopia." *Nature* 433 (2006): 296–301.

Alemseged, Z.; Wynn, J. G.; Kimbel, W. H.; Reed, D.; Geraads, D.; and Bobe, R. "A new hominin from the Basal Member of the Hadar Formation, Dikika, Ethiopia, and its geological context." *Journal of Human Evolution* 49 (2005): 499–514.

Haile-Selassie, Y.; Melillo S. M.; Vazzana, A.; Benazzi, S.; and Ryan, T. M. "A 3.8-million-year-old hominin cranium from Waronzo-Mille, Ethiopia." *Nature* 573 (2019): 214–19.

Hay, R. L., and Leakey, M. D. "The fossil footprints of Laetoli." *Scientific American* 246 (1982): 50–57.

Jablonski, N. G., and Chaplin, G. "Origin of habitual terrestrial bipedalism in the ancestor of the Hominidae." *Journal of Human Evolution* 24 (1993): 259–80.

Johanson, D. C.; Lovejoy, C. O.; Kimbel, W. H.; White, T. D.; and Ward, S. C. "Morphology of the Pliocene partial hominid skeleton (A.L. 288-1) from the Hadar formation, Ethiopia." *American Journal of Physical Anthropology* 57 (1982): 403–52.

Johanson, D. C., and Maitland, E. *Lucy: The Beginnings of Humankind.* New York: Simon and Schuster, 1981.

Johanson, D. C., and Taieb, M. "Plio-Pleistocene hominid discoveries in Hadar, Ethiopia." *Nature* 260 (1976): 293–97.

Johanson, D. C.; White, T. D.; and Coppens, Y. "A new species of the genus *Australopithecus* (Primates: Hominidae) from the Pliocene of eastern Africa." *Kirtlandia Journal* 28 (1978): 1–14.

Johanson, D. C., et al. The complete issue of the *American Journal of Physical Anthropology* 57, no. 4 (1982).

Kimbel, W. H.; Lockwood, C. A.; Ward, C. V.; Leakey, M. G.; Rak, Y.; and Johansen, D. C. "Was *Australopithecus anamensis* ancestral to *A. afarensis*? A case for anagenesis in the fossil record." *Journal of Human Evolution* 51 (2006): 134–52.

Koobi Fora Research Project field diaries, 1994–1997.

Koobi Fora Research Project field reports for the National Geographic Society, 1994–1997.

Leakey, M. D. *Disclosing the Past: An Autobiography.* London: Weidenfeld & Nicolson, 1984.

Leakey, M. D., and Harris, J. M., eds. *Laetoli: A Pliocene Site in Northern Tanzania.* Oxford: Clarendon Press, 1987.

Leakey, M. G.; Feibel, C. S.; McDougall, I.; and Walker, A. "New four-million-year-old hominid species from Kanapoi and Allia Bay, Kenya." *Nature* 376 (1995): 565–71.

Leakey, M. G.; Feibel, C. S.; McDougall, I.; Ward, C.; and Walker, A. "New specimens and confirmation of an early age for *Australopithecus anamensis.*" *Nature* 393 (1998): 62–66.

Leakey, M. G., and Walker, A. "Early hominid fossils from Africa." *Scientific American* 13 (2003): 14–19.

Leakey, R., and Morell, V. *Wildlife Wars: My Fight to Save Africa's Natural Treasures.* New York: St. Martin's Press, 2002.

Patterson, B.; Behrensmeyer, A. K.; and Sill, W. D. "Geology of a new Pliocene locality in northwestern Kenya." *Nature* 256 (1970): 279–84.

Patterson, B., and Howells, W. W. "Hominid humeral fragment from Early Pleistocene of Northwestern Kenya." *Science* 156 (1967): 64–66.

Powers, D. W. "Geology of Mio-Pliocene sediments of the lower Kerio River Valley." PhD diss., Princeton University, 1980.

Saylor, B., et al. "Age and context of mid-Pliocene hominin cranium from Woranso-Mille, Ethiopia." *Nature* 573 (2019): 220–24.

Ward, C. V.; Leakey, M. G.; and Walker, A. "Morphology of *Australopithecus anamensis* from Kanapoi and Allia Bay, Kenya." *Journal of Human Evolution* 41 (2001): 255–368.

White, T. D. "Additional fossil hominids from Laetoli, Tanzania." *American Journal of Physical Anthropology* 53 (1980): 487–504.

White, T. D. "New fossil hominids from Laetoli, Tanzania." *American Journal of Physical Anthropology* 46 (1977): 130–97.

Wynn, J. G.; Alemseged, Z.; Bobe, R.; Geraads, D.; Reed, D.; and Roman, D. C. "Geological and palaeontological context of a Pliocene juvenile hominin at Dikika, Ethiopia." *Nature* 443 (2006): 332–36.

CHAPTER 9: ANOTHER PIECE OF THE PUZZLE

Asfaw, B.; White, T.; Lovejoy, O.; Latimer, B.; Simpson, S.; and Suwa, G. "*Australopithecus garhi:* a new species of early hominid from Ethiopia." *Science* 284 (1999): 629–35.

Bush, M. E.; Lovejoy, C. O.; Johanson, D. C.; and Coppens, Y. "Hominid carpal, metacarpal and phalangeal bones recovered from the Hadar Formation: 1974–1977 collections." *American Journal of Physical Anthropology* 57 (1982): 651–78.

Coffing, K.; Feibel, C.; Leakey, M. G.; and Walker, A. "Four-million-year-old hominids from East Lake Turkana, Kenya." *American Journal of Physical Anthropology* 93 (1994): 55–65.

De Heinzelin, J., et al. "Environment and behaviour of 2.5-million-year-old Bouri hominids." *Nature* 284 (1999): 625–29.

Feibel, C. S. "Stratigraphy and depositional setting of the Pliocene Kanapoi Formation, Lower Kerio Valley, Kenya." *Contributions in Science* 498 (2003): 9–20.

Gamlin, L., and Vines, G. *The Evolution of Life.* London: Collins, 1986.

Harris, J. M., and Leakey, M. G., eds. "Geology and vertebrate paleontology of the early Pliocene site of Kanapoi, northern Kenya." *Contributions in Science* 498 (2003): 1–132.

Haile-Selassie, Y.; Melillo, S. M.; Vazzana, A.; Benazzi, S.; and Ryan, T. M. "A 3.8-million-year-old hominin cranium from Waronzo-Mille, Ethiopia." *Nature* 573 (2019): 214–19.

Jablonski, N., ed. Theropithecus: *The Rise and Fall of a Primate Genus.* Cambridge: Cambridge University Press, 1993.

Johanson, D., and Maitland, E. *Lucy: The Beginnings of Humankind.* New York: Simon and Schuster, 1981.

Johanson, D. C.; White, T. D.; and Coppens, Y. "A new species of the genus *Australopithecus* (Primates: Hominidae) from the Pliocene of eastern Africa." *Kirtlandia Journal* 28 (1978): 1–14.

Jolly, C. J. "The seed-eaters: a new model of hominid differentiation based on a baboon analogy." *Man* 5 (1970): 5–26.

Kimbel, W. H.; Lockwood, C. A.; Ward, C. V.; Leakey, M. G.; Rak, Y.; and Johansen,

D. C. "Was *Australopithecus anamensis* ancestral to *A. afarensis*? A case for ana-genesis in the fossil record." *Journal of Human Evolution* 51 (2006): 134–52.

Koobi Fora Research Project field diaries, 1994–1997.

Koobi Fora Research Project field reports for the National Geographic Society, 1994–1997.

Leakey, M. G.; Feibel, C. S.; McDougall, I.; and Walker, A. "New four-million-year-old hominid species from Kanapoi and Allia Bay, Kenya." *Nature* 376 (1995): 565–71.

Leakey, M. G.; Feibel, C. S.; McDougall, I.; Ward, C.; and Walker, A. "New speci-mens and confirmation of an early age for *Australopithecus anamensis*." *Nature* 393 (1998): 62–66.

Leakey, M. G., and Walker, A. "Early hominid fossils from Africa." *Scientific American* 13 (2003): 14–19.

Marzke, M. W., and Shakley, S. "Hominin hand use in the Pliocene and Pleistocene: evidence from experimental archaeology and comparative morphology." *Journal of Human Evolution* 15 (1986): 439–60.

Marzke, M. W.; Wullatein, K. L.; and Viegas, S. F. "Evolution of the power ('squeeze') grip and its morphological correlates in hominids." *American Journal of Physical Anthropology* 89 (1992): 283–98.

Napier, J. "The evolution of the hand." *Scientific American* 207 (1962): 56–62.

Semaw, S. "The world's oldest stone artefacts from Gona, Ethiopia." *Journal of Archae-ological Science* 27 (2000): 1197–214.

Stern, J. T., Jr. "Climbing to the top: a personal memoir of *Australopithecus afarensis*." *Evolutionary Anthropology* 9 (2000): 113–33.

Trinkhaus, E. "Evolution of Human Manipulation." In *The Cambridge Encyclopedia of Human Evolution,* edited by Jones, S.; Martin R.; and Pilbeam, D., 346–49. Cambridge: Cambridge University Press, 1992.

Ward, C. V.; Leakey, M. G.; and Walker, A. "Morphology of *Australopithecus anamen-sis* from Kanapoi and Allia Bay, Kenya." *Journal of Human Evolution* 41 (2001): 255–368.

Ward, C. V.; Walker, A.; and Leakey, M. G. "The new hominid species *Australopithecus anamensis*." *Evolutionary Anthropology* 7 (1999): 197–205.

Werdelin, L. "Carnivora from the Kanapoi hominid site, Turkana Basin, Northern Kenya." *Contributions in Science* 498 (2003): 115–32.

White, T. D. "Additional fossil hominids from Laetoli, Tanzania." *American Journal of Physical Anthropology* 53 (1980): 487–504.

White, T. D. "New fossil hominids from Laetoli, Tanzania." *American Journal of Phys-ical Anthropology* 46 (1977): 197–230.

White, T. D.; Suwa, G.; and Asfaw, B. "*Australopithecus ramidus:* a new species of early hominid from Aramis, Ethiopia." *Nature* 371 (1994): 306–12.

Wynn, J. G. "Influence of Plio-Pleistocene aridification on human evolution: evidence from paleosols of the Turkana Basin, Kenya." *American Journal of Physical An-thropology* 123 (2004): 106–18.

Wynn J. G. "Paleosols, stable carbon isotopes and paleoenvironmental interpretation of Kanapoi, northern Kenya." *Journal of Human Evolution* 39 (2000): 411–32.

CHAPTER 10: OPEN-COUNTRY SURVIVORS

Aiello, L., and Dean, C. *An Introduction to Human Evolutionary Anatomy.* London: Academic Press, 1990.

Arambourg, C., and Coppens, Y. "Sur le decouverte dans le Pleistocene inferieur de la Valle de l'Omo (Ethiopie) d'une mandibule d'australopithecien." *Comptes rendus des sciences de l'Académie des Sciences* 265 (1968): 589–90.

Asfaw, B.; White, T.; Lovejoy, O.; Latimer, B.; Simpson, S.; and Suwa, G. "*Australopithecus garhi:* a new species of early hominid from Ethiopia." *Science* 284 (1999): 629–35.

Begun, D. R. "The earliest hominins—is less more?" *Science* 303 (2004): 1478–80.

Berger L. R., et al. "*Australopithecus sediba:* a new species of *Homo*-like Australopith from South Africa." *Science* 328 (2010): 195–204.

Broom, R., and Schepers, G. W. H. "The South African fossil ape-men: the australopithecinæ." *Transvaal Museum Memoir* 2. Pretoria: Transvaal Museum, 1946, 272 + 18 plates.

Brunet, M., et al. "New hominid from the Upper Miocene of Chad, central Africa." *Nature* 418 (2002): 145–51.

Brunet, M., et al. "New material of the earliest hominid from the Upper Miocene of Chad." *Nature* 434 (2005): 752–55.

Clarke, R. J. "Discovery of complete arm and hand of the 3.3-million-year-old *Australopithecus* skeleton from Sterkfontein." *South African Journal of Science* 95 (1999): 477–80.

Clarke, R. J. "First ever discovery of a well-preserved skull and associated skeleton of *Australopithecus.*" *South African Journal of Science* 94 (1998): 460–63.

Clarke, R. J. "Newly revealed information on the Sterkfontein Member 2 *Australopithecus* skeleton." *South African Journal of Science* 98 (2002): 523–26.

Clarke, R. J. "On the unrealistic 'revised age estimates' for Sterkfontein." *South African Journal of Science* 98 (2002): 415–19.

Clarke, R. J., and Kuman, K. "The skull of StW 573, a 3.67 Ma *Australopithecus prometheus* skeleton from Sterkfontein Caves, South Africa." *Journal of Human Evolution* 134 (2019): 102634.

Cole, S. *Leakey's Luck: The Life of Louis Seymour Bazett Leakey 1903–1972.* London: Collins, 1975.

Constantino, P., and Wood, B. "The evolution of *Zinjanthropus boisei.*" *Evolutionary Anthropology* 16 (2007): 49–62.

Dalton, R. "Feel it in your bones." *Nature* 440 (2006): 1100-1101.

Dart, R. A. *Adventures with the Missing Link.* Philadelphia: The Institutes Press, 1967.

Dawkins, R. *The Ancestor's Tale: A Pilgrimage to the Dawn of Life.* London: Weidenfeld & Nicolson, 2004.

De Heinzelin, J., et al. "Environment and behaviour of 2.5-million-year-old Bouri hominids." *Nature* 284 (1999): 625–29.

Dirks, P. H. G. M., et al. "Geological setting and age of *Australopithecus sediba* from southern Africa." *Science* 328 (2010): 205–08.

Estes, R. D. *The Behaviour Guide to African Mammals.* South Africa: Russell Friedman Books, 1991.

Gibbons, A. *The First Humans.* New York: Doubleday, 2006.

Granger, D. E., et al. "New cosmogenic burial ages for Sterkfontein Member 2 *Australopithecus* and Member 5 Oldowan." *Nature* 552 (2015): 85–88.

Grine, F. E. "The diet of South African Australopithecines based on a study of dental microwear." *L'Anthropologie* 91 (1987): 467–82.

Grine, F. E.; Ungar, P. S.; and Teaford, M. F. "Was the early Pliocene hominin 'Australopithecus' anamensis a hard-object feeder?" *South African Journal of Science* 102 (2006): 301–10.

Haile-Selassie, Y. H. "Late Miocene hominids from the Middle Awash, Ethiopia." *Nature* 412 (2001): 178–81.

Haile-Selassie, Y. H.; Suwa, G.; and White, T. D. "Late Miocene teeth from Middle Awash Ethiopia, and early hominid dental evolution." *Science* 303 (2004): 1503–05.

Harmand, S., et al. "3.3-million-year-old stone tools from Lomekwi 3, West Turkana, Kenya." *Nature* 521 (2015): 310–15.

Jablonski, N., ed. Theropithecus: *The Rise and Fall of a Primate Genus.* Cambridge: Cambridge University Press, 1993.

Johanson, D. C., and Maitland, E. *Lucy: The Beginnings of Humankind.* New York: Simon and Schuster, 1981.

Johanson, D. C., et al. The complete issue of the *American Journal of Physical Anthropology* 57 (1982).

Kimbel, W. H.; Lockwood, C. A.; Ward, C. V.; Leakey, M. G.; Rak, Y.; and Johanson, D. C. "Was *Australopithecus anamensis* ancestral to *A. afarensis*? A case for anagenesis in the fossil record." *Journal of Human Evolution* 51 (2006): 134–52.

Kimbel, W. H.; Rak, Y.; and Johanson, D. C. The Skull of *Australopithecus afarensis.* Oxford: Oxford University Press, 2004.

Kingdon, J. *The Kingdon Field Guide to African Mammals.* London: Academic Press, 1997.

Kivell, T. L., et al. "*Australopithecus sediba* hand demonstrates mosaic evolution of locomotor and manipulative abilities." *Science* 333 (2011): 1411–17.

Leakey M. D. *Disclosing the Past: An Autobiography.* London: Weidenfeld & Nicolson, 1984.

Lewin, R. *Bones of Contention.* New York: Simon and Schuster, 1987.

Lebatard, A., et al. "Cosmogenic nuclide dating of *Sahelanthropus tchadensis* and *Australopithecus bahrelghazali:* Mio-Pliocene hominids from Chad." *PNAS* 105 (2008): 3226–31.

Lockwood, C. A.; Mentor, C. G.; Moggi-Cecchi, J.; and Keyser, A. W. "Extended male growth in a fossil hominin species." *Science* 318 (2007): 1443–46.

McBrearty, S., and Jablonski, N. "First fossil chimpanzee." *Nature* 437 (2005): 105–08.

Napier, J., and Napier, P. H. *A Handbook of Living Primates*. London: Academic Press, 1967.

Pickford, M.; Senut, B.; Gommercy, D.; and Treil, J. "Bipedalism in *Orrorin tugenensis* revealed by its femur." *Comptes rendus palevol* 1 (2002): 1–13.

Richmond, B. G., and Jungers, W. L. "*Orrorin tugenensis* femoral morphology and the evolution of hominin bipedalism." *Science* 319 (2008): 1662–65.

Semaw, S., et al. "Early Pliocene hominids from Gona, Ethiopia." *Nature* 433 (2005): 301–05.

Senut, B., et al. "First hominid from the Miocene (Lukeino Formation, Kenya)." *Compte rendus de l'Académie des Sciences de la terre et des planètes* 332 (2001): 137–44.

Shipman, P.; Bosler, W.; and Davis, K. L. "Butchering of giant geladas at an Acheulian site." *Current Anthropology* 22 (1981): 257–68.

Sponheimer, M., et al. "Isotopic evidence for dietary variability in the early hominin *Paranthropus robustus*." *Science* 314 (206): 980–82.

Teaford, M. F., and Ungar, P. S. "Diet and the evolution of the earliest human ancestors." *PNAS* 97 (2000): 13506–11.

Tobias, P. V. "The discovery of the Taung skull of *Australopithecus africanus:* Dart and the neglected role of Professor R. B. Young." *Transactions of the Royal Society of South Africa* 61 (2006): 131–38.

Tobias, P. V. "New researches at Sterkfontein and Taung with a note on Piltdown and its relevance to the history of palaeo-anthropology." *Transactions of the Royal Society of South Africa* 48 (1992): 1–14.

Trinkhaus, E. "Evolution of Human Manipulation." In *The Cambridge Encyclopedia of Human Evolution,* edited by Jones, S.; Martin R.; and Pilbeam D., 346–49. Cambridge: Cambridge University Press, 1992.

Walker, A.; Leakey, R. E.; Harris, J. M. H.; and Brown, F. H. "2.5-Myr *Australopithecus boisei* from West of Lake Turkana, Kenya." *Nature* 322 (1986): 519–22.

Walker, A., and Shipman, P. *Wisdom of the Bones: In Search of Human Origins.* New York: Alfred A. Knopf, 1996.

Ward, C. V.; Leakey, M. G.; and Walker, A. "Morphology of *Australopithecus anamensis* from Kanapoi and Allia Bay, Kenya." *Journal of Human Evolution* 41 (2001): 255–368.

Ward, C. V.; Leakey, M. G.; and Walker, A. "South Turkwell: a new Pliocene hominid site in Kenya." *Journal of Human Evolution* 36 (1999): 69–95.

White, T.; Suwa, G.; and Asfaw, B. "*Australopithecus ramidus:* a new species of early hominid from Aramis, Ethiopia." *Nature* 371 (1994): 306–12.

White, T. D., et al. "Asa Issie, Aramis and the origin of *Australopithecus.*" *Nature* 440 (2006): 883–89.

Wood, B. "Hominid revelations from Chad." *Nature* 418 (2002): 133–35.

Wood, B., and Constantino, P. "*Paranthropus boisei:* fifty years of evidence and analysis." *Yearbook of Physical Anthropology* 50 (2007): 106–32.

Wynn, J. G. "Paleosols, stable carbon isotopes and paleoenvironmental interpretation of Kanapoi, Northern Kenya." *Journal of Human Evolution* 39 (2000): 411–32.

Zollikofer, C. P. E., et al. "Virtual reconstruction of *Sahelanthropus tchadensis.*" *Nature* 434 (2005): 755–59.

CHAPTER 11: A FRIEND FOR LUCY?

Brown, B.; Brown, F. H.; and Walker, A. "New hominids from the Lake Turkana Basin, Kenya." *Journal of Human Evolution* 41 (2001): 29–44.

Brunet, M., et al. "*Australopithecus bahrelghazali,* une nouvelle espéce d'hominidé ancient de la région de Koro Toro (Tchad)." *Comptes rendus des sciences de l'Académie des Sciences* 322 (1996): 907–13.

Brunet, M., et al. "The first australopithecine 2,500 kilometres west of the Rift Valley (Chad)." *Nature* 378 (1995): 273–75.

Harris, J. M.; Brown, F. H.; and Leakey, M. G. "Geology and palaeontology of Plio-Pleistocene localities west of Lake Turkana, Kenya." *Contributions in Science* 399 (1988): 1–128.

Johanson, D. C.; Taieb, M.; and Coppens, Y. "Pliocene hominids from the Hadar Formation, Ethiopia (1973–1977): stratigraphic chronology and paleoenvironmental contexts, with notes on hominid morphology and systematics." *American Journal of Physical Anthropology* 57 (1982): 373–402.

Kimbel, W. H.; White, T. D.; and Johanson, D. C. "Cranial morphology of *Australopithecus afarensis:* a comparative study based on a composite reconstruction of the adult skull." *Journal of Physical Anthropology* 64 (1984): 337–88.

Kimbel, W. H., et al. "Late Pliocene *Homo* and Oldowan tools from the Hadar Formation (Kada Hada Member), Ethiopia." *Journal of Human Evolution* 31 (1996): 549–61.

Koobi Fora Research Project diaries, West Turkana, 1998.

Koobi Fora Research Project diaries, West Turkana, 1999.

Leakey, L. "Body weight estimation of Bovidae and Plio-Pleistocene faunal change, Turkana Basin, Kenya." PhD diss., University College London, 2001.

Leakey, M. G., et al. "New hominin genus from eastern Africa shows diverse middle Pliocene lineages." *Nature* 410 (2001): 433–40.

Spoor, F., et al. "Reconstructed *Homo habilis* type OH 7 suggests deep-rooted species diversity in early *Homo.*" *Nature* 519 (2015): 83–86.

White, T. "Early hominids—diversity or distortion?" *Science* 299 (2003): 1996–94.

Villmoare, B., et al. "Early *Homo* at 2.8 Ma from Ledi-Geraru, Afar, Ethiopia." *Science* 347 (2015): 1352–55.

Vrba, E. S.; Denton, G. H.; Partridge, T. C.; and Burkle, L. H. *Palaeoclimate and Evolution, with Emphasis on Human Origins.* New Haven, CT: Yale University Press, 1995.

Vrba, E. S. "Ecological and Adaptive Changes Associated with Early Hominid Evolution." In *Ancestors: The Hard Evidence,* edited by Delson, E., 63–71. New York: Alan R. Liss, 1985.

CHAPTER 12: EARLY *HOMO:* A HORRIBLE MUDDLE

Alexeev, V. P. *The Origin of the Human Race.* Moscow: Progress Publishers, 1986.

Kimbel, W. H., et al. "Late Pliocene *Homo* and Oldowan tools from the Hadar Formation (Kada Hada Member), Ethiopia." *Journal of Human Evolution* 31 (1996): 549–61.

Koobi Fora Research Project field diaries, 1972 and 1973.

Leakey, L. S. B.; Tobias, P. V.; and Napier, J. R. "A new species of the genus *Homo* from Olduvai Gorge." *Nature* 202 (1964): 7–9.

Leakey, M. D. *Disclosing the Past: An Autobiography.* London: Weidenfeld & Nicolson, 1984.

Leakey, M. D. *Olduvai Gorge: My Search for Early Man.* London: Collins, 1979.

Leakey, M. D. *Olduvai Gorge Volume 3. Excavations in Beds I and II, 1960–1963.* Cambridge: Cambridge University Press, 1971.

Leakey, M. G. Personal diaries, 1972.

Leakey, M. G., et al. "New hominin genus from eastern Africa shows diverse middle Pliocene lineages." *Nature* 410 (2001): 433–40.

Leakey, R. E. "Evidence for an advanced Plio-Pleistocene hominid from East Rudolf, Kenya." *Nature* 242 (1973): 447–50.

Leakey, R. E. *One Life: An Autobiography.* Topsfield, MA: Salem House Publishers, 1984.

Morell, V. *Ancestral Passions: The Leakey Family and the Quest for Humankind's Beginnings.* New York: Simon and Schuster, 1995.

Semaw, S. "The world's oldest stone artefacts from Gona, Ethiopia." *Journal of Archaeological Science* 27 (2000): 1197–214.

Shipman, P. *The Man Who Found the Missing Link: Eugene Dubois and His Lifelong Quest to Prove Darwin Right.* New York: Simon and Schuster, 2001.

Spoor, F., et al. "Reconstructed *Homo habilis* type OH 7 suggests deep-rooted species diversity in early *Homo.*" *Nature* 519 (2015): 83–86.

Stanford, C. B.; Allen, J. S.; and Antón, S. C. *Biological Anthropology.* Boston: Pearson, 2006.

Tobias, P. V. *Olduvai Gorge Volume IV: Skulls, Endocasts and Teeth of Homo habilis.* Cambridge: Cambridge University Press, 1991.

Villmoare, B., et al. "Early *Homo* at 2.8 Ma from Ledi-Geraru, Afar, Ethiopia." *Science* 347 (2015): 1352–55.

Walker, A., and Leakey, R. E. "The hominids of East Turkana." *Scientific American* 239 (1978): 4–64.

Walker, A., and Shipman, P. *Wisdom of the Bones: In Search of Human Origins.* New York: Alfred A. Knopf, 1996.

Wood, B. W. "Early *Homo:* How Many Species?" In *Species, Species Concepts and Primate Evolution,* edited by Kimbel, W. H., and Martin, L. B., 485–522. New York: Plenum Press, 1993.

Wood, B. W. "'*Homo rudolfensis*' Alexeev, 1986—fact or phantom." *Journal of Human Evolution* 36 (1999): 115–18.

Wood, B. W. *Koobi Fora Research Project Vol. 4. Hominid Cranial Remains.* Oxford: Clarendon Press, 1991.

Wood, B. W. "Who is the 'real' *Homo habilis?*" *Nature* 327 (1987): 187–88.

Wood, B. W., and Collard, M. "The changing face of genus *Homo.*" *Evolutionary Anthropology* 8 (1999): 195–207.

Wood, B. W., and Collard, M. "The human genus." *Science* 284 (1999): 65–71.

CHAPTER 13: BECOMING GRANDMAS

Antón, S. C. "Cranial growth in *Homo erectus:* how credible are the Ngandong juveniles?" *American Journal of Physical Anthropology* 108 (1999): 223–36.

Antón, S. C., and Leigh, S. R. "Growth and Life History in *Homo erectus.*" In *Patterns of Growth and Development in the Genus* Homo, edited by Thompson, J. L.; Krovitz, G. E.; and Nelson, A. J., 219–45. Cambridge: Cambridge University Press, 2003.

Beynon, A. D., and Dean, M. C. "Distinct dental development patterns in early fossil hominids." *Nature* 335 (1988): 509–14.

Bromage, T. G., and Dean, C. M. "Re-evaluation of the age at death of immature fossil hominids." *Nature* 317 (1985): 525–27.

Caspari, R., and Lee, S. "Older age becomes common late in human evolution." *Proceedings of the National Academy of Sciences* 101 (2004): 10895–900.

Clegg, M., and Aiello, L. C. "A comparison of the Nariokotome *Homo erectus* with juveniles from a modern human population." *American Journal of Physical Anthropology* 110 (1999): 81–93.

Coghlan, J. "With grandmothers came civilisation." *New Scientist,* July 10, 2004, 14.

Coqueugniot, H.; Hublin, J.; Veillon, F.; Houët, F.; and Jacob, T. "Early brain growth in *Homo erectus* and implications for cognitive ability." *Nature* 431 (2004): 299–302.

Dean, C., et al. "Growth processes in teeth distinguish modern humans from *Homo erectus* and earlier hominins." *Nature* 414 (2001): 628–31.

Goodall, J. *The Chimpanzees of Gombe: Patterns and Behaviour.* Cambridge, MA: Harvard University Press, 1986.

Hawkes, K.; O'Connell, J. F.; Blurton-Jones, N. G.; Alvarez, H.; and Charnov, E. L. "Grandmothering, menopause, and the evolution of human life histories." *Proceedings of the National Academy of Sciences of the United States of America* 95 (1998): 1336–39.

Huffman, O. F., et al. "Relocation of the 1936 Mojokerto skull discovery site near Perning, East Java." *Journal of Human Evolution* 50 (2006): 431–51.

Jablonski, N. G.; Whitfort, M. J.; Roberts-Smith, N.; and Quinqi, X. "The influence of life history and diet on the distribution of catarrhine primates during the Pleistocene in eastern Asia." *Journal of Human Evolution* 39 (2000): 131–57.

Lacruz, R. S., and Bromage, T. "Appositional enamel growth in molars of South African fossil hominids." *Journal of Anatomy* 209 (2006): 13–20.

Lacruz, R. S.; Rozzi, F. R.; and Bromage, T. "Variation in enamel development of South African fossil hominids." *Journal of Human Evolution* 51 (2006): 580–90.

Morwood, M. J., et al. "Revised age for Mojokerto 1, an early *Homo erectus* cranium from East Java, Indonesia." *Australian Archaeology* 57 (2003): 1–4.

O'Connell, J. F.; Hawkes, K.; and Blurton-Jones, N. G. "Grandmothering and the evolution of *Homo erectus.*" *Journal of Human Evolution* 36 (1999): 461–85.

Rosenberg, K. "Living longer: information revolution, population expansion, and modern human origins." *Proceedings of the National Academy of Sciences of the United States of America* 101 (2004): 10847–48.

Rosenberg, K. R., and Trevathan, W. R. "The evolution of human birth." *Scientific American* 13 (2003): 80–85.

Smith, H. "Dental development in *Australopithecus* and early *Homo.*" *Nature* 323 (1986): 327–30.

Smith, H. "Patterns of dental development in *Homo, Australopithecus, Pan* and *Gorilla.*" *American Journal of Physical Anthropology* 94 (1994): 307–25.

Smith, S. L. "Skeletal age, dental age and the maturation of KNM-WT 15000." *American Journal of Physical Anthropology* 125 (2004): 105–20.

Walker, A., and Leakey, R. E, eds. *The Nariokotome* Homo erectus *Skeleton.* Cambridge, MA: Harvard University Press, 1993.

Walker A., and Shipman, P. *Wisdom of the Bones: In Search of Human Origins.* New York: Alfred A. Knopf, 1996.

CHAPTER 14: GROWING BRAINS

Acredolo, L., and Goodwyn, S. *Baby Signs: How to Talk with Your Baby Before Your Baby Can Talk.* Chicago: Contemporary Books, 2002.

Aiello, L., and Dean, C. *An Introduction to Human Evolutionary Anatomy.* London: Academic Press, 1990.

Allman, J. M. *Evolving Brains.* New York: W. H. Freeman & Co., 2000.

Bramble, D. M., and Lieberman, D. E. "Endurance running and the evolution of *Homo.*" *Nature* 432 (2004): 345–52.

Cole, S. *Leakey's Luck: The Life of Louis Seymour Bazett Leakey, 1903–1972.* London: Collins, 1975.

Falk, D. *Braindance: New Discoveries About Human Origins and Brain Evolution.* New York: Henry Holt & Co., 1992.

Leakey, L. S. B. Lecture, Yale University, October 1971.

Leonard, W. R. "Food for thought. Dietary change was a driving force in human evolution." *Scientific American* 13 (2003): 62–71.

Martin, R. D. *Primate Origins and Evolution: A Phylogenetic Reconstruction.* London: Chapman and Hall, 1990.

McNutt, J., and Boggs, L. *Running Wild: Dispelling the Myths of the African Wild Dog.* Washington, D.C.: Smithsonian Books, 1997.

Milton, K. "Diet and primate evolution." *Scientific American* 16 (2006): 22–29.

Savage-Rumbaugh, S., and Lewin, R. *Kanzi: The Ape at the Brink of the Human Mind.* Somerset, NJ: John Wiley & Sons, 1994.

Spoor, F.; Wood, B.; and Zonneveld, F. "Implications of early hominid labyrinthine morphology for evolution of human bipedal locomotion." *Nature* 369 (1994): 645–48.

Stanford, C. B.; Allen, J. S.; and Antón, S. C. *Biological Anthropology*. Boston: Pearson, 2006.

Turner, A., and Anton, M. *The Big Cats and Their Fossil Relatives*. New York: Columbia University Press, 1997.

Walker, A., and Leakey, R. E., eds. *The Nariokotome* Homo erectus *Skeleton*. Cambridge, MA: Harvard University Press, 1993.

Walker, A., and Shipman P. *Wisdom of the Bones: In Search of Human Origins*. New York: Alfred A. Knopf, 1996.

CHAPTER 15: THE ICEHOUSE

Bowen, M. *Thin Ice: Unlocking the Secrets of Climate in the World's Highest Mountains*. New York: Henry Holt & Co., 2005.

Campisano, C. J. "Milankovitch cycles, paleoclimatic change, and hominin evolution." *Nature Education Knowledge* 4 (2012): 5.

Flannery, T. *The Weather Makers*. New York: Atlantic Monthly Press, 2005.

Imbrie, J., and Palmer Imbrie, K. *Ice Ages: Solving the Mystery*. Cambridge, MA: Harvard University Press, 1979.

Lepre, C. J.; Quinn, R. L.; Joordens, J. C. A.; Swisher III, C. C.; and Feibel, C. S. "Plio-Pleistocene facies environments from the KBS Member, Koobi Fora Formation: implications for climate controls on the development of lake margin hominin habitats in the northeast Turkana Basin (northwest Kenya)." *Journal of Human Evolution* 53 (2007): 504–14.

Macdougall, D. *Frozen Earth: The Once and Future Story of Ice Ages*. Berkeley: University of California Press, 2004.

Maslin, M. A., and Christensen, B. "Tectonics, orbital forcing, global climate change, and human evolution in Africa: introduction to the African paleoclimate special volume." *Journal of Human Evolution* 53 (2007): 443–64.

McDougall, I.; Brown, F. H.; and Fleagle, J. G. "Stratigraphic placement and age of modern humans from Kibish, Ethiopia." *Nature* 433 (2005): 733–36.

Turner, A., and Anton, M. *The Big Cats and Their Fossil Relatives*. New York: Columbia University Press, 1997.

Vrba, E. S.; Denton, G. H.; Partridge, T. C.; and Burkle, L. H. *Palaeoclimate and Evolution, with Emphasis on Human Origins*. New Haven, CT: Yale University Press, 1995.

Ward, P., and Brownlee, D. *The Life and Death of Planet Earth: How the New Science of Astrobiology Charts the Ultimate Fate of Our World*. New York: Henry Holt & Co., 2003.

Zachos, J.; Pagani, M.; Sloan, L.; Thomas, E.; and Billups, K. "Trends, Rhythms and Aberrations in Global Climate 65 Ma to Present." *Science* 292 (2001): 686–93.

CHAPTER 16: THE FIRST EXPLORERS

Antón, S. C. "Developmental age and taxonomic affinity of the Mojokerto child, Java, Indonesia." *American Journal of Physical Anthropology* 102 (1997): 497–514.

Antón, S. C. "Natural history of *Homo erectus.*" *Yearbook of Physical Anthropology* 46 (2003): 126–70.

Antón, S. C.; Leonard, W. R.; and Robertson, M. L. "An ecomorphological model of the initial hominid dispersal from Africa." *Journal of Human Evolution* 43 (2002): 773–85.

Balter, M. "Skeptics question whether Flores hominid is a new species." *Science* 306 (2004): 1116.

Bar-Yosef, O., and Belfer-Cohen, A. "From Africa to Eurasia—early dispersals." *Quaternary International* 75 (2001): 19–28.

Bobe, R., and Behrensmeyer, A. K. "The expansion of grassland ecosystems in Africa in relation to mammalian evolution and the origin of the genus *Homo.*" *Palaeogeography, Palaeoclimatology, Palaeoecology* 207 (2004): 399–420.

Bobe, R.; Behrensmeyer, A. K.; and Chapman, R. E. "Faunal change, environmental variability and late Pliocene hominin evolution." *Journal of Human Evolution* 42 (2002): 475–97.

Bobe, R.; Behrensmeyer, A. K.; Eck, G. G.; and Harris, J. M. "Patterns of Abundance and Diversity on Late Cenozoic Bovids from the Turkana and Hadar Basins, Kenya and Ethiopia." In *Hominin Environments in the East African Pliocene: An Assessment of the Faunal Evidence,* edited by Bobe, R.; Alemseged, Z.; and Behrensmeyer, A. K., 129–58. Dordrecht, Netherlands: Springer, 2007.

Bobe, R., and Eck, G. G. "Responses of African bovids to Pliocene climatic change." *Palaeobiology Memoirs* 27 (2001): 1–47.

Brown, F. "Some Considerations on Early Movements Out of Africa." Lecture, Human Evolution Symposium "Out of Africa," Stony Brook University.

Brown, P., et al. "A new small-bodied hominin from the Late Pleistocene of Flores, Indonesia." *Nature* 431 (2004): 1055–61.

Bunn, H. T. "The Bone Assemblages from the Excavated Sites." In *Koobi Fora Research Project Vol. 5: Plio-Pleistocene Archaeology,* edited by Isaac, G. L., and Isaac, B., 402–44. Oxford: Clarendon Press, 1997.

Burroughs, W. J. *Climate Change in Prehistory: The End of the Reign of Chaos.* Cambridge: Cambridge University Press, 2005.

Culotta, E. "When Hobbits (slowly) walked the earth." *Science* 320 (2008): 433–35.

Dalton, R. "Hobbit was 'a cretin.'" *Nature* 452 (2008), doi.org/10.1038/news.2008.643.

Falk, D., et al. "The brain of LB1, *Homo floresiensis.*" *Science* 308 (2005): 242–45.

Flannery, T. *The Weather Makers.* New York: Atlantic Monthly Press, 2005.

Gabunia, L.; Antón, S. C.; Vekua, A.; Justus, A.; and Swisher III, C. C. "Dmanisi and dispersal." *Evolutionary Anthropology* 10 (2001): 158–70.

Gabunia, L., and Vekua, A. "The environmental contexts of early human occupation of Georgia (Transcaucasia)." *Journal of Human Evolution* 38 (2000): 785–802.

Gabunia, L., et al. "Earliest Pleistocene hominid cranial remains from Dmanisi, Republic of Georgia: taxonomy, geological setting, and age." *Science* 288 (2000): 1019–25.

Gibbons, A. "A new body of evidence fleshes out *Homo erectus*." *Science* 317 (2007): 1664.

Gibert, J.; Gibert, L.; Ferrandez-Canyadell, A.; and Gonzales F. "Venta Micena, Barranco León-5 and Fuenteneuva-3. Three Archeological Sites in the Early Pleistocene Deposits of Orce, Southeast Spain." In *The Human Evolution Source Book*, edited by Ciochon, R. C., and Fleagle, J. G., 327–35. New York: Routledge, 2006.

Hoberg, E. P.; Alkire, N. L.; De Queiroz, A.; and Jones, A. "Out of Africa: origins of the *Taenia* tapeworms in humans." *Proceedings of the Royal Society London B* 268 (2001): 781–87.

Holmes, R. "Hobbit hand waives away doubters." *New Scientist* 195 (2007): 14.

Jashasvili, T., et al. "Postcranial evidence from early *Homo* from Dmanisi, Georgia." *Nature* 449 (2007): 305–10.

Larson, S. G. "Evolutionary transformation of the human shoulder." *Evolutionary Anthropology* 16 (2007): 172–87.

Larson, S. G., et al. "*Homo floresiensis* and the evolution of the hominin shoulder." *Journal of Human Evolution* 53 (2008): 718–31.

Leakey, L. "Body weight estimation of Bovidae and Plio-Pleistocene faunal change, Turkana Basin, Kenya." PhD diss., University College London, 2001.

Leakey, M. D. "List of identified faunal remains from known stratigraphic horizons in Beds I and II." Appendix B in *Excavations in Beds I &II 1960–1963* by Leakey, M. D. Cambridge: Cambridge University Press, 1971.

Leonard, W. R. "Food for thought. Dietary change was a driving force in human evolution." *Scientific American* 13 (2003): 62–71.

Lewis, M. E., and Werdelin, L. "Patterns of Change in the Plio-Pleistocene Carnivorans of Eastern Africa: Implications for Hominin Evolution." In *Hominin Environments in the East African Pliocene: An Assessment of the Faunal Evidence*, edited by Bobe, R.; Alemseged, Z.; and Behrensmeyer, A. K., 77–105. Dordrecht, Netherlands: Springer, 2007.

Lieberman, D. E. "Homing in on early *Homo*." *Nature* 449 (2007): 291–92.

Mirazón Lahr, M. "Saharan Corridors and Their Role in the Evolutionary Geography of 'Out of Africa 1.'" In *Out of Africa 1: The First Hominin Colonization of Eurasia,* edited by Fleagle, J.; Shea, J.; Grine, F.; Baden, A.; and Leakey, R. E. Dordrecht, Netherlands: Springer, 2010.

Mirazón Lahr, M., and Foley, R. "Human evolution writ small." *Nature* 431 (2004): 1043–44.

Morwood, M. J.; O'Sullivan, P. B.; Aziz, F.; and Raza, A. "Fission-track ages of stone tools and fossils on the east Indonesian island of Flores." *Nature* 392 (1998): 173–76.

Morwood, M. J., et al. "Archaeology and age of a new hominin from Flores in eastern Indonesia." *Nature* 431 (2004): 1087–91.

Obendorf, P. J.; Oxnard, C. E.; and Kefford, B. J. "Are the small human-like fossils found on Flores human endemic cretins?" *Proceedings of the Royal Society B: Biological Sciences* (2008): 1287–96.

Sahnouni, M., et al. "Further research at the Olduvai site of Aïn Hanech, northeastern Algeria." *Journal of Human Evolution* 43 (2002): 925–37.

Shipman, P.; Bosler, W.; and Davis, K. L. "Butchering of giant gelada at an Acheulian site." *Current Anthropology* 22 (1981): 257–68.

Stanford, C. B.; Allen, J. S.; and Antón, S. C. *Biological Anthropology.* Boston: Pearson, 2006.

Stringer, C., and McKie, R. *African Exodus: The Origins of Modern Humanity.* London: Jonathan Cape, 1997.

Sutikna, T., et al. "Revised stratigraphy and chronology for *Homo floresiensis* at Liang Bua in Indonesia." *Nature* 532 (2016): 366–69.

Swisher III, C. C., et al. "Age of the earliest known hominids in Java, Indonesia." *Science* 263 (1994): 1118–21.

Swisher III, C. C., et al. "Latest *Homo erectus* in Java: potential contemporaneity with *Homo sapiens* in Southeast Asia." *Science* 274 (1996): 1870–74.

Templeton, A. R. "Out of Africa again and again." *Nature* 416 (2002): 45–51.

Tocheri, M. W., et al. "The primitive wrist of *Homo floresiensis* and its implications for hominin evolution." *Science* 317 (2007): 1743–45.

Van der Bergh, G. D., et al. "*Homo floresiensis*-like fossils from the early Middle Pleistocene of Flores." *Nature* 534 (2016): 245–48.

Vekua, A., et al. "A new skull of early *Homo* from Dmanisi, Georgia." *Science* 297 (2002): 85–89.

Vrba, E. S.; Denton, G. H.; Partridge, T. C.; and Burkle, L. H. *Palaeoclimate and Evolution, with Emphasis on Human Origins.* New Haven, CT: Yale University Press, 1995.

Walker, A., and Shipman, P. *Wisdom of the Bones: In Search of Human Origins.* New York: Alfred A. Knopf, 1996.

Werdelin, L., and Lewis, M. E. "Plio-Pleistocene Carnivora of eastern Africa: species richness and turnover patterns." *Zoological Journal of the Linnean Society* 144 (2005): 121–44.

Wong, K. "The littlest human." *Scientific American* 292, February 2005, 40–49.

Wong, K. "Stranger in a new land." *Scientific American* 16, June 2006, 38–47.

Wynn, J. G. "Influence of Plio-Pleistocene aridification on human evolution: evidence from paleosols of the Turkana Basin, Kenya." *American Journal of Physical Anthropology* 123 (2004): 106–18.

Zachos, J.; Pagani, M.; Sloan, L.; Thomas, E.; and Billups, K. "Trends, rhythms and aberrations in global climate 65 Ma to present." *Science* 292 (2001): 686–93.

Zhu, R. X., et al. "Earliest human in northeast Asia." *Nature* 413 (2001): 413–17.

Zhu, R. X., et al. "New evidence on the earliest human presence at high northern latitudes in northeast Asia." *Nature* 431 (2004): 559–62.

CHAPTER 17: A VERY GOOD HOMININ

Antón, S. C. "Developmental age and taxonomic affinity of the Mojokerto child, Java, Indonesia." *American Journal of Physical Anthropology* 102 (1997): 497–514.

Antón, S. C. "Evolutionary significance of cranial variation in Asian *Homo erectus*." *American Journal of Physical Anthropology* 118 (2002): 301–23.

Gabunia, L., et al. "Dmanisi and dispersal." *Evolutionary Anthropology* 10 (2001): 158–70.

Gabunia, L., et al. "Earliest Pleistocene hominid cranial remains from Dmanisi, Republic of Georgia: taxonomy, geological setting, and age." *Science* 288 (2000): 1019–25.

Gray, H. *Anatomy, Descriptive and Surgical.* Edited by Pickering, T., and Howden, R. New York: Bounty Books, 1997.

Groves, C. P., and Mazák, V. "An approach to the taxonomy of the Hominidae: Gracile Villafranchian hominids of Africa." *Casopsis pro Mineralogii a Geologie* 20 (1975): 225–47.

Leakey, R. E. "Further evidence of Lower Pleistocene hominids from East Rudolf, North Kenya." *Nature* 237 (1972): 264–69.

Leakey, R. E. *One Life: An Autobiography.* Topsfield, MA: Salem House Publishers, 1984.

Marquez, S.; Mowbray, K.; Sawyer, G. J.; Jacob, T.; and Silvers, A. "New fossil hominid calvaria from Indonesia—Sambungmacan 3." *The Anatomical Record* 262 (2001): 344–68.

Spoor, F., et al. "Implications of new early *Homo* fossils from Ileret, east of Lake Turkana, Kenya." *Nature* 448 (2007): 688–91.

Stanford, C. B.; Allen, J. S.; and Antón, S. C. *Biological Anthropology.* Boston: Pearson, 2006.

Swisher III, C. C., et al. "Age of the earliest known hominids in Java, Indonesia." *Science* 263 (1994): 1118–21.

Swisher III, C. C., et al. "Latest *Homo erectus* in Java: potential contemporaneity with *Homo sapiens* in Southeast Asia." *Science* 274 (1996): 1870–74.

Vekua, A., et al. "A new skull of early Homo from Dmanisi, Georgia." *Science* 297 (2002): 85–89.

Walker, A., and Shipman, P. *Wisdom of the Bones: In Search of Human Origins.* New York: Alfred A. Knopf, 1996.

CHAPTER 18: THROUGH THICK AND THIN

Abbate, E., et al. "A one-million-year-old *Homo* cranium from Danakil (Afar) depression of Eritrea." *Nature* 393 (1998): 458–60.

Asfaw, B., et al. "Remains of *Homo erectus* from Bouri, Middle Awash, Ethiopia." *Nature* 416 (2002): 317–20.

Bowen, M. *Thin Ice: Unlocking the Secrets of Climate in the World's Highest Mountains.* New York: Henry Holt & Co., 2005.

Bräuer, G. "The ES-11693 Hominid from West Turkana and *Homo sapiens* Evolution in East Africa." In *Hominidae,* edited by Giacobini, G., 241–45. Milan: Jaca Book, 1989.

Bräuer, G. "The KNM-ER 3884 Hominid and the Emergence of Modern Anatomy in Africa." In *Humanity from African Naissance to Coming Millennia,* edited by Tobias, P. V.; Raath, M. A.; Moggi-Cecchi, J.; and Doyle, G. A., 191–97. Johannesburg: Witwatersrand University Press, 2001.

Bräuer, G., and Leakey, R. E. "The ES-11693 cranium from Eliye Springs, West Turkana, Kenya." *Journal of Human Evolution* 15 (1986): 289–312.

Bräuer, G., and Mabulla, A. Z. P. "New hominid fossil from Lake Eyasi, Tanzania." *Anthropologie* 34 (1996): 47–53.

Bräuer, G., and Mbua, E. "*Homo erectus* features used in cladistics and their variability in Asian and African hominids." *Journal of Human Evolution* 22 (1992): 79–108.

Dean, C., et al. "Growth processes in teeth distinguish modern humans from *Homo erectus* and earlier hominins." *Nature* 414 (2001): 628–31.

EPICA. "Eight glacial cycles from an Antarctic ice core." *Nature* 429 (2004): 623–28.

Flannery, T. *The Weather Makers.* New York: Atlantic Monthly Press, 2005.

Hublin, J. "Climatic Changes, Paleogeography and the Evolution of the Neandertals." In *Neandertals and Modern Humans in Western Asia,* edited by Akazawa, T.; Aoki, K.; and Bar-Yosef, O., 295–310. New York: Plenum Press, 1998.

Imbrie, J., and Palmer-Imbrie, K. *Ice Ages: Solving the Mystery.* Cambridge, MA: Harvard University Press, 1979.

Jones, D. "The Neanderthal within." *New Scientist* 193 (2007): 28–32.

Klein, R. G., and Edgar, B. *The Dawn of Human Culture: A Bold New Theory of What Sparked the "Big Bang" of Human Consciousness.* New York: John Wiley & Sons, 2002.

Lambert, F., et al. "Dust-climate couplings over the past 800,000 years from the EPICA Dome C ice core." *Nature* 452 (2008): 616–19.

Leakey, R. E.; Butzer, K. W.; and Day, M. H. "Early *Homo sapiens* remains from the Omo River region of Southwest Ethiopia." *Nature* 222 (1967): 1132–38.

Macdougall, D. *Frozen Earth: The Once and Future Story of Ice Ages.* Berkeley: University of California Press, 2004.

McBrearty, S., and Brooks, A. S. "The revolution that wasn't: a new interpretation of the origins of modern human behaviour." *Journal of Human Evolution* 39 (2000): 453–56.

McDougall, I.; Brown, F. H.; and Fleagle, J. G. "Stratigraphic placement of modern humans from Kibish, Ethiopia." *Nature* 433 (2005): 733–36.

Olson, S. *Mapping Human History: Genes, Race, and Our Common Origins.* Boston: Houghton Mifflin, 2002.

Ponce de Leon, M. S., and Zollikofer, C. P. E. "Neanderthal cranial ontogeny and its implications for late hominid diversity." *Nature* 412 (2001): 534–38.

Rightmire, G. P. "Human Evolution in the Middle Pleistocene, the Role of *Homo heidelbergensis.*" *Evolutionary Anthropology* 6 (1998): 218–27.

Rosas, A., et al. "The growth pattern of Neandertals reconstructed from a juvenile skeleton from El Sidron (Spain)." *Science* 357 (2017): 1282–87.

Ruff, C. B. "Climate and body shape in human evolution." *Journal of Human Evolution* 21 (1991): 81–105.

Ruff, C. B., and Walker, A. "Body Size and Body Shape." In *The Nariokotome* Homo erectus *Skeleton*, edited by Walker, A., and Leakey, R. E., 234–65. Cambridge, MA: Harvard University Press, 1993.

Stanford, C.; Allen, J. S.; and Antón, S. C. *Biological Anthropology*. Boston: Pearson, 2006.

Stringer, C., and Gamble, C. *In Search of the Neanderthals*. London: Thames and Hudson, 1993.

Stringer, C., and McKie, R. *African Exodus: The Origins of Modern Humanity*. New York: Henry Holt & Co., 1997.

Tattersall, I. *The Monkey in the Mirror: Essays on the Science of What Makes Us Human*. New York: Harcourt, 2002.

Templeton, A. R. "Out of Africa again and again." *Nature* 416 (2002): 45–51.

Trinkaus, E., and Shipman, P. *The Neandertals: Changing the Image of Mankind*. London: Jonathan Cape, 1993.

Vrba, E. S.; Denton, G. H.; Partridge, T. C.; and Burkle, L. H. *Palaeoclimate and Evolution, with Emphasis on Human Origins*. New Haven, CT: Yale University Press, 1995.

Walker, A., and Leakey, R. E., eds. *The Nariokotome* Homo erectus *Skeleton*. Cambridge, MA: Harvard University Press, 1993.

Walker, A., and Shipman, P. *Wisdom of the Bones: In Search of Human Origins*. New York: Alfred A. Knopf, 1996.

White, T. D., et al. "Pleistocene *Homo sapiens* from Middle Awash, Ethiopia." *Nature* 423 (2003): 742–47.

Wong, K. "Who Were the Neanderthals?" *Scientific American* 13, April 2003, 28–37.

CHAPTER 19: MIGRATING MUTANTS

Bar-Yosef, O., and Belfer-Cohen, A. "From Africa to Eurasia—early dispersals." *Quarternary International* 75 (2001): 19–28.

Behar, D. M., et al. "The dawn of human matrilineal diversity." *American Journal of Human Genetics* 82 (2008): 1–11.

Bowen, M. *Thin Ice: Unlocking the Secrets of Climate in the World's Highest Mountains*. New York: Henry Holt & Co., 2005.

Burroughs, W. J. *Climate Change in Prehistory: The End of the Reign of Chaos*. Cambridge: Cambridge University Press, 2005.

Cann, R. L.; Stoneking, M.; and Wilson, A. C. "Mitochondrial DNA and human evolution." *Nature* 325 (1987): 31–36.

Cann, R. L., and Wilson, A. C. "The Recent African Genesis of Humans." *Scientific American*, April 1992, updated in 2003, 54–61.

Chou, H., et al. "Inactivation of CMP-N-acetylneuraminic acid hydroxylase occurred prior to brain expansion during human evolution." *Proceedings of the National Academy of Sciences* 99 (2002): 11736–41.

Dalton, R. "Neanderthal genome sees first light." *Nature* 444 (2006): 254.

Díaz, S., et al. Summary for policymakers of the global assessment report on biodiversity and ecosystem services for the Intergovernmental Science-Policy Platform on Biodiversity and Ecosystem Services. Advanced Unedited Version, May 6, 2019. https://www.ipbes.net/sites/default/files/downloads/spm_unedited_advance_for _posting_htn.pdf/.

Evans, P. D.; Mekel-Bobrov, N.; Vallender, E. J.; Hudson, R. R.; and Lahn, B. T. "Evidence that the adaptive allele of the brain size gene *microcephalin* introgressed into *Homo sapiens* from an archaic *Homo* lineage." *Proceedings of the National Academy of Sciences* 103 (2006): 18178–83.

Green, R. E., et al. "Analysis of one million base pairs of Neanderthal DNA." *Nature* 444 (2006): 330–36.

Harris, E. E., and Hey, J. "X chromosome evidence for ancient human histories." *Proceedings of the National Academy of Sciences* 96 (1999): 3320–24.

Heinz, T., et al. "Updating the African human mitochondrial DNA tree: relevance to forensic and population genetics." *Forensic Science International: Genetics* 27 (2017): 156–59.

Jablonski, N. G. *Skin: A Natural History.* Berkeley: University of California Press, 2006.

Jablonski, N. G., and Chaplin G. "The evolution of skin coloration." *Journal of Human Evolution* 39 (2000): 57–106.

Jablonski, N. G., and Chaplin G. "Skin deep." *Scientific American,* October 2000, 72–79.

Jones, D. "The Neanderthal within." *New Scientist* 193 (2007): 28–32.

Kittler, R.; Kayser, M.; and Stoneking, M. "Molecular evolution of *Pediculus humanus* and the origin of clothing." *Current Biology* 13 (2003): 1414–17.

Macaulay, V., et al. "Single, rapid coastal settlement of Asia revealed by analysis of complete mitochondrial genomes." *Science* 308 (2005): 1034–36.

Moorjani, P., et al. "Variation in the molecular clock of primates." *PNAS* 113 (2026): 10607–12.

Noonan, J. P, et al. "Sequencing and analysis of Neanderthal genomic DNA." *Science* 314 (2006): 1113–18.

Olson, S. *Mapping Human History: Genes, Race, and Our Common Origins.* Boston: Houghton Mifflin, 2002.

Pennisi, E. "The dawn of Stone Age genomics." *Science* 31 (2006): 1068–71.

Poznik, G. D., et al. "Sequencing Y chromosomes resolves discrepancy in time of common ancestor of males versus females." *Science* 341 (2013): 562–65.

Rosas, A., et al. "The growth pattern of Neandertals reconstructed from a juvenile skeleton from El Sidron (Spain)." *Science* 357 (2017): 1282–87.

Shreeve, J. "The greatest journey." *National Geographic Magazine,* March 2006.

Soodyall, H., ed. *The Prehistory of Africa: Tracing the Lineage of Modern Man.* Cape Town: Jonathan Ball Publishers, 2006.

Stanford, C. B.; Allen, J. S.; and Antón, S. C. *Biological Anthropology*. Boston: Pearson, 2006.

Stringer, C., and Gamble, C. *In Search of the Neanderthals*. London: Thames and Hudson, 1993.

Tattersall, I. *The Monkey in the Mirror. Essays on the Science of What Makes Us Human*. New York: Harcourt, 2002.

Templeton, A. R. "Out of Africa again and again." *Nature* 416 (2002): 45–51.

Torroni, A.; Achilli, A.; Macaulay, V.; Richards, M.; and Bandelt, H. "Harvesting the fruit of the human mtDNA tree." *Trends in Genetics* 22 (2006): 339–45.

Trinkaus, E., and Shipman, P. *The Neandertals. Changing the Image of Mankind*. London: Jonathan Cape, 1993.

Weiner, J. *Time, Love, Memory*. New York: Vintage, 1999.

Wells, S. *Deep Ancestry: Inside the Genographic Project*. Washington, D.C.: National Geographic, 2007.

Wells, S. *The Journey of Man: A Genetic Odyssey*. London: Penguin Books, 2002.

EPILOGUE

Gore, A. *An Inconvenient Truth: The Planetary Emergency of Global Warming and What We Can Do About It*. New York: Rodale, 2006.

Leakey, R. E., and Lewin, R. *The Sixth Extinction: Biodiversity and Its Survival*. London: Weidenfeld & Nicolson, 1996.

Index